A CASE FOR NUCLEAR-GENERATED ELECTRICITY

(or why I think nuclear power is cool and why it is important that you think so too)

SCOTT W. HEABERLIN

Battelle Press
Columbus • Richland

Library of Congress Cataloging-in-Publication Data

Heaberlin, S. W.
A case for nuclear-generated electricity, or, Why I think nuclear power is cool and why it is important that you think so too / Scott W. Heaberlin
p. cm.
Includes bibliographical references and index.
ISBN 1-57477-136-1 (alk. paper)
1. Nuclear engineering—Popular works. I. Title: Case for nuclear-generated electricity. II. Title: Why I think nuclear power is cool and why it is important that you think so too. III. Title.

TK9146.H43 2003
333.792'4—dc21

2003052499

Printed in the United States of America

Battelle Press
505 King Avenue
Columbus, Ohio 43201-2693, USA
614-424-6393 or 1-800-451-3543
Fax: 614-424-3819
E-mail: press@battelle.org
Website: www.battelle.org/bookstore

Cover photo courtesy of Energy Northwest, Norris Clark, photographer.

CONTENTS

ACKNOWLEDGEMENTS

First, I need to acknowledge the extraordinary efforts of my editor and long-time friend, Nancy Burleigh. While dealing with a number of other challenges, she undertook the mind-numbing task of examining every single character in the book. She was faced with finding all those thousands of creative ways I have of generating errors that the most sophisticated of word processing software will not find. In that, she has reintroduced the concept of punctuation and gotten me to write in actual complete sentences most of the time. Not always. Done good, Nancy.

I would also like to thank Dr. Andy Klein, Professor and Head, Department of Nuclear Engineering and Radiation Health Physics, Oregon State University, for his review. His advice encouraged me to work harder to remove redundancies and streamline the effort. If the reader finds this work not streamlined, I offer the teenager's defense, "Mom, you should have seen my room before I cleaned it up!"

Ann Doherty, Manager Nuclear Materials and Engineering Analysis, Pacific Northwest National Laboratory, also has my gratitude for her patient and thorough reading of the work and her many insightful suggestions. Like Nancy, she touched every page. Her efforts have enhanced my confidence in the value of the book.

Diana Love deserves a big thank you for several years' worth of capturing items and articles that support my message. Ritchie Jensen has my gratitude for alerting me to a number of things, most notably Dr. Deffeyes' book.

I would also like to thank Gayle McNeece for her layperson's review of the section on nuclear waste in Chapter 6. She was a good enough friend to tell me I missed the mark, and to point out all the places I failed. If the discussion as it now stands does convince you, then Gayle gets most of the credit.

Joe Sheldrick and his team at Battelle Press deserve a big thanks for all their efforts to turn all this into the physical thing you hold in your hands. Kristin Manke and Susan Vianna stand out in that team in making magic happen.

Finally, I would like to acknowledge all the help and support I got from my wife, Dr. Joan O. Heaberlin. She has patiently excused me for many hours as I disappeared into this thing. She has been the sounding board for a lot of ideas, and she has gently corrected more than one errant tangent. As I have often observed, she is not only the most beautiful nuclear engineer I ever met, she is also the most competent.

A CASE FOR NUCLEAR-GENERATED ELECTRICITY

1
SUMMARY

There is a lot of negative feeling about nuclear energy. I think that feeling is based on a lack of understanding of nuclear energy and the failure to see nuclear energy, energy in general, and all of human technology in a wider, more appropriate context. My goal in this book is to explain nuclear energy to you and to give you some of that wider context. I believe that if you have some relevant facts and see those in a valid context you will conclude that the negative feelings about nuclear energy are largely unfounded and in fact harmful to the future success of humanity.

In this summary we will hit the high points. The rest of the book will take you on a number of interesting journeys that explore the history of time, humanity's place in the cosmos, and why we must have technology to survive as humans. I will tell you how nuclear energy works so that you will understand both the basics of the science and how those basics are applied to make nuclear generated electricity. I will help you understand some of the "bad" things you have heard about nuclear power, and we will look at the overall energy needs of humanity and the choices we have to make about energy.

It is a fascinating story and one that everyone in the modern world needs to understand. Along the way you are going to learn a lot, and together we are going to have some fun. Ready? Here we go!

First, it is technology that allows 6 billion humans to live on planet earth. Without technology a very large fraction of those people would have to get off the planet. A critical part of that technology is the energy required to drive it, and a critical part of energy is electrical energy. Electrical power is essential to modern life. Unless you are ready to let a large portion of the current human population die off, you need electricity. It does a lot more than run hair dryers. It is the mobile, apply it where it is needed, convenient energy source that makes our industrial society work. Without it our industry doesn't work, our society doesn't work, no one has jobs, and even if food would show up at your local grocery store, which it wouldn't, none of us could buy it.

Second, nuclear generation of electricity is a wholly remarkable way to create electricity. Without getting too technical, here is the deal: most ways of making heat come from burning something, like coal, oil, natural gas, or wood. This is a chemical reaction, which is all about changes in electrons that swirl around at the edge of atoms. Nuclear energy is about changing the core or nucleus of an atom. So what? The energy obtained from individual chemical reactions is measured in electron volts. Nuclear reactions are measured in **millions** of electron volts. We aren't working with wimpy electrons out at the distant outskirts of the atom. We are in the heart of it, where the real energy is.

Why is that cool?

A coal-fired electrical plant producing 1000 million watts of power (enough to run about 800,000 hair dryers all at once) requires 2,600,000 metric tons of coal (a metric ton is 2200 pounds) each year. A nuclear power plant producing 1000 million watts brings in fuel every 18 months. At each refueling about a third of the fuel load is replaced. The total load of fuel in a nuclear reactor of this size is about 120 metric tons. This means you use about 30 metric tons of fuel each year. Not 2,600,000 metric tons, 30.

Look at it another way. One kilogram (2.2 pounds) of coal will make 3 kilowatt-hours of electricity. That is enough to run one of those hair dryers for about 2.5 hours. One kilogram of nuclear fuel[1] produces 240,000 kilowatt-hours. That is 80 thousand times more energy per kilogram of fuel.

Let's look at it yet another way. Assume that you have the choice of a nuclear power plant or a coal-powered plant. However, the deal is you personally have to bring the fuel to the plant. Let's also assume that you share this duty with the other people served by the power plant. For round numbers we will assume the 1000 million watt plant provides power to a million people, which is in round numbers about average for the United States. Because the coal plant requires 2.6 million metric tons per year, your share is 2.6 metric tons. That is about 70 cubic feet, or if you are want to carry your

[1]This assumes a 4% enriched light water reactor uranium dioxide fuel, a 33% thermal efficiency of the power plant, and 20,500 kilowatt-hours of thermal energy per gram of uranium-235. Chapters 4 and 5 will explain what all that means, but the point is that this is pretty typical and measures the energy output in terms of the nuclear fuel loaded to the reactor.

coal in one gallon buckets, like those that paint comes in, you get to carry about 500 buckets for your yearly contribution. For the proper mental image, be sure you picture 999,999 of your friends and neighbors joining you, each trooping back and forth to carry their 500 buckets of coal to the power plant each year. By the way, this works out coincidentally that, if you carry two buckets at a time and make one trip a day, you have to carry coal 5 days a week every week, allowing for 10 holidays off, but no vacations, and no getting sick. If you fail to get your coal to the plant on any of your assigned 250 days a year, the lights go out, the supermarket closes, and the kids get sent home from school.

If you chose the nuclear plant, your share is 1 millionth of the 30 metric tons. This works out to 30 grams a year or slightly less than the weight of six U.S. quarter-dollar coins. I bet you could get that down to the plant in one trip.[2] Cool, huh?

These numbers are approximate, but they are real. That is the magic of nuclear energy.

That is why the Navy uses nuclear reactors for aircraft carriers. They fuel the ship when it is built, then they run it for 10 years before it needs new fuel. They are working on a new design in which the fuel lasts as long as the ship. They will build it and never fuel it again. No fuel tankers to mess with; no question of finding friendly harbors to refuel in.

Nuclear energy has a much higher energy density than any chemical energy source.

Why is this amazing energy density so cool?

The amount of waste produced in nuclear-generated electricity is vastly less than in fossil fuels. All the coal that is burned in a coal-fired power plant turns into combustion products, greenhouse or smog producing gases, and ash. Remember, 2.6 million metric tons of coal per year are needed for a 1000 million watt plant versus 30 metric tons for a nuclear plant of the same size. The 2.6 million metric tons turns into over 2 million tons of gases released to the air you breathe and into 500,000 metric tons of solid waste products. The nuclear power plant produces just 30 metric tons of spent fuel to produce the same energy. It is also an interesting and widely unknown fact that because uranium is commonly found in trace amounts in anything you dig out of the ground including coal, that in burning coal, more radiation is released from a coal-fired electrical generating plant than from a nuclear plant.

So, why so much concern about nuclear waste? There is a story in that and it takes a little telling. I am giving you just a taste in this summary. You will get the chance to learn more in later chapters. The short answer is that nuclear waste is pretty hazardous stuff. It is nothing you want to curl up next to, but

[2]And, yes, you could carry it safely in your hand, just please don't eat it. That would be about as smart as swallowing a lead sinker.

when seen in scale, the concerns are pretty much overrated nonsense. There is not that much of it, and it can be stored safely away from people and other living things.

It is important that we all understand these things, so we can see nuclear power in a fair light and allow it to have what should be a valuable role in helping provide the energy we need to allow humans to survive. Right now we are getting by with a mix of energy sources mainly based on fossil fuels (coal, oil, and natural gas). In the near term much of our energy will have to come from fossil fuels. However, in the long run this can't continue. While globally there are enough fossil fuels to last quite a while, the geographic concentration of those resources (e.g., oil in the Middle East) is going to cause increasing international tensions and conflicts. And the problem will get a lot worse as developing countries increase industrialization required to raise the standard of living of their populations. That increase requires more energy and at least for now more fossil fuels. And the uneven concentration of fossil fuels is only part of the problem. Burning fossil fuels produces, in addition to other pollutants, greenhouse gases that cause the atmospheric temperature to increase and can lead to severe impacts on climate and in turn all biological life on the planet.

If we don't kill each other trying to get fossil fuel resources away from each other, the greenhouse gases we produce burning the stuff will render the planet uninhabitable anyway. We need to have near-term options that allow us to reduce the use of fossil fuels. The single most significant option we have is nuclear energy. What I am all about here is trying to ensure nuclear energy gets a fair hearing and the chance to do its part toward saving the planet and the people on it.

So why is "everybody" so against nuclear power? Well strangely enough according to a number of polls about 60% of the American people are in favor of nuclear power, twice as many as oppose it. However, if you ask those same people if the majority of Americans favor nuclear energy, only 20% will say yes. So the majority of Americans favor nuclear energy, but a majority of those think that they are a minority. How can that be?

The answer is that nuclear energy makes a great bogeyman. It is new, powerful, invisible, and science fiction thrilling. The hype put out against nuclear power is unbalanced and is "spun" to give the most sensational story. Negatives are more sensational than positives. In fairness there are probably a number of anti-nuclear people who honestly believe they are saving the planet. I hope to be able to show you that they are in fact putting the future of all humanity at terrible risk. If you have the patience to read more, I would like to try to give you some balance. In advance, I give you fair warning, I am a fan of nuclear power. However, what I will give you are facts, real numbers, not "biased" words chosen for "spin." For example, I once read where a politician referred to what I knew was about 150 gallons of radioactive waste per year as "tremendous volumes." I am an engineer. I work with numbers. I will give you the numbers and some comparisons; you can supply the adjectives of your choice.

Lastly, why is it important that you think that nuclear-generated electricity is cool? It is because decisions are being made all around you every day that affect your life. Those making these decisions are influenced by what you think. The decisions are about what government programs to fund, what laws are going to be passed, how those laws are going to be interpreted and implemented, what projects various industries are going to take on, and what laws, projects, and all those things that are **not** going to happen. All these affect how your taxes are spent, how much your taxes are, how much you pay for essentially everything you buy, how clean the air you breathe is, how much acid rain happens hundreds of miles away from you, and what kind of a world is left for our children's grandchildren. And, let's be very clear here, these decisions are extremely important to the future of humanity. Business as usual, the mass reliance on fossil fuels (coal, oil, and natural gas), with an unrealistic belief that alternative sources of energy such as solar or wind power are going to save the day, is condemning the legacy of humanity to a new stone age or worse.

The people making the critical decisions, the politicians, government officials, bureaucrats, company presidents, industry executives, all care what you think.

They care because everyone on that list works for you. There is an old joke in business circles that if you ever get confused about who your boss is, just ask who provides your salary and who has the power to fire you. As voters, taxpayers, electrical ratepayers, and individual investors and buying public, you are the people who provide the money to all of those on that list above, and collectively you can cause every one of them to lose their job if they do not follow your wishes. And here is the big secret that most folks don't realize. Anyone on that list who is going to be successful knows that their job depends on what the Constitution so powerfully proclaims as "We the people . . ." think and do. Friends, that is you and me. The key here is that those decision makers only respond to those of "We the people" who make their views known. If you are silent, you get what others clamor for whether that is good for you or not.

At the end of the book, we will talk more about how much weight each of you has, if you will use it. Here I want to get you thinking about how much you can influence the actions of all those folks who want to do what you want them to do, if you help them know what it is that you actually want. If they keep thinking that only a minority of the American population supports nuclear-generated electricity because a small percentage of the population including the sensationalists in media and government see the nuclear bogeyman as a great cause to keep their names in the press and to sell media, then the rest of us are going to get what they want, not what we want.

So, what I am going to try to do is to show you why technology and specifically electrical energy is vital to humanity and why I think nuclear-generated electricity is a really cool way to make electricity. I want to get you as excited about it as I am and offer you a few easy things you can do to show all those decision makers what it is you would like them to do. The point is if you don't understand the importance of energy, or the issues and the choices being made about how energy is produced in this country and around the world, someone else will make those decisions for you. You and your children's grandchildren will live with what you do or don't do.

Let me give you a quick summary of what you will find in this book. The order of presentation is intended to give you a grounding step by step, but feel free to jump to a chapter that sounds more interesting or skip one for now until you see later how it would help explain things.

In Chapter 2, I will make a case for technology. As I said above, technology is essential to allow the current human population to exist. In this chapter, I will show you what I mean by that and hopefully change your perspective on our point in history.

In the third chapter, I will continue describing the importance of technology on human life, with a much sharper focus on today's world and on electricity in particular.

In the fourth chapter, we will detour into a semi-technical description of what nuclear energy is all about. There is a bit of science in it but nothing any of you can't handle. If you are to understand the issues and the rhetoric on nuclear energy, it will help to understand the basic science behind it.

In Chapter 5, I will provide a little more technical stuff about how nuclear reactors work and what some of the most common reactor types are like. This chapter builds on the basic nuclear science you will learn in Chapter 4 and makes you a real reactor physicist; well, almost.

Chapter 6 will see us struggle with the giant atomic mutant ants and all the other "bad stuff" we hear about things nuclear. We will look at what the real scary stuff is and put that into some context by comparisons to other more commonplace things.

In the seventh chapter, we will look at alternatives to nuclear power. It is important that we do so because in the near term nuclear power alone is not going to keep us going. I will explain a bit about our current mix of energy sources, what each is used for, how much of that resource we think there is, and what the impacts of using that source are. I will also provide a realistic assessment of "green alternative" energy sources.

In Chapters 8 and 9, respectively, I will give you an overview of nuclear power today and what future nuclear energy options might entail.

In Chapter 10, I will return to the issue of why all this is important to you and offer some suggestions to help get your views heard and get you engaged in the critical energy decisions that are being made.

2
THE CASE FOR TECHNOLOGY

The "Technology is Bad" Message

Why do I need to make a case for technology?[1] Because unless you have been on Mars for the last 30 years or so, you have been under a continuous assault—being told in many ways, by many sources that technology is bad. Somewhere, in the sixties and seventies we started to lose faith in big institutions. Government was mired in Vietnam and Watergate. Air and water pollution were apparent to anyone living in an industrial area. There were plenty of real examples. Air quality in major cities was awful and getting worse. Rivers and streams had become devoid of life. We had huge scandals like Love Canal and Times Beach. We learned that just about everything made by man caused cancer. Asbestos in our buildings and artificial sweeteners in our soft drinks were poised to kill us.

All this was "their" fault. It was caused by "their" technology. The "they" was somehow a conspiracy of this untrustable government and greedy,

[1] I need to make clear what I mean by "technology." It is increasingly common to use this term to mean information technology. While information technology is certainly included in my meaning, I am using the original definition that would include all human application of science and machines. This ranges from chipped flint spear points to nuclear medicine. It encompasses mechanized agriculture, transportation, buildings, and manufactured goods. My Webster's calls it, "the totality of the means employed to provide objects necessary for human sustenance and comfort."

unscrupulous industrialists. This message got hammered into us in hundreds of ways. We saw many films in which the villains were evil corporation tyrants. The cartoon Captain Planet showed children how the fight against these evil folks must be fought. The toxic avenger suffered for our sins. The message appeared in textbooks, billboards, popular songs, just about everywhere. In this message was the axiom that the fundamental evil was technology. Technology causes pollution. Technology destroys the natural world. Technology serves the evil government and corporations interests.

So we rushed to a new environmental paradigm. We needed to save Mother Nature. It was the cause of the day. It was a new kind of global patriotism. In truth we did some pretty good things. A current model car produces only a fraction of the smog causing pollution of a 1960s model. Air and water quality standards have had real impacts, and industrial pollution has been greatly reduced. We also did some really silly things along the way. At the peak of this fervor, Congress gave us the Delaney Amendment that stated any substance shown to cause cancer in laboratory animals must be banned from human food. It was a great law for the providers of laboratory rats, although tough on the rats. The problem is the science was terrible. It is very hard to connect animal cancers to humans. It is also very hard to show the effects of small doses. The result is that experimenters would give massive overdoses to rats, who then would show signs of cancer, and that product would then be banned. This may or may not have anything to do with the real potential for cancer in humans at realistic doses. In many cases, perfectly good products were denied access to the market. Industry and, as a result, consumers paid huge costs. The Delaney Amendment was retired a few years ago to wide acclaim within the scientific world.

One very important question about the environmental movement is why it was allowed to become such an encompassing force in our society. While there were legitimate reasons for improving our environmental housekeeping, the main reason why it took on such force is because we could. We could afford it. Even in the tough economic times of the 1970s, the modern industrial countries had sufficient excess capacity that we could add environmental controls to our industrial complex and still produce goods at affordable prices. Our technology-based system was strong enough that we could accept the additional burden. No one starved because we wanted clean air. No one had to go a winter without heat because we wanted to have clean water. A price was certainly paid and paid by all of us, but no real pain was inflicted. Therefore, we could all feel good about doing the right thing. The honest truth was it didn't hurt much.

However, this was not globally true then, and it is not globally true now. Developing nations sometimes face much more difficult choices. It is fine for citizens of industrial nations to call for saving the rainforest as they read about its destruction in an article nearly lost among the ad-filled newspaper, but to the struggling third world farmer who needs more land to feed his family, clearing an acre of forest might be the difference between life and death to one of his children. If pollution controls on a plant producing pesticides reduce production investments in a developing country, how many tons of crops go to the bugs instead of going to feed hungry people? In the United States we have a problem of agriculture over-production and depressed prices. Developing countries do not have that problem.

I once was working on a project with a major international company. Talking to one of their engineers I was told this interesting little story. It seems this engineer's job was to sell an acid recovery system for a factory to a group of Indian businessmen. He was having a very tough time of it. They kept asking him why they should buy it. He kept explaining that if they had the recovery system, the acid would not be discharged from the plant. They understood that, but why should they want that? The engineer was getting more and more frustrated. Clearly, not discharging acid was a good thing. Couldn't they see that? He even was citing the religious implications of the discharge to the holy Ganges but was just not getting any reception. Finally one of the Indian businessmen saw how to ask their question another way. Can we make money selling the recovered acid and would that be enough to pay for the recovery system? The environmental consequence was not in their calculation. They were interested only in the economics. It wasn't that they were "bad" people. It was that they could not afford to consider anything other than the economics, because that meant people were either fed or they went hungry.

The mindset that technology is bad, and the wish to protect us from **all** harm are far from gone. While the Environmental Protection Agency has been moving toward performance-based standards that really measure effect, there are still proscriptive standards in force that set limits far beyond any real impact. There are stories of clean-up criteria so strong that a child could eat cups of the dirt at an industrial site for years without impact. This is great for clean-up but way beyond any real need and only achievable at great cost, and that is cost passed on to consumers. It is sometimes ruefully observed that it is a shame that science can detect pollutants at levels far below harm. There is a perverse feature of human nature that if you can see the spot of filth, you should clean it up. This is a particular problem for nuclear energy related areas where our detection methods are very, very good, and we can detect extremely small amounts of radioactive material.

If you doubt the message "technology is bad" is still out there, I challenge you to watch one night of television or walk past one display of popular magazines and not see one or more subtle message touting that technology is bad. This has become a cornerstone of popular belief.

The message tells us that the opposite of technology is nature and that nature is a very good thing. The message is so pervasive and persuasive that now we all "know" that natural products are better than artificial products. Smallpox is a naturally occurring biological organism, but I don't want any in my breakfast cereal, please. Roses occur in nature, so does cow dung, so does arsenic. What does "natural" mean? My point is that it has turned into a biased word. It no longer means "occurring in nature." It means "good." We have come to an Orwellian state where technology equals bad, and nature equals good. Technology is in fact neither good nor bad. You can use a hammer to build a house or beat your brother Abel's head in. Pollution is not good, but building harvesters to bring you wheat or producing chemical fertilizers so that we can feed everyone is not bad.

I once saw a painting of a pioneer family. Their Conestoga wagon was cresting a hill looking down into a wide river valley. A broad river ran through the valley, and with one exception the entire scene was unbroken forest. The exception was a little clearing on the river with a few log cabin structures with a column of black smoke rising from the settlement. The buckskin clad leader mounted on a horse along side the wagon team was pointing with obvious excitement at the little village below. Seen through the eyes of a properly calibrated, environmentally aware modern person, my first response was, "How sad that those people were destroying that lovely natural valley." But the look of delight on the scout's face made me think a little more. At the time the painting was done, I am guessing about 1820, "nature" was not good. Nature was a great big terrible thing trying its best to kill you. That little village meant that this pioneer family probably would not starve or die of exposure this night where the night before that was significantly less certain. The smoke wasn't pollution. It was warmth, food, and life.

I will return to this idea in a bit, but first I want to take you on a detour. The purpose of the detour is to alter your perspective on humanity's place in time and space. You need this perspective shift to escape the programming we have all suffered. All you need do is sit back, relax, and loosen up your brain a bit to be ready to look at things a little differently. Let's step back a bit from the current world and all its complexities and see if we can't find a true perspective.

A Little Detour for Perspective

About 4.5 billion years ago, the earth was formed.

I told you we were going to step back a bit. Don't worry, it goes pretty fast, and I think you will enjoy the ride.

As I said, about 4.5 billion years ago, the earth was formed. (If you want to follow along, Table 2.1 gives you a time line summary of this history.) Quite surprisingly life appeared just about as soon as it possibly could. There is still some controversy as to just when, but it was somewhere between

TABLE 2.1 A brief history of earth.

Event	Years Before Present
Earth is formed	4.5 billion
Comet/meteor rain ends	3.8 billion
Earliest indication of life	3.8 billion
Earliest prokaryote fossils	3.5 billion
Continental land masses begin to form	2.5 billion
Snowball event	2.5 billion
First eukaryote life	2.5–2 billion
Oxygen from plants starts to go to seas and atmosphere	1.8 billion
First land plants	1.2 billion–800 million
First animals	1 billion to 600 million (there is significant controversy on the timing)
Snowball event	800–600 million
Continental drift	600–500 million (happens over 10–15 million years)
Cambrian explosion	550–500 million
First land animals	500–425 million
Carboniferous period	345–280 million
Permo-Triassic extinction	250 million
Mesozoic era	225–65 million
K-T extinction	65 million
"Recent" ice ages	40 million–present
First of genus Homo	2.6 million
Genus Homo invents fire	1.5 million
First Homo sapiens	400 thousand
First modern humans	100 thousand

3.8 and 3.5 billion years ago. The planet was pretty much a mess at this time. The heavy bombardment of comets and meteors was just finishing up. There was no oxygen in the atmosphere and little dry land. The life that sprang up was pretty simple stuff. These were the prokaryotes that were the forerunners of bacteria among other things. For the next billion years, things were pretty boring. The prokaryotes adjusted their internal chemistry to adapt to new conditions rather than change their outward form. It worked pretty well for them because some modern forms have not changed much for billions of years.

About 2.5 billion years ago the continental landmasses started to form. They had not settled into their current locations. That came later. Somewhere, between that time and 2 billion years ago, the eukaryotes appeared. This is the branch of life that led to all plants and animals. While eukaryotes were much more complex than prokaryotes, they were still pretty simple stuff at this stage. Microscopic in size, they had very simple structures. Some of these little guys were however capable of photosynthesis, which you will remember from your high school biology as the process that plants use to capture the energy from sunlight. In that process the "plants" breathe carbon dioxide and exhale oxygen. Because there was a lot of carbon dioxide in the atmosphere at that time, these plants had it pretty good. In fact they did very, very well and in the process produced a lot of oxygen. Until about 1.8 billion years ago this was fine because the exhaled oxygen was being bound up in free iron. Oxygen is a very reactive chemical and likes to combine with lots of things. Sand is silicon oxide, water is hydrogen oxide, and most corrosion is some form of oxide. Good old iron rust is iron oxide and, for hundreds of millions of years, the oxygen released by the primitive plants went into converting free iron into rusty iron oxide ore. But by 1.8 billion years ago, the iron was essentially used up. At that point, free oxygen started to appear in the water and in the atmosphere. While great news for what would become oxygen-breathing animals, this was death for a lot of eukaryote species. They had evolved to live in a carbon dioxide rich atmosphere and were poisoned by their waste product of oxygen.

For the next billion or so years, things get boring again, at least from the perspective of the fossil record. Somewhere between 1.2 billion to 800 million years ago simple plants started to be found on land. However, up to about 600 million years ago the majority of life was confined to vast planktonic algae filled seas. There is a lot of controversy about when the first animals appeared. Genetic projections suggest as far back as one billion years, but the geological record cannot support anything beyond 600 million years. So, during this time span, there was plankton soup in the oceans, few plants on land, simple animals toward the end of the period, and no organisms with any developed structure. Then something pretty amazing happened. The fossil record went from microscopic dots to fully developed multi-parted animals.[2] This was the Cambrian explosion.

In a period of only about 50 million years, evolution went nuts and spit out very complex animals with legs, arms, feet, claws, eyes, and shells. It was not just the complexity of individual animal types, it was the near instantaneous invention of whole groups of animals. The largest division biologists use to classify life is the phylum. In animals this can be thought of as body plans. The phylum Chordata, to which we belong, are animals that use a central stiffening rod to hang the rest of the animal on. We use our spines. The phylum Arthropoda includes insects and spiders. During the Cambrian explosion up to 100 phyla burst on the scene. About 40 made it through those exciting times, and **none** have been added in the 500 million years since then.

With that great start, nature got moving. Land animals first appeared between 500 and 425 million years ago. The Carboniferous period is marked as that period between 345 and 280 million years ago. This is significant in that the now more complex land and sea plants formed vast jungles and living oceans that, lucky for us, fossilized and turned into coal, petroleum, and natural gas. Other geological spans contributed to our current fossil fuel reserves, but they began during this period. The Mesozoic era[3] saw the rise of the giant reptiles, the first flowering plants, a vast evolution of insects, and the first birds and mammals. The Mesozoic era ended with the famous K-T extinction event in which a meteor or comet smashed into the earth.

The last 40 million years of the planet's history is a bit odd in that we went through an extended series of ice ages. While there have been retreats, called inter-glacials, lasting from 10,000 to 15,000 years, things have been cooler than "normal." While the earth's climate is, from a solar system's perspective, very constant, 95% of the earth's history has seen temperatures warmer than what we have now. During those normal warm times, humid tropical forests go all the way into the arctic regions. We are actually still in an ice age albeit at the warmer end of one.

Getting a little more focused here, let's start looking for humanity. The genus Homo shows up about 2.6 million years ago. You might not recognize one of these guys as your uncle, but they used tools and moved with bipedal locomotion. They figured out how to make fire about 1.5 million years ago,

---

[2]There is some controversy about this. Some suggested that complex animals evolved over hundreds of millions of years. They don't show up in the fossil record because they didn't have shells. When the oceans reached the right state of dissolved calcium, shells were possible and fossils were possible. However, others point to the "soft" fossils of simple worms as impressions on mud and ask where the other "complex" animal soft fossils are.

[3]An "era" is a collection of "periods." The history of the earth is divided into four eras—Precambrian 4.5 billion to 600 million, Paleozoic 600 million to 225 million, Mesozoic 225 million to 65 million, and Cenozoic 65 million to the present.

although there is some controversy as to how much use was made of fire until relatively recent times. The species Homo sapiens shows up about 400 thousand years ago. Modern humans, folks that you could put in a three-piece suit and walk down a Manhattan street and no one would blink, arrived just 100 thousand years ago.

We will take up the time line again in a bit, but let's do a check here. What does all this have to do with nuclear power or energy or much of anything?

Remember, I am trying to adjust your perspective. With this super-compressed history of the earth, you ought to get the picture that nature and life are not static things. While there are very long spans where things just coast along, there are also periods where very dramatic things happen very fast. I did not mention them above, but in the table you will find two "snowball events," one at 2.5 billion years ago and one from 800 million to 600 million years ago. These are super ice-age events. The oceans froze almost to the equator. O.K., so what?

The problem with ice ages is that there is a lot of ice. But you see the thing about ice is that it reflects the sun's heat back into space. That causes the planet to be cooler, which causes more ice to form, which reflects more of the sun's heat, which cools the planet more, which causes more ice, which...., you get the point. So why didn't we freeze? The answer is that during these two events, we nearly did. The only thing that saved us was the production of the very greenhouse gases that everyone is so concerned about now. With ice covering up most of the land and oceans, the normal, natural means of collecting carbon dioxide, dissolving it in ocean water and binding with chemicals in rocks, was lost. Natural processes like volcanoes continued to produce carbon dioxide. With the collecting mechanisms curtailed and the production side working business as usual, the atmospheric fraction of carbon dioxide built up.

Greenhouse gases, like carbon dioxide, have the wonderful property of letting the energy of the sun's light through, but will not let the re-radiated earth's heat back out. That is because the frequency of light is proportional to the temperature of the radiating surface. The sun is very hot; hence it has a higher frequency of light. The earth is heated but to a much lower temperature than the sun, so its re-radiated heat is a much lower frequency of light. Carbon dioxide is transparent to the sun's frequency of light, but absorbs light at the frequency of earth's re-radiated heat. Glass has this same property, which is why greenhouses work, and where the name greenhouse gases comes from.

So, what happened is carbon dioxide buildup trapped enough heat to allow the planet to thaw out. In the case of the event 800–600 million years ago, it set the stage for the Cambrian explosion, which set off the rush to life forms as we know them now. Think of it, 4 billion years and all that existed were tiny dots of life and perhaps some simple worms. Then a big freeze followed by a thaw, and, bang, all the phyla were created and we were off and running. But it was not a smooth, gradual climb to more advanced creatures.

The Permo-Triassic extinction 250 million years ago killed 80–90% of all the species on the planet. The famous K-T extinction that killed the dinosaurs only killed 50% of all the species. Here is the scary part. We are not sure what caused the Permo-Triassic extinction. Our best guess is that there was a sudden and large degassing of carbon dioxide from ocean sediments combined with unusually large volcanic events. The resulting flood of carbon dioxide poisoned much of the sea life and caused severe global warming lasting 10,000 to 100,000 years. There have been recent suggestions that the Permo-Triassic extinction was caused by a meteor or comet impact. We just don't know. The point is we are not the only ones who can trash the neighborhood. The planet can do it all by itself. If that fails, the solar system can throw monster rocks at us. In fact, there are up to 15 mass extinction events in the fossil record, although the Permo-Triassic seems to be the worst.

Here is another interesting bit of information. In his book *Extinction—Bad Genes or Bad Luck?* David Raup looks at the geological record of ocean reefs. Reefs are good things to study because, as they are built up by the activity of a large group of interacting life forms and they leave good fossils, they give a good record of what ocean life is doing at any given time. In his study he finds that active reef building has gone on for about half the time since the Cambrian explosion. He notes six different periods in which reef building was completely stopped. Each of these followed a major extinction event. Here is the interesting part. After each cessation, reef building started up again. However, each time the species composition is different from the pre-extinction communities. Nature tries again, but not necessarily in just the same way.

The cosmos is a tough place. It is wild and capricious. In their excellent book *Rare Earth—Why Complex Life Is Uncommon in the Universe,* scientists Peter Ward and Don Brownlee make the compelling point that, while there may be billions and billions of stars out there, the set of very unique circumstances to allow the development of complex animals might be extremely rare. Life, they argue, might be pretty easy. On the earth, life appeared just about as soon as liquid water was stable; that is, the planet cooled enough and very big things from the sky stopped slamming into us so often. But complex life took a very long time to happen. If comets or meteors of the K-T extinction event size or larger were more common, complex life would get wiped out every time it tried to start. Having the gas giant planet Jupiter out there sucking up most of the rocks and comets has saved our bacon. But how common is that in the universe? How common is it to get the right mix of chemicals that allows the balancing act of carbon dioxide to prevent either runaway snowball events or runaway greenhouse events. Our sister planet Venus lost to the greenhouse, and lead will melt on its surface.

And where did we leave the narrative? With modern humans showing up 100 thousand years ago. How odd was that? It would seem, very odd.

In his book *On Giants' Shoulders*, Melvyn Bragg recounts a discussion with Richard Dawkins, an eminent Oxford professor and a specialist in evolution.

> "I asked Richard Dawkins whether, if we razed the planet clean and started again, *Homo sapiens* would inevitably turn up or would it be purely a matter of luck.
>
> "*Homo sapiens* would not turn up, that is just too implausible. If you rephrase the question, would a brainy animal turn up? Would language turn up? Would a bipedal brainy animal turn up? Then it is less uncertain. However, I would say that although bipedalism has turned up—dinosaurs, birds, kangaroos—extreme braininess appears really only to have arisen once; language has certainly only arisen once, unlike, say, the eye which has arisen forty or fifty times independently in the animal kingdom. It looks as though eyes evolve at the drop of a hat, contrary to many of Darwin's critics who thought they were very difficult things to evolve. But brainy things and, especially language, in the face of all this time, only evolved once. So I should have thought it is highly unlikely that it would evolve again."[4]

The thing is that evolution was having a tough time. These persistent ice ages over the last 40 million years with the odd short warm spells coming in-between made it hard to evolve successful creatures, as conditions kept changing. The conditions were right to allow something different to have a go at it. What got spit out of the random mutations was a smart monkey that could use its hands. These genus Homo creatures could use tools, figured out fire, and therefore had a pretty good chance of dealing with changing conditions. Once started down the road to braininess, evolution continued producing a variety of hominids along the way to modern humans. Now here is the really weird part. While the original members of genus Homo were very smart monkeys, the modern humans of 100 thousand years ago were just as smart as you or me. They were not as knowledgeable, not as technologically advanced, but they were just as brainy. What evolution needed for success was something smart enough to figure out which berries were not poisonous, how to find them again next season, how to tie chipped flint to a stick and poke a deer with it, and how to work together to go after big game. What it got was something smart enough to do calculus, invent quantum mechanics, build airplanes, and take themselves to the moon and back. That appears to be a serious overshoot. For our ancestors rambling around out there on the African savanna, that much braininess was really wasted. But we got it.

---

[4]*On Giants' Shoulders*, Melvyn Bragg, John Wiley & Sons, 1998, pp. 176-177.

Here is another interesting note from our history. We, that is to say modern humans, were not the only attempt by nature to work with smart monkeys. In their book *Extinct Humans,* authors Ian Tattersall and Jeffrey Schwartz explore the fossil record and uncover as many as 17 different hominid species. Many of which co-existed. Neanderthals shared Europe with "modern" humans as late as 30,000 years ago. From DNA research it has been extrapolated that about 100,000 years ago things got very tough and the total number of Homo sapiens was down to about 10,000 individuals. The path to "us" was not a long continuous climb. It was a series of experiments, all of which failed other than us. But for a little different luck, perhaps, we would have failed as well. You see Neanderthals were pretty successful. Tattersall and Schwartz characterize them as highly adaptable and flexible whose success in a range of habitats was, in their words, "partly, at least, based on technology." If the Neanderthals seem unsophisticated to us, it might be humbling to recognize they existed, if not flourished, for twice as long, perhaps three times as long as we have so far.

Although modern humans have been around since about 100,000 years ago, something happened about 50,000 years ago.[5] Tattersall and Schwartz speculate it might be the development of language. Whatever it was, it turned loose all the braininess the Homo sapiens had been lugging around for the last 50,000 years. These guys started using much more innovative and varied stone tools. They invented sewing, used bone and antlers, had art and music, and just generally had better ways of doing everything. The Neanderthals had none of these things. It took about 15,000 years, and there is some question of just how hot the competition was, but the Neanderthals are gone and we are still here. It appears that a similar competition happened in Asia with modern humans winning out over the indigenous Home erectus of Java Man fame. Tattersall and Schwartz end their book with the following observation.

> "There was nothing inevitable about how we got to where we are today, and in the truly unique and important aspects of our being we are not simply the result of steady improvements in a linear progression. The emergence of behaviorally modern H. sapiens was emphatically not just an extrapolation of earlier trends; and cognitively we are not merely a refinement of what went before. For reasons, and through mechanisms, that we still don't fully understand, something truly unprecedented happened in one of the terminal branches of our bushy family tree, and we are still learning how to live with the consequences."

---

[5]Please recognize there is much speculation about these dates. You can find other estimates that vary some thousands of years. However, this should give you a feel for the time spans involved.

The point to all this is that we have a very unique situation on earth with its complex animal life. The hold any complex life has is very precarious. The fossil record indicates that the average life span of a species is 5 million years. Some longer, some shorter. The most famous dinosaur of them all Tyrannosaurus rex, the king of the dinosaurs, got two million years. Rise to glory and two million years later, splat, a big rock from space takes you out. Evolution, nature, and the cosmos are not patient. Try your genes out. If you cannot be successful with that mix, pass your organic chemicals on to something else and nature will try that out instead. There is no ultimate goal. There are lots of other ways of being alive rather than being human. Insects or bacteria could give us a good run. In *Journey of the Ants*, Bert Holldobler and Edward Wilson estimate that in terms of shear weight, ants just about match that of humans. In individual numbers, they estimate the ants have us outnumbered 16 million to one, and there are a lot more insects than just ants.

We are not the chosen, **unless we choose us**, more on that later.

I will return now to the time line, focusing on humanity, that very brainy if improbable monkey. In that I hope to show you how our brainy ancestors used those brains to capture a hold on our spot on the planet and how important technology is for us to hold on to it within our little window in time in this very tough neighborhood.

We left the story with modern humans showing up 100 thousand years ago. Once again, we get a span of coasting, without much going on; that is, not much changes for a while. About 50,000 years ago there appears a technological leap with a series of improved stone tools. Not a big deal, just better blades and more specialized implements. However, this was accompanied with a rapid expansion in human populations and locations. Humans spilled out of Africa, across Asia, and into Australia. Within another 10,000 years they had spread to Europe and were living next door to the Neanderthals who were already there. There are indications that these modern humans organized large hunting parties and had some semi-permanent communities. In puzzling the question of why the modern humans succeeded over the Neanderthals, one theory focuses on the ability to cooperate in groups. This theory looks at the skeletons of Neanderthals and concludes their vocal apparatus was inferior to modern humans. That is, they could not make as many and as varied sounds as we can. This, the theory goes, gave us an advantage in communication. We could talk to one another, which allowed us to plan and work together better than the Neanderthals. That means we could figure out better tools by joining our skills and ideas better, and we could plan and execute the hunt better. All because we could talk to each other better.

Another leap in time. About 10,000 years ago another inter-glacial started, and the ice sheets over North America and Europe retreated. One might assume this was good news for the humans, but actually it introduced new challenges. With wide scale glaciation we also had large herds of big relatively slow mammals, mammoths, bison, and the like. With the warming, the windswept grasslands were replaced with forests populated with the more difficult to hunt elk and moose. Human hunters rose to this challenge with new technologies of snares, traps, and the bow.

About this time human technology made its biggest leap, perhaps only exceeded by that of the development of language. This was the domestication of plants and animals. The fertile crescent of the Tigris and Euphrates Rivers is often cited as the birthplace of this revolution. It appears that this technological leap arose independently in the Americas, Southeast Asia, China, and Africa in times ranging from 8500 B.C. in the Fertile Crescent to 2500 B.C. in the Eastern United States. The importance of the domestication of plants and animals was not just that it gave humans a more reliable supply of food. It also allowed some fraction of the population to be assigned to tasks other than the procurement of dinner. While in any subsistence agriculture society, even today, the majority of the population are farmers, not everyone must be committed to getting food. Also they tended to stay in one place. This allowed them to collect stuff. Anyone who has ever cleaned out the hall closet knows humans are good at collecting stuff. But the key here is when humans stopped following the game herds, they could have more stuff than they could carry. This meant they could make more and bigger tools to help them make other things. This leads to pottery, improved weaving, and metallurgy. This improved technology allowed them, and now us, to have a better quality of life.

With a society in which not everyone has to be getting food, you can organize and conduct joint projects. This can be communal farming, clearing fields, building granaries, building mills for flour, useful things that are beyond the resources of one individual or family. This leads to villages, towns, organized societies, and governments.

Governments lead then to leaders, some of which will have big plans. Some of those were extremely useful such as Egyptian irrigation projects, Grecian public buildings, and Roman water and sewage systems. Some were just impressive public works projects, such as the pyramids or the Coliseum in Rome. While we can marvel at these works of antiquity, there is also a dark side to all of them. When we look at these great construction projects, we have to recognize that the technology that built them required energy. The only form of energy available was human muscle. And the source of that human muscle was usually conscripted labor, the labor of slaves. We can admire the great works of the Greeks and Romans in advancing the human condition, but we must also recognize that it took a great deal of human energy in the form of slave labor to support those societies.

The next great leap for humans was the development of machine sources of energy to replace human muscle power. While some animal energy was used to replace human muscle power, the big leap didn't happen until we harnessed thermal energy. Wind and water powered mills provided the technological bases for this leap, but by themselves were insufficient to power the full rush of the industrial revolution. Once we figured out how to turn the heat of burning coal into steam and steam into rotary motion of machines, we were off and running.

With the mechanization of farming, we vastly increased the productivity of each farmer. With the development of nitrogen-based fertilizers and pesticides, we greatly increased the productivity of each acre of farm. This freed more farmers to become factory workers, scientists, clerks, and all the diverse mix of occupations that populate the developed nations.

At this point it is instructive to take a look at some differences between nations in the world today. By comparing developed and developing countries, we can see where the developed countries have come from and where the developing countries would like to go. Table 2.2 takes a look at a set of nations.

What is most striking about these numbers is the strong correlation between the fraction of the labor force in agriculture and the per capita Gross

TABLE 2.2 Developed vs. developing nations.[6]

Nation	Per Capita GNP (US$/person)	% Labor Force in Agriculture	Daily Food Consumption (kcal/person)
Burundi	$220	91.5%	2283
Uganda	$250	81.4%	2113
Bangladesh	$180	69.2%	1963
China	$350	68.2%	2622
Mexico	$2010	30.6%	3118
Costa Rica	$1780	24.5%	2757
Canada	$19,030	3.5%	3400
United States	$20,910	2.4%	3595

GNP = Gross National Product.

[6]These values and those of Table 2.3 come from "Food, Land, Population and the U.S. Economy" by David Pinentel and Mrio Giampietro, Carrying Capacity Network, 2000 P Street, NW Suite 240, Washington, D.C., November 21, 1994.

National Product (GNP). While money alone should not be directly equated with standard of living or quality of life, GNP per person does give one a feel for what life must be like for the average person. At $180 per year, life is probably pretty tough for someone in Bangladesh.

If dollars seems too venal a measure for quality of life, then look at the last column. That is the number of calories in the average diet of people in these countries. You might say, gee, they don't vary all that much, what is the big deal? The thing is that a human being needs about 2000 kcal per day to be healthy and functional. Even at 2283 kcal per day, the average person in Burundi probably spends some of the time hungry and their health suffers from that. Recognize also that this is the average. This means many people will be below the average within any given country. The correlations of food consumption and fraction of labor in agriculture are even more telling on quality of life. The fraction of the work force committed to agriculture is probably a pretty good measure for the degree technology is integrated into the social structure.

My point here is application of technology in the society is directly related to quality of life. It is not just whether you have your own personal stereo system or not. It is whether your job is mucking about in rice paddies or in an air-conditioned office. It is whether you and your children go to bed hungry or you join a health club to try to work off those excess calories.

Table 2.3 gives us a deeper look into the nature of agriculture in these sample countries.

O.K., there are a lot of numbers, but let's look at the story these numbers tell. For a surprise, look at the numbers for the amount of fertilizer used in Costa Rica and China and compare those to the U.S. and Canada. Are you as

TABLE 2.3 Agriculture intensities.

Country	Arable Land (hectare/capital)	Arable Land (hectare/farmer)	Fertilizer (kilograms nitrogen/ hectare)	Output (million kcal/ hectare)	Output (kcal/farmer)
Burundi	0.20	0.44	1.9	4.43	1.95
Uganda	0.28	0.52	0.05	3.17	1.64
Bangladesh	0.08	0.4	69.7	7.50	2.98
China	0.08	0.2	210	13.02	2.65
Mexico	0.27	2.50	55.2	4.52	11.3
Costa Rica	0.1	1.14	217	10.35	11.75
Canada	1.75	99.5	25.8	3.36	335
United States	0.76	64	53.3	6.03	386

surprised as I was that those nations use much more than we do? After China, the next biggest user of fertilizer per hectare[7] is Bangladesh. Looking at the available arable land, either as per capita or per farmer, shows you why this is. Those countries have a relatively small amount of farmable land available per person, so they farm that land intensely; e.g., they put a lot of fertilizer on it. The result is seen in the output per hectare in which China gets over twice as much food value out per hectare as we do in the United States. Costa Rica is nearly as productive, and Bangladesh also exceeds us. But the most remarkable variation in the numbers is the output per farmer. There we see that despite only moderate amounts of fertilizer used, U.S. and Canadian farmers are much more productive than their counterparts elsewhere. The U.S. farmer produces 145 times the food value of a Chinese farmer. That is the application of technology. Large tractors, harvesters, and other farm machinery allow 2.4% of our labor force to not only overfeed us but also to provide a large supply of food for export to foreign nations.

Human Population and Technology

I will come back to the interplay of technology in the current world, but let's go back to history for just a bit and look at some population numbers.

Joel E. Cohen wrote a book with the wonderful title *How Many People Can the Earth Support?* With that eye-catching title he then goes on in 400 pages to explain why the answer isn't all that easy to come up with. We will come back to the question shortly, but first I want to share with you the historical information Dr. Cohen compiled. He looked at several historical population estimates and tabulated them in an appendix. One must accept that these are estimates at best, but especially as we get to later historical times, the various estimates start to agree fairly closely. Taking a rough average of the various estimates, I came up with the numbers in Table 2.4.

TABLE 2.4

Estimates of historical human populations.

Date	Estimated Population (millions)
100,000 B.C.	2.6
25,000 B.C.	3.3
10,000 B.C.	4
5000 B.C.	5
3000 B.C.	14
1000 B.C.	65
1 A.D.	240
1000	265
1300	360
1500	425
1700	610
1800	900
1900	1625
1925	2000
1950	2516
1970	3698
1990	5292
2000	6000

[7]A hectare is the metric measure for large areas. It is 10,000 square meters, or equal to a square 100 meters on a side. If like me, you are hopelessly stuck in English units, one hectare is 2.471 acres.

The numbers alone don't really give you the flare for what is happening here. Let me give you those numbers as a series of graphs in Figures 2.1 through 2.4. First, let's see the whole thing, the population as a function of time starting at 100,000 B.C.

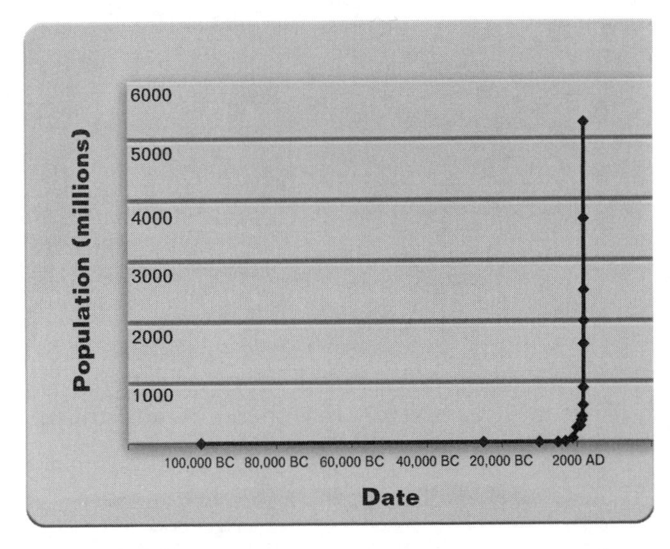

Figure 2.1. Estimated human population 100,000 B.C. to 2000 A.D.

Well, that wasn't too useful was it? Let's ignore the first 90% of the time period and just look at 10,000 B.C. to the present.

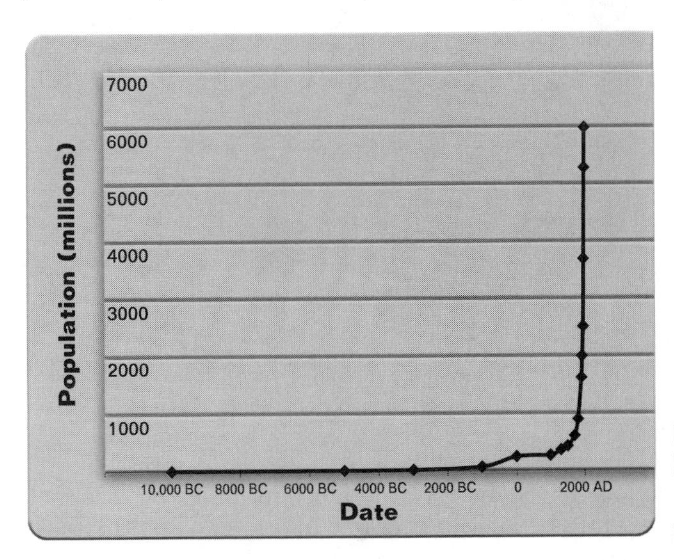

Figure 2.2. Estimated human population 10,000 B.C. to 2000 A.D.

That isn't much better. It still looks like a straight line across the bottom of the scale, then zip, straight up. Let's pitch out all the B.C. time and look at just the last 2000 years, which is just the last 2% of the history of modern humans.

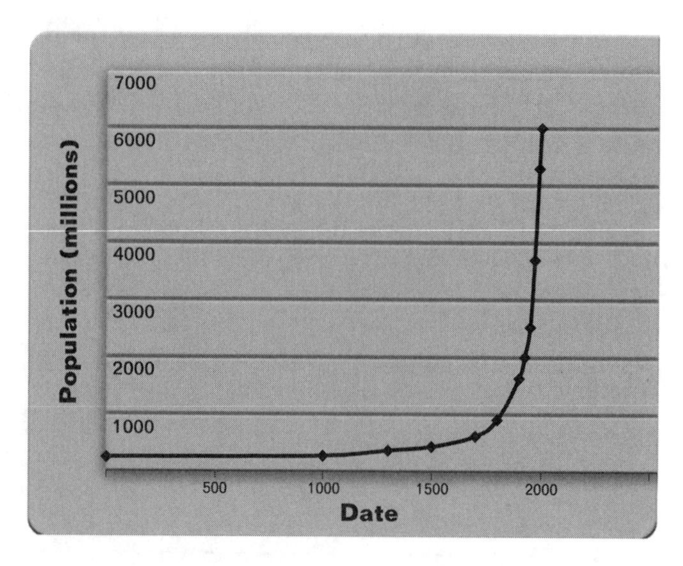

Figure 2.3. Estimated human population 1 A.D. to 2000 A.D.

This is better, but still it is almost completely flat, then zips straight up. Let's try just the last 300 years, 0.3% of our history.

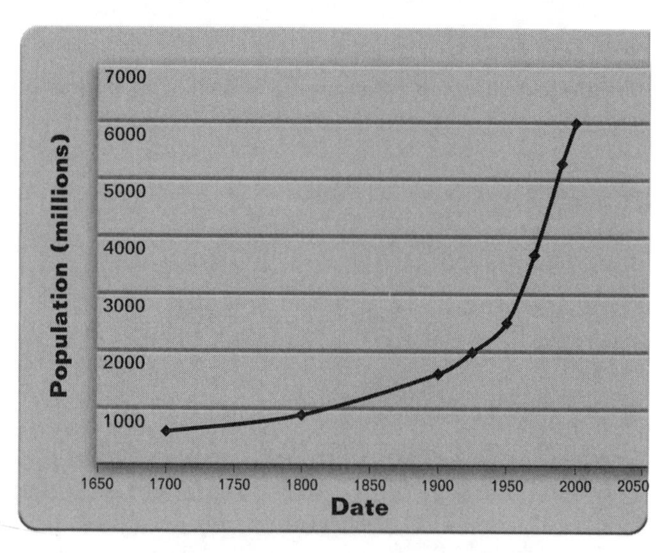

Figure 2.4. Estimated human population 1700 A.D. to 2000 A.D.

Finally, we have some shape you can see.

My point is that if you look at the history of modern humans, you see that except for the very, very last little bit of time, there weren't many of us. Out of all of the 100,000 years of modern humans, only the last 100 years sees the fantastic growth in our numbers. It is not much of a leap to see that it is technology that has allowed this growth. We didn't all of a sudden start to have more children. What happened is we got technology to push back all those things good old kindly Mother Nature has been trying to kill us with. Things like cold, damp, excessive heat, famine, and disease.

Cohen suggests four periods for changes in the growth rate of human populations. These are shown in Table 2.5.

TABLE 2.5 Evolutions in human population growth.

Evolution	Date	Doubling Time Before (years)	Doubling Time After (years)
Local agriculture	8000 B.C.	40,000–300,000	1400–3000
Global agriculture	1750 A.D.	750–1800	100–130
Public health	1950	87	36
Fertility	1970	34	40

Cohen uses the concept of doubling time to show the rate of population growth. This is simply the time it took for the population to get twice as big.

Before the domestication of plants and animals, it took a very long time, up to 300,000 years, to double the population. After that technological leap, the doubling time shrank to 1400 to 3000 years, which was perhaps as much as 100 times greater population growth.

The global agriculture he refers to is the introduction of crops found in the New World (Americans) to Europe and eventually to Africa and Asia. The big actors here were maize (corn), potatoes, and sweet potatoes. These crops could produce 130 to 180% of the calorie value per hectare of the grains they replaced. They also expanded what was considered arable land by growing these new crops in places that the traditional grains and rice would not grow. The result was another factor of 10 increase in the rate of population growth.

It is important to recognize that the technological leap here wasn't just in agriculture although that is where the payoff was felt. You can't get new crops from the New World unless you had traffic with the New World. That didn't happen until you got a technological leap in navigation, shipbuilding, and all the industries that went along with them.

The next evolution in Cohen's list is public health. It would seem to me that there should be an intermediate step having to do with the Industrial Revolution that gave us the concept of a developed country and all the quality of life benefits that comes with that. However, that may have a lot more to do with how well you live and less with how long you live. Clearly, the numbers support Cohen in showing that something very remarkable happened to population growth around 1950. What Cohen cites as the thing that happened is that improvements in administration and transportation allowed modern medicine, hygiene, and agriculture to reach the developing countries preventing the epidemics and famines that previously, through massive deaths, have been holding back their numbers.

The introduction of antibiotics developed during the Second World War, the widespread use of inexpensive insecticides to control insect-carried diseases; vaccines against smallpox, measles, and diphtheria; drugs to treat tuberculosis and malaria; and improvement in clean water supplies gave rise to an astonishing reduction in the death rate. The overall population growth rate jumped by a factor of 3 to 4. In developing countries such as India, it jumped by a factor of 6.

For some perspective let me make one small aside here. Environmentalists correctly exposed the hazards of the pesticide DDT. The really bad thing about DDT is that it doesn't break down rapidly and, therefore, builds up in plants and animals over time. This was killing a lot of birds as well as bugs. However, it was, and is still, a very cheap way to kill mosquitoes that carry malaria. So, although DDT killed birds, it probably allowed many millions of people to live. If you look at the grandchildren who wouldn't exist if their grandparents had died of malaria, it might well be hundreds of millions.

In the April 2001 issue of *Scientific American*, their "50, 100, 150 Years Ago" page had the following reprint from April 1951:

> DDT Shortage—"The World Health Organization last month reported a developing shortage of DDT so serious that it threatens the breakdown of the campaign against insect-borne disease, which since the end of the war has wiped out malaria in many parts of the world. The shortage is due to increasing use of the insecticide by farmers and by the armed forces for the defense program, and shortages of the ingredients. Roberto Caceres Bustamente, Under Sectary of Public Health in El Salvador, declared: 'DDT is for us a problem of living or dying. In a population of 2,500,000 there are more than 200,000 cases of malaria.'"

There was a National Public Radio item on December 4, 2000, in which it was explained that the chief environmental group that had convinced the South African government to ban DDT was completely reversing that position and endorsing the use of DDT. The more expensive substitute had proved to be ineffective in controlling mosquito-carried malaria and a significant increase in human fatalities was occurring.

On April 19, 2001, President Bush announced that the United States was signing an international treaty to ban a number of environmentally hazardous chemicals including DDT; however, he explicitly stated that we would assist developing countries in obtaining effective, affordable pesticides.

The point is that one has to consider all the impacts of technology, both good and bad before adopting, rejecting, or modifying that technology.

While we are on the subject of haves and have-nots, it would be instructive to get a sense of where that leaves us today to start thinking about the gap between developed and developing nations. Cohen points out that in about 1990, there were 800 million people with average annual incomes of $22,000 and there were 800 million people who chronically failed to get enough calories to grow normally and be active.

Now, let's come back to our discussion about the rapid population increase in the middle 1900s as a result of the step change improvement in public health. Looking at Figure 2.4 you can really see this effect. Between 1900 and 1950 the rate of growth and hence the total population really takes off. The thing that happened, once again, is that technology pushed back on Mother Nature and stopped her from killing so many of us. Our rate of reproduction didn't change, so if deaths are less and births are the same, you get more people, a lot more people.

If you look at the life expectancy rates in the United States shown in Table 2.6, you will see a similar story, if shifted a bit forward in time.

TABLE 2.6

U.S. life expectancy.[8]

Year	Years of Expected Life at Birth
1900	47.3
1910	50
1920	54.1
1930	59.7
1940	62.9
1950	68.2
1960	69.7
1970	70.8
1980	73.7
1990	75.4

This shows that in the United States, a very similar improvement in human health happened, albeit we can see the impact coming in a little earlier than 1950. The numbers are pretty dramatic. There is a 44% improvement between 1900 and 1950. That is nearly half again as long a life expectancy.

This is personal to me. My father was born in Colorado in 1916. He had six siblings. Of these, three died in childhood. Of the four children that made it to adulthood, two died in early adulthood and only my father and my aunt lived long enough to have children. All the deaths were due to disease. When I look back on my childhood, I can't recall a single classmate dying of disease, not one. In my father's case, the odds were 2 in 7 for success. But in my generation, my little sample, I can't count one out of the hundreds of kids I knew being taken by disease. I recall two accidental deaths, but no one lost to disease. If my father and his siblings were born 30 years later, would they have had the benefit of technology that would have allowed them to live?

[8]World Almanac Books, *1960 World Almanac and Book of Facts 2001*, New York: Press Pub. Co.

Where does energy play in all this change in the human condition? Well, it is pretty important. As we have seen, the ancestors of modern humans already had fire, but over the history of humans, most work was done by human muscle. We didn't figure out how to turn fire into useful work until very recently. Cohen provides the estimate that one person can, for a short period, produce about 800 watts of power. Hold on to that number for a bit.

Let's think a little about the energy available to humans. Animals helped some after their domestication some time after 8000 B.C. Water and windmills were a big boost but didn't see wide use until the Middle Ages. The really big boost comes in during 1800s as the English were running out of forests, they discovered large deposits of coal and that clever Scot James Watt developed practical steam engines. Cohen provides data, given in Table 2.7, to show what happened in terms of human use of energy after that.

TABLE 2.7 Human energy use 1860 to 1987.

Year	World Population (millions)	Inanimate Energy From All Sources (MW-hr per year)	Energy Per Person (MW-hr per year per person)
1860	1200	1078	0.9
1900	1625	6089	3.7
1950	2500	20,556	8.2
1980	4488	68,354	15.2
1987	5000	77,229	15.5

MW-hr = megawatt-hours.

If we extrapolate the energy use to the year 2000 when the population reached 6 billion, we get 16.3 megawatt-hours per year per person.

Now recall that even as we admired the achievements of the Greeks and Romans we recognized that the dark side of those achievements was that they were achieved on the backs of slaves. If one human working flat out can produce 800 watts, that would be equivalent to $800 \times 24 \times 365 = 7$ megawatt-hours per year if they could keep that rate up indefinitely. Clearly, no one could. Cohen estimates one full-time slave is effectively one megawatt-hour per year. So in 1987, each of us had effectively 15.5 full-time slaves in terms of inanimate energy working to provide our wants and needs. In 2000 this might have increased to just over 16.

Those are global numbers. That is, this applies to every one of the 6 billion people on the planet. The 6 billion had energy working for them equal to the work of 96 billion people.

To stress how much that energy use is concentrated in the developed countries, Cohen offers that in low-income economics other than China and India, energy consumption per person is 10% of the world average. While in the developed countries the per person energy consumption was 3.3 times the world average and in the United States and Canada, it was higher yet.

According to the Energy Information Administration in the United States, we used 73.35 quadrillion BTU of energy in 1999. That is 21,491 million megawatt-hours. There were about 280 million of us, so that is 76.8 megawatt-hours for each of us, or the equivalent of about 77 humans energy worth for the average American. Your energy servants do the work of 77 people, if they had to rely on muscle power only.

Cohen's fourth evolution is a glimmer of hope for any of you clever enough to look at any of the population graphs and figure out that we are going to have a serious problem if this keeps up. Clearly, if our numbers continue to shoot straight up before long we won't be able to find a place to stand.

The fourth evolution is the apparent reduction in fertility rates. We have pushed back the forces that kill us. If we don't reduce the rate of births, we will eat up the planet with our numbers. The good news is that there is a reduction in fertility and the reduction appears to be increasing. If you look carefully at Figure 2.4, you will see just a bit of reduction in the steepness of the curve. The bad news is we really don't know yet just how great that reduction in fertility will be and if it will be enough to prevent major disasters in the future.

This brings us back to the wonderful title of Dr. Cohen's book, *How Many People Can the Earth Support?* As I said earlier, Dr. Cohen takes us on a very interesting study to show that this is a very difficult question to answer. The simplest, most straightforward answer to the question is "Well, it depends." It depends on what assumptions you want to make, who does the analysis, and just what kind of a life will you accept and still call that living.

Cohen walks us through a long set of analyses that over the years have sought to answer his question. The analyses have ranged from extremely simplistic to extremely sophisticated, and the numbers have ranged very widely from one billion to 1000 billion. The biggest number was calculated from maximum photosynthesis production under optimum conditions from all the earth's land. No allowance was given for a place for the people, i.e., cities or factories. If you allowed 750 square meters (a square 90 feet on a side) each for people to actually live in, it brought the estimate down to 146 billion.

Most assessments looked at some aspect of resource needs and focused mainly on agriculture. The most optimistic looked at the most productive farmland on the planet, assumed we could farm all arable land like that, estimated the total arable land on the planet, and worked the math. These numbers come out very encouraging especially if you base your math on a subsistence diet for all those folks. If you will only accept a modern European diet, the numbers get roughly reduced to half or even a third.

However, this gets increasingly complicated when you start looking at the competing demands on agricultural land. Wood for construction and land for non-food crops such as wine grapes take up a surprising fraction of arable land. Furthermore, there is the ratio of meat to plants in the diet. Grazing animals as a source of calories are about 10% as efficient as growing human edible plants on the same land. Therefore, as you increase the ratio of meat to plants in a diet, you decrease the number of people you can feed on the same amount of land. If there is good news in that observation, it is that in developed countries we have shifted our diet too far toward the meat end of the ratio to be optimally healthy. Therefore, if we in the developed countries were to shift back toward plants instead of meat, not only would there be more room at the global dinner table, we would be healthier as well.

And that brings up another point, the haves versus the have-nots, and who gets to choose who is which. In estimating how many people you can feed, a big assumption is how you are going to organize the society you are trying to feed. If you set up an Orwellian state in which everything is carefully controlled and evenly distributed, you can eke out every morsel and ensure it gets well used. If you are willing to let the current state continue, the rich and poor divide will go on with the rich investing in weight reduction spas while the poor don't get enough calories to maintain normal vigor and health. On the other hand, getting to the equal division, albeit allowing a greater total population, would be pretty ugly in terms of the social upheavals required to make that adjustment in our social structure, not to mention the violence done to individual liberty in the process.

Then there is also the question of can you farm all arable land as intensively and productively as you can the most productive land on the planet? The answer to that it turns out is also "Well, it depends." It depends on how long you plan to do that and what technology you will accept.

We saw above that if you use 4 times the amount of fertilizer that we use on average in the United States, you could get over twice the food value per hectare. However, that is a little complicated by the fact that in the comparison between say China and the United States, the Chinese also used 140 times as much human labor per unit food value gained from farms than we do in the United States. The point in either comparison, food per hectare or food per farmer, is that to get more food you need more energy driven technology. United States farmers' extensive use of machines requires petroleum, lots of it. Chinese extensive use of fertilizers requires fossil fuel sources, lots of them.

Let's look at fertilizers just a bit. The key thing we need in fertilizers is nitrogen. Sure, you also need phosphorus and potassium. You also might need secondary elements such as calcium, magnesium, and sulfur along with a number of trace elements depending on the particular soil you are working with. But the big thing is nitrogen. The basis of most nitrogen fertilizers is ammonia. Ammonia is one nitrogen atom combined with three hydrogen atoms. Well, that ought to be pretty easy, huh? If you recall any of your grade school science you know that nitrogen makes up about 80% of the atmosphere. That means there is plenty of it and it is everywhere. And hydrogen, gee, that ought to be a snap. It is the single most common element in the universe and available on earth in many forms, the most common of which is plain old water. So how do we make ammonia? Well, we use the Haber-Bosch process in which you do get nitrogen from the air, but you get hydrogen from natural gas.

Yep, that's right. To get ammonia we burn natural gas. A lot of natural gas. The thing is that making ammonia takes energy. The nitrogen and hydrogen won't combine to make ammonia if you just bring them together. You have to add energy. Natural gas is great in that regard; you get not only the supply of hydrogen you need but also the energy source. Now natural gas is actually a mixture of gases, but mainly it is methane, which is one carbon with four hydrogen atoms. This means for each four molecules of ammonia you make you also have three carbon atoms left over. The process of using the methane turns the carbon atoms into carbon dioxide and expels them into the air as a waste product. Carbon dioxide is one of the chief greenhouse gases leading to global warming. So not only do you use up natural gas in making ammonia, you also produce greenhouse gas.

The annual production of nitrogen-based fertilizers takes 62 billion kilograms of nitrogen from the atmosphere. Doing a little bit of chemistry would tell you that you get three carbon dioxide molecules for each four ammonia molecules or 36 grams of carbon for each 56 grams of nitrogen in the ammonia. So the 62 billion kilograms of nitrogen being made into fertilizers causes the release of 40 billion kilograms of carbon to the atmosphere each year.

Well, then let's not do that, huh? Well, no, not hardly, because without that use of fertilizers we couldn't produce the food to feed the population. We just couldn't do it. Here are some comparisons.[9]

If you used no fertilizers or pesticides you could get 500 kilograms of grain from a hectare in a dry climate and as much as 1000 kilograms in a humid climate. If you got organic and used animal manure as fertilizer, assuming you could find enough, you might get as much as 2000 kilograms per hectare. For a sense of scale, the average in the United States, where recall we only get half the food value to hectare as the intensively farmed Chinese crop land, we get

[9]From "Food, Land, Population and the U.S. Economy" by David Pinentel and Mrio Giampietro, Carrying Capacity Network, 2000 P Street, NW Suite 240, Washington, D.C., November 21, 1994.

about 4500 kilograms per hectare on the average. In serious cornfields with fertilizer, irrigation, and pesticides, the value is 7000 kilograms per hectare.

Modern mechanized, chemically supported agriculture produces 7 to 14 times the food that you would get without those advantages. Even the best organic farming would produce only 30 to 45% of the food value you would get from the same sized chemically fertilized farm, and that is assuming you could get the manure you needed to make it work.

In very stark terms, without the chemically enhanced farming we would have probably something like one-fifth the food supply we have now. That means four-fifths the population would not be fed, at least as we are organized now. So, no, just giving up on fertilizers is not in the deal.

However, we could get the hydrogen and energy from sources other than natural gas. Nuclear energy could be used to provide electricity to extract hydrogen from water and produce the process heat required to combine the hydrogen and nitrogen from the air. That is just a thought to stick in your mind.

While we are looking at energy use in agriculture, here are a few more numbers for you.[10] If you look at the energy input into agriculture and the energy you get out, you see some interesting facts. By combining the energy used to make fertilizers and pesticides, power irrigation, and run the farm machinery in the United States, we use about 0.7 kcal of fossil fuel energy for each 1 kcal of food we make. This doesn't include the energy needed to process and transport the food. In Europe where they farm more intensely, the amount of energy out is just about the same as energy in. In Germany and Italy the numbers are 1.4 and 1.7 kcal energy input to each 1 kcal output respectively. **The point is you need energy to feed people, well at least a lot of people.**

Which gets us back to Cohen and his question. One of the studies he examined looked at a "self-sustaining solar energy system." For the United States, this would replace all fossil energy and provide one-fifth to one-half the current energy use. The conclusion of the study was that this would either produce "a significant reduction in our standard of living . . . even if all the energy conservation measures known today were adopted" or if set at the current standard of living, "then the ideal U.S. population should be targeted at 40–100 million people." The authors of that study then cheerfully go on to point out that we do have enough fossil fuel to last a least a century, as long as we can work out the pesky environmental problems. So, you can go to a "self-sustaining" energy economy as long as you are willing to shoot between 2 out of 3 and 6 out of 7 of your neighbors.

---

[10]Also from "Food, Land, Population and the U.S. Economy" by David Pinentel and Mrio Giampietro, Carrying Capacity Network, 2000 P Street, NW Suite 240, Washington, D.C., November 21, 1994.

And this is a real question. The massive use of fossil fuel driven agriculture to provide the fertilizers and pesticides, and power the farm equipment, is a) vitally important to feed everyone, and b) something we just can't keep up in a business-as-usual fashion. Sustainable means you can keep doing it. Fossil energy supplies are finite; you will run out some time. Massive use of fossil energy and the greenhouse gases they produce also may very well tip the planet into one of those extinction events in which a lot of very bad things happen to a lot of the life on the earth.

O.K. to Cohen's big question, how many people can the earth support? What it comes down to is that the "Well, it depends" answer depends on

- what quality of life you will accept,
- what level of technology you will use, and
- what level of social integration you will accept.

We have seen some of the numbers regarding quality of life. Clearly if you are willing to accept the Bangladesh diet, you can feed 1.8 times more people than if you chose the United States diet.

If you choose the back-to-nature, live like our hearty forefathers, level of technology, you can feed perhaps one-fifth as many people as you can with modern chemical fertilized agriculture. The rest have to go.

And here is the tough one. You can do a lot better, get a lot more people on the planet, if you just force a few things. Like, no more land wasted in growing grapes for wine or grains for whiskey and beer. No cropland used for tobacco. No more grain wasted on animals for meat, just grain for people. No more rich diets for the rich countries, share equally for everyone. No more trade barriers; too bad for the farmers in Japan and France, those countries would just have to accept their dependence on other countries for their food. It is easy to see that at least some of those might actually be a pretty good thing; however, the kicker is how do you get them to happen? After all, Mussolini[11] did make the trains run on time. How could you force these things without a totalitarian state? Are you willing to give up your ability to choose for yourself for the common good? It is not pretty, is it?

Cohen looked at all the various population estimates and concluded that most fell into the range of 4 to 16 billion. Taking the highest value when researchers offered a range, Cohen calculated a high median of 12 billion and taking the lower part of the range a low median of 7.7 billion. The good news in this is 12 billion is twice as many people as we have now. The bad news is

[11]Mussolini was the fascist dictator of Italy during World War II and led that nation to disaster, but it was said that he restored an order of sorts to the country, often summarized as "he made the trains run on time."

that the projections for world population for 2050 are between 7.8 and 12.5 billion. That means we have got no more than 50 years before we exceed the nominal carrying capacity of the earth. Cohen also offers a qualifying observation by stating the "First Law of Information," which asserts that 97.6% of all statistics are made up. This helps us appreciate that application of these numbers to real life is subject to a lot of assumptions and insufficiencies in our understanding of the processes and data.

However, we can draw some insights from all of this. What it comes down to is that if you choose the fully sustainable, non-fossil fuel, long-term options with only limited social integration, the various estimates Cohen looked at give you a number like 1 billion or less people that the earth can support. That means 5 out of 6 of us have got to go, plus no new babies without an offsetting death.

On the other hand, if you let technology continue to do its thing and perhaps get even better, the picture need not be so bleak. We haven't made all our farmland as productive as it can be. Remember, the Chinese get twice the food value per hectare as we do in the United States. There is also a lot of land that would become arable if we could get water to it. And, of course, in case you need to go back and check the title of this book, there are alternatives to fossil fuels to provide the energy to power that technology.

So given a positive and perhaps optimistic view of technology, we can look to some of the high technology assumption based studies from Cohen's review. From the semi-credible set of these, we can find estimates from 19 to 157 billion as the number of people the earth could support with a rough average coming in about 60 billion. This is a good time to be reminded of the First Law of Information. The middle to lower end of this range, however, might be done without wholesale social reprogramming. Hopefully we would see the improvement in the quality of life in the developing countries as they industrialize and increase their use of energy. Hopefully, also this would lead to a matching of the reduction in fertility rates that has been observed in the developed countries, which in turn would lead to an eventual balancing of the human population.

The point to all this is the near-term future of the human race depends on technology. If we turn away from technology, a very large fraction of the current and future human race will starve. If we just keep on as we are, with our current level of technology and dependence on fossil fuel resources, in the near term it will be a race between fertility decrease and our ability to feed ourselves, with, frankly, disaster the slight odds-on bet. In a slightly longer term, dependence on fossil fuels has got to lead to either social chaos or environmental disaster. There are no other end points to that road. It doesn't go anywhere else.

However, if we accept that it is technology that makes us human, that technology uniquely identifies us as the only animal that can choose its future, we can choose to live, choose to make it a better world for everyone and all life. This means more and better technology. It means more efficient technology that is kinder to the planet but also allows humans to support large numbers in a high quality of life. That road is not easy and has a number of ways to screw up. However, it is a road that can lead to a happier place, a better place.

Two Concluding Thoughts on the Case for Technology

Two more points and I will end my defense of technology. First, I want to bring you back from all the historical tour and all the numbers about population to something more directly personal. Let me ask you two questions.

What do you do for a living?

What did you have for breakfast?

Don't see any connection between these questions or of their connection to the subject of technology? Don't worry, the point will come out shortly. I am just trying to bring the idea of technology back from this grand vision to its impact on your daily life.

Just as a wild guess, your answer to the first question was something that, say 500 years ago, didn't even exist. If we look 20,000 years ago, the only job was "get food." Even if you have a really directly socially valuable job like a medical doctor, 20,000 years ago you would have been extraneous. That is, the tribe couldn't afford you. What, no way! A doctor could save lives, surely a tribe would value such a skill. Well, sure, but the tribe could not afford taking one of their members out of the productive "getting the food" job for 20 years while that individual learned all those doctor skills.

If you examine the "what you do for a living" just a bit I think you will see a grand interconnectedness of all things. I personally find it pretty remarkable that we have a society that values nuclear engineers enough that I can make a living at it. Think about it. Somehow what I have done has been of enough value that, through various taxpayer and utility ratepayers, society has given me enough money for food and shelter. The tribe 20,000 years ago wouldn't have put up with me for a day.

You see, that is why we as humans are successful, wildly successful in fact. We work together. "Yeah, sure we do," you reply, " read a newspaper lately?" Well, O.K., we fuss and fight a good deal and some of us do some pretty stupid and pretty mean things. But the degree of cooperation is amazing if you just step back a bit.

O.K., what did you have for breakfast: orange juice, coffee, toast, maybe some cereal and milk? Where do these things come from? Orange juice came from Florida or California. Coffee came from South America. Bread for the toast came perhaps from Kansas; cereal, from the Mid-West somewhere. The jam on the toast may have come from Oregon, or maybe Chile. Milk is probably the only thing that came from within a hundred miles of your breakfast table. Think about it. There were hundreds of people involved in your breakfast. Farmers, food-processing workers, packaging manufacturers, transportation people, energy producers, wholesale and retail people. Perhaps each one only spent a second on their personal contribution to your personal breakfast, but they touch thousands of other people's breakfasts as well. In turn, you buying the various components of your breakfast supported, in your part, all those people. They in turn, in some way or another, bought whatever you provide to society that allowed you to buy breakfast. Pretty amazing, don't you think?

Now when you look at all that, think about what ties all the planetwide interconnection. Yep, you guessed it: technology. Without technology, you get what is available within your personal reach, and what you produce is available only to those who are near enough that you can personally carry it to them on your own two feet. Technology makes our world work. It gives you personally a productive and socially valuable way to make both a living and to provide your contribution to the rest of us.

I want you to stop a minute and really think about that. What would your life be like without technology? Could you do what you currently do? Would anyone be able to use what you do? Would anyone pay you for that? "But I am a school teacher," you say, "of course, they would pay me!" Are you sure? Why do you need schools if there is no technology? All I need is to teach the kid how to farm and how to hunt. Sons and daughters can learn that by working in the fields along with their parents. See what I mean?

Now, I have hopefully reset your brain. Sure, you are still going to be hit with daily **"technology is bad"** messages. Hopefully, you are a bit more shielded against that din, and you have been given some perspective to balance that message and are prepared to see the true critical value of technology to human existence. The point is that technology is what makes us human. Without it, we are just slightly smarter monkeys.

You may feel that 6 billion of us are too many, and that may very well be. I personally don't know how to make that value decision. Which particular person does one select as being one of the excess ones?

However, the fact is that there are 6 billion of us, and it looks like we are headed for 10 to 12 billion in the next 50 years. Without not only the technology we have, but significantly better and more environmentally friendly technology, the world is going to get ugly as we approach these numbers.

On the other hand, with the right technologies we can not only support those numbers, we can do it while we close the gap between the haves and have-nots. We can make it a better place for everyone. It takes technology and the energy to drive it. Choosing technology is what we have to do to secure the evolutionary selection of us as a successful species. Remember, some pages back in discussing the unlikely evolutionary path to us, I said we are not the chosen, unless. Unless we choose us. This is what I meant. We are totally unique in all of evolutionary history. We humans have the unique ability and opportunity to choose either our evolutionary success or failure. A choice of technology gives us a chance. A choice rejecting technology dooms us as a species and gives the cockroaches the chance in our place. Nature doesn't care what survives, algae seas, dinosaurs, humans, cockroaches, or whatever is successful. If we care, we have to choose correctly.

As an aside, let me address a point of philosophy here. If any of this offends your personal theology, I offer this for your consideration. Genesis tells us God gave all the Earth to humanity and charged us with the stewardship thereof. So it is ours to use as well as we can. That insightful social philosopher Niccolo Machiavelli put it this way in 1501:

> "What remains to be done must be done by you; since in order not to deprive us of our free will and such share of glory as belongs to us, God will not do everything Himself."

O.K., you are saying. I give. You have beaten the socks off me. Technology is good; technology is the identifying human trait and our only hope. But what is this stuff about choosing technology or not? Technology just happens doesn't it? I mean, technology always advances, it always has, so why the big deal?

Well, that is my last point on technology. It doesn't always just happen, and people have chosen to turn away from technology. In what might have seemed at the time to be a practical social decision, huge future implications were imposed on many generations to come. It has happened. Let me take you on one more trip through history. I think you will find it enlightening.

In *Guns, Germs, and Steel,* Jared Diamond explores the question of why the European societies came to be dominate over all the other human cultures on earth. It is a fascinating story and provides a lot of insight into how modern societies evolved. In moving through history, he comes across a very odd discontinuity. He observes that if you came to earth from space in the year 1400 A.D., looked around, and went home to write your research paper on the probable future of the earth, you would clearly conclude the Chinese would run the entire planet shortly. Furthermore, you could conclude they would do it pretty darn well. If those same extraterrestrial researchers were to pop into their time machine and come back to earth in any year from say 1800 to now, they would be totally amazed to see China as a large, but relatively backward, country, struggling to catch up with their European and American peers.

To understand the significance of this, you have to go on that research trip with the extraterrestrials and look at China before 1400. In *The Lever of Riches*, Joel Mokyr dedicates one chapter looking at the comparisons of technology development in China to that in Europe. He lists the following as technology advantages China had in the centuries before 1400:

- Extensive water control projects, alternately draining and irrigating land, significantly boosting agricultural production
- Sophisticated iron plow introduced sixth century B.C.
- Seed drills and other farm tools, introduced around 1000 A.D.
- Chemical and organic fertilizers and pesticides used
- Blast furnaces and casting of iron as early as 200 B.C., not known in Europe until fourteenth century
- Advanced use of power sources in textile production, not seen in Europe until the Industrial Revolution
- Invention of compass around 960 A.D.
- Major advances in maritime technology (more in a bit on this)
- Invention of paper around 100 A.D. (application as toilet paper by 590 A.D.)

In the year 1400 A.D., China was a world power, perhaps the only true world power. Their technology in agriculture, textiles, metallurgy, and maritime transportation were far in advance of any other country. They had a strong central government and a very healthy economy.

Their naval strength provides a real insight into the degree of this dominance. Dr. Diamond sends us to an extremely readable book *When China Ruled the Seas—The Treasure Fleet of the Dragon Throne 1405–1433* by Dr. Louise Levathes. Dr. Levathes takes us on an inside tour of the Chinese empire during these years. She focuses on the great treasure fleets that China set forth in these early years of the fifteenth century. In her book she has a wonderful graphic that overlays a Chinese vessel of the treasure fleet (~1410) with Columbus's *St. Maria* (1492). At 85 feet in length and three masts, the *St. Maria* is dwarfed by the nine-masted, 400-foot-long Chinese vessel.

The Chinese sailed fleets of these magnificent vessels throughout oceans of South Asia, to India, and even as far as the eastern coast of Africa. With this naval domination China claimed tribute from Japan, Korea, the nations of the Malay Archipelago, and various states within what is now India. Through both trade and the occasional application of military force, China provided an enlightened and progressive direction for all the nations within this sphere of influence. If two princes in India were fighting over a throne, it was the recognition, or lack thereof, from the Chinese emperor that decided who would rule. Setting a policy of religious inclusion and tolerance, the Chinese engaged the Arabian traders and calmed religious disputes within Asia.

With applications of power sources in textiles and advanced metallurgy, the Chinese were in the same position in 1400 as the British were in 1750, ready to launch into the Industrial Revolution. They traded with nations thousands of miles from home with vast, sophisticated shipping fleets. They were poised to extend this trade all the way to Europe and perhaps find the New World by going east instead of the European's going west in search of the rich Chinese markets.

But if we pop into that extraterrestrial time machine and drop into China in 1800, we find a technologically backward nation, humbled by a relatively small force of Europeans with "modern" military technology who wantonly imposed their will on the Chinese. The Chinese have been struggling to catch up with European and American technology ever since and so far not quite being able to do that. The domination of China by the Japanese during World War II shows how complete the turnaround was. In 1400 Japan was but one of many vassal states huddled about the feet of the Imperial Chinese throne. In 1940 the Japanese military crushed the Chinese government while marching on to control much of South Asia.

What could have happened to turn this clear champion of technology, trade, enlightened leadership with all its advantages over both its neighbors and yet-distant foreign competitors into such a weak, backward giant?

Mokyr goes through a pretty complete list of potential causes. He looks at diet, climate, and inherent philosophical mindset rejecting each as a credible actor mainly on the bases that all of these conditions were present during the period of technological and economic growth as well as the subsequent stagnation. Therefore, these were not determining factors in the turnabout. In the end he concludes, as does Diamond and Levathes, that it was just politics.

Yep, that is right. It was good, old human politics. Dr. Levathes gives us a delightful insider's view of the personalities and politics of Imperial progressions during this critical time period. To make a short story of it, the party that had been in control during the expansionist period supported the great treasure fleets, commerce with foreign nations, use and expansion of technology, and a rather harsh control of the rival party. The rival party was based on Confucian philosophy that preached a rigid, inward-looking, controlled existence.

When the Confucian party gained control of the throne, they had their opportunity to push back on the prior ruling party that had oppressed them so harshly for so long. And they did. They wanted nothing to do with foreigners; we have all we need at home, here in China, they said. The fleet was disbanded and the making of ocean-going vessels forbidden. Technology was no longer "encouraged." Again, their position was what we have is good enough, stop with all this new nonsense. Over a period of just a few years, the course of the entire nation was shifted from what would have appeared to be a bright future as the leading power in the world to a large, but relatively insignificant, backwater, rich in history and culture, but all backward looking to a former glory.

That was it. A shift in the political agenda. At the time, to the leaders in control, one that made sense. Focus at home, use what you have now, create order, discipline, control. In 50 years Japanese pirates controlled the coast of China, and the former ruler of the seas from Asia to Africa could not get out of their harbors safely.

So, you see if the "technology is bad" message gets incorporated into too many of our daily decisions, we can turn from our bright future into something else. The difference is that this time the stakes are much higher than they were in fifteenth century China. If we, in the developed nations, make the wrong choices, we doom all of humanity by our folly. It is not just that we miss the potential bright future, we miss the chance to avoid the combined human population growth and resources exhaustion disaster coming at us like a runaway train. Technology is the only way to prevent that train wreck. We can hear the siren's call of anti-technology, come back to nature and let the train run us down in a bloody mess, or we can try our best to use technology wisely and win free to make a better life for everyone.

3
THE CASE FOR ELECTRICAL POWER

The great science fiction writer Arthur C. Clarke once observed that any technology sufficiently advanced is indistinguishable from magic. If you think about electrical power for a bit, you will see it meets this definition of magic pretty well.

About 120 miles from where I am sitting right now, tons of water per second are rushing down a very large pipe. Along with all the other falling water being turned into electricity, about 1.5 ounces per second is turned into a particular bit of electricity and sent zooming a heartbeat later to my house where it lights up my laptop and lets me record these words. Magic? What else could you call that? Falling water at about a one-week walking distance from me gets essentially instantly turned into the energy that makes this magic little box capture my words and correct my spelling.

In the next second it may be that the particular bit of electricity zipping into my laptop came from the neighborhood nuclear power plant about 20 miles from where I am sitting. In that case about a trillion uranium atoms fissioned to give me that second's bit of power. Does that sound like a lot? That is about 40 billionths of a gram. That means I could run my laptop for 78 years on one gram of uranium or for more than 35 thousand years on a pound of uranium.

Only in the last hundred years or so have we been able to make energy in one place and turn it into useful work at a distant location. Before electricity, the best we could do was transmit the power of a water wheel or a steam boiler throughout one factory using spinning shafts and belts or rushing steam. We could heat water in the basement of a building and pump it to heat several floors above, maybe even heat a portion of a town. However, before commercial electrical generation and distribution, there was no way to produce vast amounts of energy at one place and distribute it to thousands, even millions of users tens and hundreds of miles away.

Even more magical are all the ways we can produce electricity. We can tickle electricity out of light with photovoltaic panels. Or much more commonly, anything that can produce motion can be used to make electricity. Wind turning windmills and water turning water turbines works. Anything that produces heat can be turned into motion, e.g., flowing steam. So you can get electricity by burning coal, natural gas, wood, and even garbage. You can also get it from nuclear fission, as we will learn about in Chapters 4 and 5.

More amazing yet is all the things we make electrical energy do. With motors we can turn it into motion. With resistance or induction heaters, we can turn it into heat. By using motors to compress and then expand a refrigerant, we can turn electricity into cold. We can make light and sound. We can make coded electromagnetic waves to broadcast radio and television signals. We can decode those electromagnetic waves to guide electron beams to precisely strike a phosphorescent TV screen and to activate moving magnets in speakers and turn the waves into sound and moving pictures in our homes.

These are perhaps all very common things to most of us, and we accept them as a matter of course. If you can put yourself in the mind of one of your ancestors three or four generations back, or if you have ever had to experience several days without electricity, perhaps you can appreciate just how wondrous electrical energy production and application are.

If you have gone through a multi-day blackout, you also understand how key electrical energy has become to our society.[1] Probably you couldn't report to work because your place of business wasn't functioning. You may have begun to get concerned about the quality of your water supply, if you had one, because the water purification system wasn't running. Unless the local gas station had a standby generator, you couldn't even buy gasoline for your gasoline-powered generator if you had one, because the gas pumps wouldn't work. The perishables in the grocery store, well, perished. Lots of frozen food became inedible and very smelly. As the folks in California discovered in the

[1]My personal experience was the Columbus Day storm that hit the Pacific Northwest in 1962. We were without power for 5 days. Fortunately for us, we had a gas-powered generator, as did the local gas station. We could run the electric pump on our well and run our freezer from time to time. We got very popular with our friends and neighbors. Our generator ran a lot during those 5 days.

spring of 2001, even an hour without electricity to operate the traffic signals makes a pretty bad day.

O.K., electricity is magic, and we are hooked, so you say. You need to tell me more than that to convince me that electricity is not part of that "bad" technology.

Challenge accepted.

Let's go at this from the start. Let me first tell you a little about just what electricity is. Then I will give you a little history. In completing the history I will try to show you how electricity came to be so important in developed countries. I will finish with some more detail on how we use electricity today. Ready? Here we go.

What Is Electricity?

Electricity as we commonly use it is the flow of electrons.

Well, great. Thanks a lot. You just defined one thing I don't understand with another thing I don't understand. Very helpful.

Sorry, I will try to make it more clear. If it helps at all, no one, really, at the heart of it, understands exactly just what an electron is. However, that doesn't prevent us from having some models about electrons and using these to make electrons do a lot of useful things for us. Things like light the room you are in, make factories operate, make steel, drive railroad engines, look inside people's bodies to find diseases or broken bones, pasteurize milk, and cook food.

Let's start again. All stuff is made up of atoms. In Chapter 4 we will talk about atoms a good deal more, but for now realize atoms are very, very small, and they are the tiny bits that make up all matter. Electrons are the outside part of atoms. They are the smallest part of an atom by weight but take up most of the space. The important thing about electrons is that they carry electrical charge. You are just going to have to trust me on that. Electrical charge is a fundament property of matter, and the electrical force it generates is one of the primary forces of nature. They just are.

The thing that is important about electrical charge is that it can be used to create electrical force, and we can use electrical force to do all the neat things we do with electricity. It also drives all chemical reactions. If you heat coal in air, the electrons in the carbon within the coal and the electrons in the oxygen within the air reconfigure themselves and we get a carbon dioxide molecule and heat. If you want to get metal aluminum, you have to put a lot of energy into aluminum oxide to get the electrons in the aluminum oxide to let go of each other to give back the separate atoms of aluminum and oxygen. All of chemistry is manipulating electrons. Energy is added to or released from a chemical reaction because the chemical reaction is changing the nature of the electrons and using the electrical forces of the electrons.

We have all played with magnets. You know that if you align the south pole of one magnet with the north pole of another they will be pulled together. If you align like poles and push them together there is a resisting force. This is what electrons are doing at tiny, tiny levels.

If you can create an abundance of electrons in one place and a relative lack of electrons some place else, then you create an electrical force between the places, just like a south magnetic pole attracted to a north magnetic pole. If the force is great enough and you have a path that allows electrons to flow, the electrical force will try to create a balance, and electrons will flow from the abundance to the lacking place.[2] This is electric current. There are two really cool things about such a flow of electrons. First, they can carry energy. Second, they can carry information. The information is pretty limited: on, off, varying degrees of intensity, and the timing of the on, off, and intensity changes. This may not seem like a lot, but it is enough to transmit information to deconstruct and reconstruct the zillions of possible combinations of sound, pictures, and all the other hosts of electronic information.

So, electricity is the flow of electrons. The flow of electrons is caused by the imbalance of electrons in one place relative to another. This causes an electric force that moves the electrons. O.K., how do you make that happen?

Some of the first observations of electrical force and electricity were recorded by the ancient Greeks. They noticed that if you rubbed an amber rod with a piece of fur, the rod would then magically attract bits of straw or feathers. This observation is preserved in that "elektron" is the Greek word for amber. In rubbing the amber rod, electrons were being transferred between the rod and the fur. They were creating a site of electron abundance and lack. They might have noticed that the fur was acting a bit odd as well, with the hairs standing up and trying to push away from each other.

Remember, we said that chemical reactions were all about electrons and electrical force? Another way to get electrical current is to use chemical reactions. If you stack copper and zinc plates together and put them in a moist bath, electrons want to form a new chemical compound. To do this, the electrons have to flow from one metal to the other. If you give them a conducting path, like a copper wire, they will move from one metal to the other causing a flow of electrons which we can use to do useful things. If you doubt that, go turn on your flashlight.

However, the most useful way to create a flow of electrons is with a generator. Here we turn back to the observation that electrical force is a lot like the force we see in magnets. Others observed this and examined what

[2]When nature pushes air and water vapor around inside a thunderstorm, it creates an imbalance of electrons between clouds or between clouds and the ground. The result is lightning, which is a very impressive flow of electrons.

happens when you mix the two ideas together. It was discovered that if you pushed a magnet into a loop of a conductor, like a copper wire, electric current flowed in the wire. If you pulled the magnet back out of the loop, the electric current flowed the other way. If you held the magnet still inside the loop, nothing happened. From this folks figured out that around a magnet is something called a magnetic field. You can't see it, but you can see its effects. Somewhere you have seen the experiment where they put a sprinkle of iron filings on a piece of glass then put a strong magnet underneath. The filings arrange themselves in loops about the point of the magnet. This is a picture of the magnetic field that the magnet is putting out.

Now, if we move the magnet, we move the magnetic field. If we expose a conducting material to a changing magnetic field, electrons are induced to flow. This is electromagnetic induction. It doesn't matter if you move the magnet or the conductor. They really don't know or care which is being moved. All they care about is that there is relative motion.

You can think of it as the magnetic field pulls at electrons as it moves by. If the field is not changing, the electrons just sit where they are and don't move. But if you move the field, the electrons are tugged on and move a bit. If the field moves some more, the electrons move some more. So if you keep the field moving, you keep the electrons moving.

A few quick words on conductors. Conductors are usually metals. In metals, electrons are not uniquely bound to individual atoms but rather shared among a lattice of atoms. This means electrons are free to move about with limited resistance in a metal. Some metals are better conductors than others are. Copper is just about the best there is. It is about 5.5 times better than iron. Silver is a little bit better than copper as a conductor, but considering the relative cost, copper is obviously more commonly used.

So, we now know that we can induce electrons to flow in a conductor if we either move a magnetic field relative to a stationary conductor or move a conductor relative to a stationary magnetic field. If we do the former, we have built an alternator; if the latter, we have built a generator. The best way to get a constant relative motion is to spin one around the other. So to get a generator, we get a big stationary magnet then build a loop of conductor and rotate that loop through the magnetic field of the generator. This causes electric current, a stream of electrons, to flow continuously through the conductor as long as we continue to turn it.

We know a lot of ways to get things to turn. You can crank it by hand, have a horse go around a big wheel, use the wind to spin a fan, let falling water turn a waterwheel, or let expanding steam from water you have boiled with any heat source either expand inside a cylinder driving a piston and crank or turn a fan blade. So, you connect your turn-y thingy to the loop of a conductor that is going through the magnetic field from the stationary magnet, and you have made electricity.

There are mechanical inefficiencies in this process, and you never get quite as much electrical energy out as mechanical energy you put in, but it is pretty close. However, there is a bit of a problem in getting the mechanical energy, especially if you are going to get it from heat. If you use a heat engine to convert heat to mechanical energy, you lose quite a bit of the energy from the heat. It isn't wastefulness, it is just physics.[3] The amount of mechanical energy you can get from a heat engine is called the thermal efficiency. In the history of electrical power discussion, which follows, we will see how far we have come in improving thermal efficiency.

A few words on thermal efficiency: as noted above, the fact that you lose a lot of the heat energy in making mechanical energy is not wastefulness, it is physics. In 1824, the French engineer Nicolas Leonard Sadi Carnot worked out the limits of heat engines. A heat engine is a mechanical device that uses the difference in two reservoirs of heat to produce mechanical energy. In other words, a heat engine uses heat to make something move. But it is important to recognize that it can do that only if there is a cold place as well as a hot place. There are no heat engines in Hell. You have to have a hot place and a cold place to make a heat engine work. If everything is hot, you can't extract any mechanical energy out of the heat.

So, what Carnot figured out is that the fraction of the heat of the hot place that you can turn into mechanical energy depends on the ratios of the temperatures of the hot and cold places of your heat engine. Furthermore, the greater the difference, the greater the thermal efficiency. His formula is pretty simple, fraction = 1 - cold/hot, with cold and hot being the temperatures of the cold and hot places of your heat engine. The key is that you have to use absolute temperatures for the hot and cold temperature numbers.

We are used to Fahrenheit and Centigrade temperature scales. In those scales, room temperature is 72 and 22 degrees, respectively. But zero on either scale isn't as cold as you can get. If you cool things down as low as you can get, you get to minus 460°F and minus 273°C. Because starting zero at the lowest possible and counting up makes more sense than the semi-arbitrary points in the Fahrenheit and Centigrade scales, the "absolute" temperature scales of Rankine and Kelvin do that. The Rankine scale uses the same spacing as Fahrenheit, and the Kelvin scale uses the same as Centigrade. So, 0°C is 273°K, and 0°F is 460°R, and 0°R equals 0°K, equals, well, zero, as cold as there is. Got it?

---

[3]Recognize that this isn't just a problem for electrical generation. The reason your car has a radiator is that your gasoline engine is a heat engine processing heat between the flame of burning gasoline and the outside air. That systems operates with something like 20% overall efficiency.

So, if we have a steam boiler that heats water to 600°F or 1060°R (600+460) and uses the outside air at 60°F or 520°R as a heat sink, the limit on thermal efficiency is 1- 520/1060 = 0.51, or 51%.

What a disappointment, only half the energy of the hot water can be turned into mechanical energy. But what if I raised the hot side or lowered the cold side? Well, that would help. In fact, if you lower the cold side to absolute zero, you would get all the energy of the heat source. But there are no lakes or other heat sinks on earth available at absolute zero, so maybe 32°F is about as good as it ever gets on the cold side. For the 600°F steam using a 32°F heat sink would get you up to 53.5% in Mr. Carnot's perfect world.

On the hot side, you can go a little higher, but as you get much higher you run into problems with materials getting soft and corrosion rates increasing, so there are limits.

Now, the really bad news. Carnot gave us the ideal limits. In the real world there is friction in mechanical systems and leakage of heat from our heat engine. You can never get the ideal limit. Our 600°F steam boiler probably is going to max out at 30% or so efficiency. Modern combined-cycle gas turbine plants can get thermal efficiencies over 40%, but that is about all we can do for now.

The History of Electrical Power[4]

The history of electrical power is really the story of a number of ideas coming together. As we saw above, the original observations of electricity go back to the ancient Greeks. However, the exciting stuff happens in the 1800s. In fact we can start the story in 1800.

It was 1800 when Alessandro Volta put together a set of copper and zinc plates within a salt solution. He connected the ends of the copper and zinc with a conductor and electricity flowed. He had invented the battery. At that time this was the only way to "make" electricity. It would remain one of the major sources of electricity for many years to come.

So, we had a source of electricity, albeit pretty limited. What could you do with it? In 1801 an Englishman, Humphry Davy, managed to make light with electricity, but only as a spark between two carbon rods. It wasn't a sustained light, so while very bright, it wasn't very useful. In 1821 Michael Faraday showed that electrical currents working with magnetic fields could be used to create motion, the principle that would lead to electric motors.

---

[4]While taken from a number of sources, much of this section, especially the early history, comes from *Engineering in History,* by Kirby, Withington, Darling, and Kilgour. With its initial publication date of 1956, this book is a bit dated for current perspectives; however, it is an excellent reference for anyone interested in the impact of engineering and technology across the centuries.

But electricity was still a scientific toy. Faraday starts the ball really rolling in 1831 when he demonstrates the principle of electromagnetic induction. Remember, that is inducing electrical current by moving a magnetic field relative to a conductor. This was really important, because now you could turn mechanical energy into electrical energy. This opened the door to making vast amounts of electricity.

However, while the principle had been demonstrated, there wasn't a big need for a practical means of generating electricity because there weren't any practical uses for a lot of electricity yet. It took a number of things to come together to get the supply and demand for electricity synergistically supporting one another. For example, Francis Walton developed an electric motor, one that would turn a shaft if you supplied electric current, in 1835. However, the Faraday generator was only a toy, and, therefore, without any economical supply of electricity, Walton's motor was a toy as well.

The pace picked up in 1863. In that year Antonio Pacinotti developed a much-improved generator. He also showed that if you supplied electricity to the generator, it became a motor, which gave us two for one. Cool!

At this same time, the carbon arc illumination business was showing a little life. Two French lighthouses were given carbon arc lights. Demand was calling on supply. Supply answered.

The Belgian Zenobe Gramme developed an improved Pacinotti-like generator in 1871 and enhanced that in 1873. These were driven with a reciprocating steam engine. Gramme continued to improve these generators and also developed a line of electric motors, providing both a source of and demand for electric power. He did a good business in the 1880s.

The illumination business was getting interesting at this time as well. A gentleman by the name of Jablochkoff, who worked for Gramme, came up with an improved carbon arc light, called an arc candle in 1876. Charles Brush, an American, made a better one that burned twice as long in 1879. Brush did pretty well with this. He designed a generator system to go with his arc light and sold it as a street lighting system to several American and European cities.

But, of course, we are all waiting for that great American inventory Thomas Edison to enter the game. He started his search for an electric light in 1877. Edison wanted to make a practical incandescent light. That is a light that gets light from a filament that is heated until in glows. The arc lights produced a very bright, but harsh, light and, despite Brush's clever self-adjustment features, were difficult to set right and burnt out rapidly.

The idea of incandescent light went back to as early as 1820, but it had not worked in any form that could be economically produced for mass use.

The problem was that it was very hard to get anything hot enough to glow enough to give off decent light without burning up. Which leads to Edison's famous search through a myriad of potential filaments. Along the way he had successes with platinum wire, carbonized thread, and split bamboo.

Several things were happening at once now, but let's follow lighting for just a bit.

The carbon filament light of 1881 produced about 1.7 lumens per watt of electricity. The units are not important, but for a sense of scale, a modern 100-watt light bulb produces about 2200 lumen. That means the 1881 bulb would have to be powered by about 1300 watts to make the same light.

By 1905, the carbon filament lights had improved up to 3.4 lumens per watt. A metallized carbon filament replaced these in 1905, which brought the bar up to 4.25 lumens per watt. These were produced up to 1918 but were being given a run by tungsten filament lamps that started out at 8 lumens per watt in 1902. By 1918, gas-filled tungsten filament lamps were the way to go and are represented by the modern light bulb. Further improvements have brought the efficiency up to 22 lumens per watt, 13 times better than the original carbon filament bulb.

However, that is not the end of the story on illumination. Fluorescent lights first appeared in the 1930s arriving in the United States in 1938. In a way this returns to the idea of the old arc lights, with an arc being generated down the length of the tube. The difference is the arc is used to excite or fluoresce material in the inside coating of the tube. The result is a cool white light at 65 lumens per watt, 38 times as efficient as the original carbon filament and nearly 3 times as efficient as the modern incandescent lamp.

Now to get back to our buddy Thomas Edison. We left him just as he was getting an economically producible incandescent light working.

His first application of the new lamp was fitting the arctic exploration steamer *Jeannette* with lamps and a generator in 1879. He fitted out another steamship, the *Columbia,* in 1880. It ran between San Francisco and Portland, Oregon, with an electric light in each stateroom.

With these modest beginnings, Edison went on to install over 150 generator and lighting systems in individual buildings and ships by 1882. The most famous of these was the Pearl Street Station in New York, completed in 1882. It had six generators producing about 900 horsepower or about 670 kW, enough for 7200 lamps. He started up the system with 60 customers and powered 1300 lamps in the vicinity.

And that was the problem; it was available only in the vicinity. Although Edison's systems were allowing energy to be produced in a central station and transmitted to distant users, the distant user had to be, in fact, pretty close. Edison doggedly stood by direct current. The problem with direct current at the time was that it was limited in transmission voltage. Voltage is like pressure in hydraulic systems. If the pressure is low, you can't move much fluid. At a low voltage, you can't move much electricity, at least not very far. The resistance in the lines uses up all the pressure after you go just a short way. A competing system, alternating current, could be easily stepped up in voltage for transmission and stepped down for use at the other end. Higher voltage makes for much more efficient transmission.

Alternating current systems were being developed in Europe at this time. In 1885 George Westinghouse bought the rights to one of the more promising systems. Westinghouse built a generating station in Buffalo, New York, in 1886. It also helps that a practical electric motor that could use alternating current was developed in 1888.

In 1891 Westinghouse built the first hydroelectric plant at Willamette Falls and used his alternating current transmission system to ship the power 13 miles to Portland, Oregon. Later in that same year, a 109-mile transmission line was built in Germany.

While there was a protracted battle between Edison and Westinghouse, it was the practicality of shipping power over long distances, demonstrated by the Willamette Falls and German installations, that gave the victory to Westinghouse.

We need not feel bad for Edison. Through the development of many other inventions and good business management, his company General Electric is one of the largest and most successful in the world today.

The story of parallel development isn't complete just yet. With long distance transmission working and electric lights and motors creating demands, the stage was set for large central station power plants. In 1900 these were either hydro plants or steam plants. The steam plants were boilers driving large reciprocating engines. The largest of the reciprocating engines were built in 1904 producing 7500 horsepower. These were monsters. If you ever get the chance to go to the Smithsonian Museum of American History in Washington, D.C., look for the display of one of these. You get to stand on the top of one of the pistons where there is room for a dozen of your friends, then go downstairs to see the bottom of the connecting rod. For a sense of scale, the connecting rod in your car is maybe 4-inches long. These beauties were perhaps 20 feet.

While a very impressive massive hunk of steel, these engines were very inefficient. But there was a better way to get steam to make something turn: the turbine. The ancient Greeks recognized expanding steam could be used to spin a wheel. A number of designs were developed in the mid-1880s, but none were very efficient. Carl de Laval produced a practical steam turbine in 1889. Note the serendipitous date. Just as demand for electricity is growing, an improved means of converting heat energy into mechanical energy at much higher efficiencies comes along. A number of improvements followed. By 1930, the 7500-horsepower reciprocating steam engines had been replaced by 240,000-horsepower turbines.

Turbines weren't the only way to improve on the efficiency of making electrical power. More efficient burning of fuels in the steam boilers and better materials that allowed operation at higher temperatures and steam pressures helped. Figure 3.1 shows how the thermal efficiency of the production of electricity has changed over time.

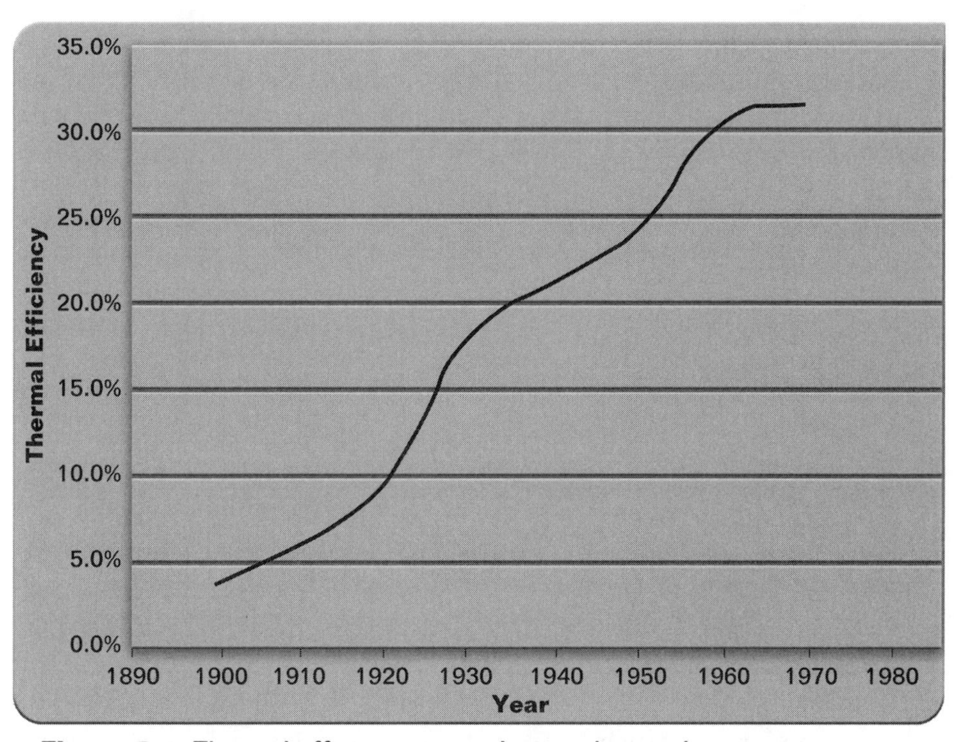

Figure 3.1. Thermal efficiency in producing electrical power.

Remember that thermal efficiency is a measure of how much of the heat energy gets eventually turned into electrical energy. From this figure, you can see we have come a very long way. Because a lot of the cost of producing electricity, at least for fossil fueled electricity generation plants, is in the fuel, it was very important to get as much electricity out of each unit of fuel you used. Therefore, this improvement in thermal efficiency has done a lot to reduce the effective price of electricity. Figure 3.2 shows you how the price of electricity has changed over the last 75 years.

Except for price increases at the start of the Great Depression and following the oil shocks that began in 1973, the cost of electric power has gone down steadily such that the 1996 rate is just about one-quarter the 1926 rate.

A lot of that is due to the improvements in thermal efficiency, but a lot of it is also due to the economies of scale. That is, it is sometimes cheaper to do something big than to do it small. For instance, if you built a standard family

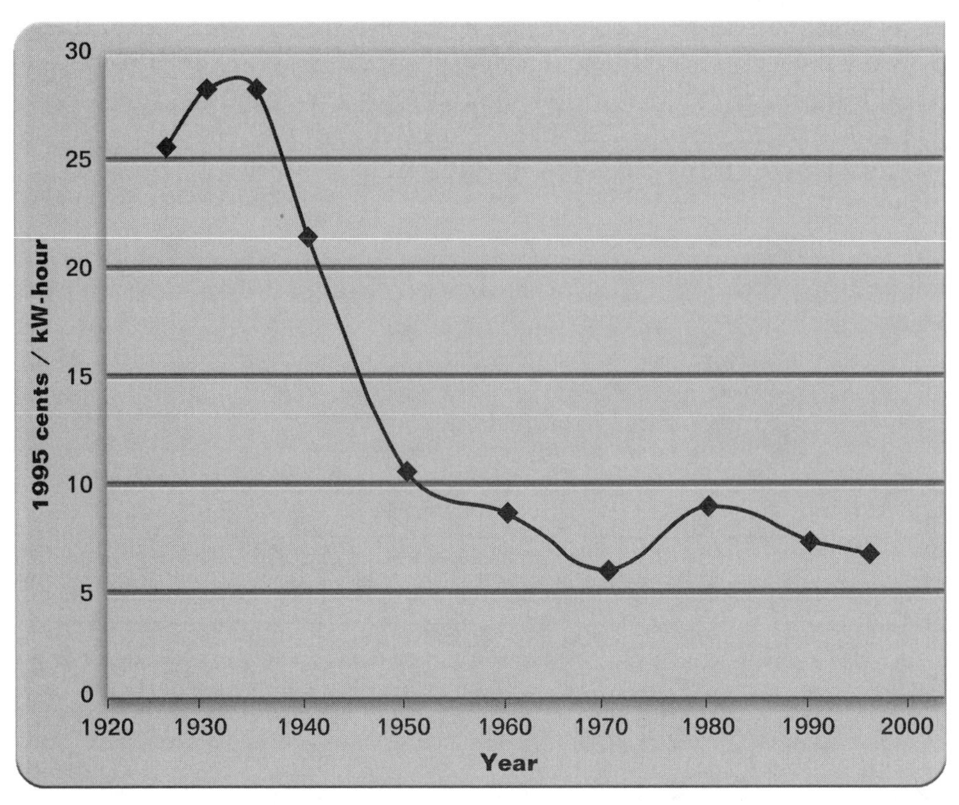

Figure 3.2. Cost of electrical power in the United States.

Taken from: *Electricity Prices in a Competitive Environment-Marginal Cost Pricing of Generation Services and Financial Status of Electrical Utilities*, DOE/EIA-0614. U.S. Department of Energy, Energy Information Administration, August 1997.

sedan by hand, from scratch, it might cost you a million dollars versus perhaps $20,000 if you make a million of them a year.

Making electricity on a grand scale didn't happen overnight. Let's go back to the story and pick it up around 1900.

The number of utilities had spread rapidly in the 1890s, but these were still fairly small and local. After 1900 more of these small utilities sprang up, and a number of them merged into larger firms, spanning state lines. Steam turbines began to replace reciprocating steam engines providing both greater efficiencies and much larger capacities. The federal government began to invest in large hydroelectric projects. Overall the electrical generation in the United States grew at a 12% annual rate from 1901 to 1932. That means it doubled about every 6 years despite significant, real drops as the Depression hit in 1929.

Where was all this new electrical power going? Just about everywhere. In 1907 only 8% of the dwellings were electrified. By 1932 that was up to 67%, two-thirds. In the cities the value was higher yet at 80%, while the rural farms were only 11% electrified.

In recognition of the limited expansion of the benefits of electricity to the rural communities and as part of the Depression recovery efforts, the federal government launched a series of programs to support rural electrification. With loans and other assistance, getting power out to the farms became a priority. By 1941 rural electrification was up to 35%. Perhaps a modest increase, but it was an impressive achievement to triple the number of farms with electric power in only 9 years.

During the 1930s the federal government also invested heavily in hydroelectric power. The Tennessee Valley Authority was created in 1933, and the Bonneville Power Administration in 1937. These large federal agencies created vast hydroelectric systems that still serve us today. The total addition of new electric generation from 1932 to 1941 grew at about 8% per year. Although a bit off the 6-year doubling time from 1901 to 1932, this is still doubling every 9 years. Besides increasing total capacity, the equipment and systems for distribution of electrical power also were significantly improved during this period. The basis of a national infrastructure for electrical power was being set.

When fortress America responded to the Second World War, the industrial might of the nation was given a massive challenge and answered it. The capacity of the federal electrical generation systems doubled during the war. Industrial applications of electricity such as the smelting of aluminum and the fabrication of aircraft, tanks, and all manner of munitions blossomed. The rural electrification was not left behind, with over half of all farms electrified by the war's end.

It is also important to recognize the amazing drop in the real price of electricity over this time. From 1935 to 1950, the price of power dropped more than a factor of 2.5. Such a reduction fostered even greater acceptance and use. More and more ways were found to use electricity.

The post-war economic boom supported continued rapid growth in electric power generation and use. The generation growth rate was an annual 8.5% in the 1950s and 7.5% in the 1960s. The 1950s saw a major increase in residential electric use and essentially the completion of rural electrification. In the 1960s electricity became increasingly important as a means of providing convenience, quality of life improvements, as well as continuing the amplification of manufacturing productivity. The price of electricity continued to fall but more slowly. The improvement in thermal efficiency was getting nearer its limit. Additional improvements would be increasingly difficult to obtain. We also started to impose environmental controls. The passage of the National Environmental Policy Act of 1969 shepherded in a new era of government oversight.

The 1970s were tough on the entire energy business. The oil shocks that began in 1973 set off massive fuel price swings, worldwide economic recessions, and crippling inflation. Electricity prices rose for the first time since the Great Depression. While generation still grew at 4% per year, capacity had grown faster than demand.[5] This put even further strain on the firms generating electricity.

During the early part of the 1980s, the overall economy declined. As the recovery kicked in around 1984, electrical power enjoyed a rapid growth. The commercial sector became more important and led the growth in new electrical demand. During this period most of the oil-fired electrical generation plants were replaced by coal-, nuclear-, or natural gas-fired electricity.

In the 1990s we saw an increasing number of non-utility generators. These took two forms. One was large industrial users generating electricity as part of their process. They either produce electricity as a by-product of their process or, while producing electricity for themselves, produce extra that they can sell to other users. The second type of non-utility generator is the independent power producer. Over the years the U.S. and state governments allowed firms to hold monopolies on producing and distributing electric power. Early on it was recognized that this was the more efficient way of getting electricity to users. It didn't make any sense for company ABC to string wires right next to company XYZ. As long as the government made sure the rates were reasonable and the monopolies didn't abuse their customers, individual companies were granted sole "ownership" of given areas. Similar systems were set up for telephones, natural gas, and cable TV, for all the same reasons.

Now we are seeing increasing deregulation of electrical power. The first stage is to allow non-utilities to produce power. A regulated utility may still be the distributor of the power, but more independent power producers are appearing to sell their power to the utilities.

In 1998 non-utility generators produced about 11% of all electricity.

While there have been major gains in environmental performance of fossil fuel-fired power plants, the 1990s saw increasing costly restrictions. With oil-fired plants almost out of the picture and natural gas being relatively clean, greenhouse gases not withstanding,[6] most of the new restrictions have fallen on coal-fired plants, which still provide about half of all our electricity.

---

[5]Capacity is how much power the system is capable of making if it runs full out. Generation is how much energy the system actually does produce in any given period. Because the amount of electricity we use changes with the time of day and time of year, the electrical system must always have more capacity than average need. To equalize the distribution you need something in excess of the maximum demand. The bad news is that the power producers then have some generating capacity sitting idle all the time and, therefore, during that time making no money for them.

[6]Natural gas is free of a lot of the pollutants that come in coal or oil. However, all fossil fuels produce carbon dioxide. Carbon dioxide is the greenhouse gas of primary concern related to global warming caused by greenhouse gases. The good news is that per kilowatt-hour generated, natural gas produces about one-third less carbon dioxide than coal.

As we look back over the history of electrical power, it is important to understand the impact it has made on our lives. Let me introduce the concept of productivity. This is an economic measure of how much output we get per labor hour expended. It is how much you can get done per unit time per worker. Well, so what? Why should I care? Sounds like something a stuffy economist would put into long boring tables of financial statistics. Well, economists, stuffy or otherwise, do worry a lot about productivity, and you should too.

If one worker can build one chair a week, that chair is going to be pretty expensive and you are probably going to need about a week's pay to buy one. If, however, a worker can build 10 chairs of equal quality in an hour, those chairs are going to be a lot more affordable for everyone.

So, how can you increase the rate of chair production? Well, electrical saws, drills, sanders, and paint sprayers will be a big help. If you add in computer-aided design, parts inventory, and product distribution, you get even more means of reducing the ultimate price of that chair you want to buy.

The point is: the more stuff you can make per labor hour the less expensive that stuff is, the more things you can afford, and the quality of your life gets better.

I have toured both Mount Vernon and Monticello. Now, in 1800, George Washington and Thomas Jefferson lived about as well as you could at that time. In addition to a lot of slaves to work the farms, both had an extensive household staff to prepare the food, wash the clothes, make the soap, and on and on. The average Joe and Jane today are rich beyond the wildest dreams of George and Tom. Not only do we have electric servants to wash clothes and dishes, get the fuel to cook the food, and a hundred other tasks that would have been drudgery in 1800, we have a vast electrical-powered economy that makes an amazing array of products for us that George and Tom couldn't begin to imagine. And due to the high productivity driven in part by the universal availability of cheap electrical power, the average person can afford what would have been a king's ransom of those products.

Let's look just a bit at the relation between productivity and electrical energy use. Figure 3.3 shows how the manufacturing productivity index and the use of electricity in the industrial sector has changed with time.

Now, to be clear, I did fiddle with the data just a bit. I plotted the electrical use in units of 10 billions of kilowatt-hours per year. This conveniently made the units of the Census Department's electrical energy use for the industrial sector match their manufacturing productivity index. However, regardless of how one might normalize the data, it is an impressive match.

For scientific honesty, you have to understand the difference between correlation and causation. Correlation means two parameters are behaving the same. Causation means the reason they are correlated is that one is causing

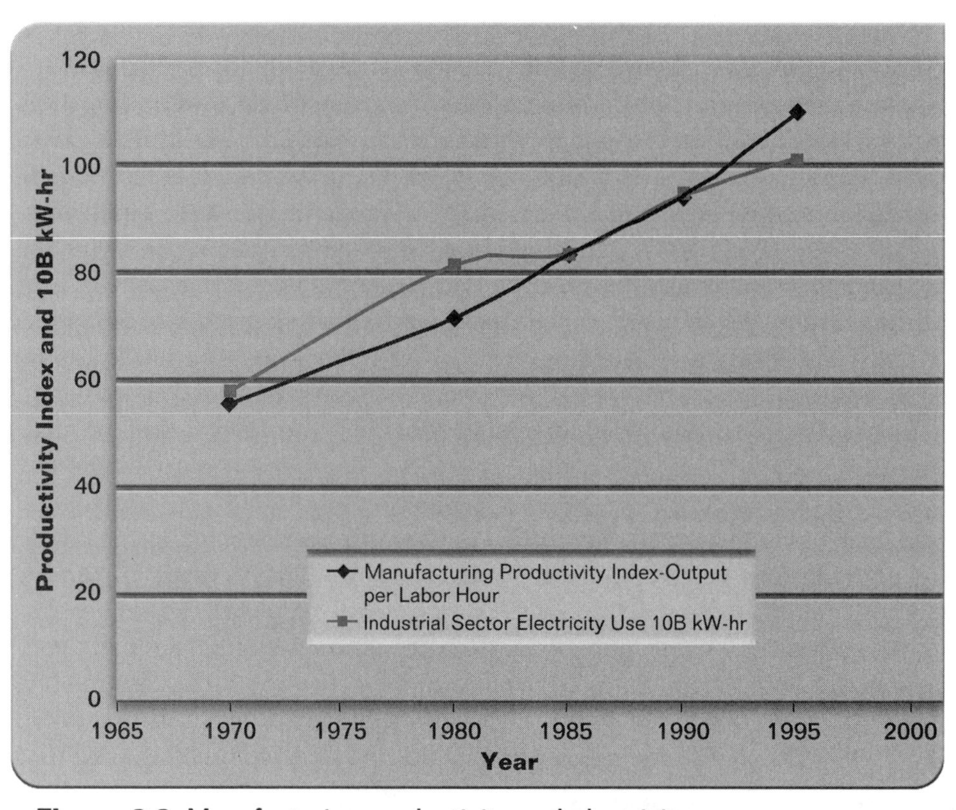

Figure 3.3. Manufacturing productivity and electricity use.

the other. One might plot the sales of bubble gum and the rate of communication satellites launches. Choosing suitable normalization factors you might very well show both have been growing at about the same rate over time. That does not mean bubble gum leads to satellites, nor do satellites lead to bubble gum.

So, you should not take Figure 3.3 to prove that electricity use leads to productivity. Having said that for the sake of honesty, I think you can postulate that because electricity has the ability to replace and augment human energy, increasing electrical use could enhance productivity. And that the data shown in Figure 3.3 strongly support that theory.

Which is why the price of electricity is so very important. Our standard of living can increase if our income increases faster than our cost of living. If I get raises faster than prices go up, I can buy more stuff, of course. But how can that be? If everyone gets a raise, that money has to come from the people who buy what all of us produced. So clearly on the global scale of things, you can never win. If we get paid more, everything we buy costs more, so we don't get anywhere, at least not as a group overall.

But wait! If productivity goes up, then you can produce more for the same cost, which means either you can afford to pay the worker more while keeping the cost of the product the same, or you can reduce the price of the thing produced. Great, let's all get more productive. Well, there are two ways to do that. One, get smarter. Find a better, more efficient way to produce whatever it is you make. Two, use more energy to allow the worker to make more of the product per unit time, which is fine as long as the additional energy costs less than the labor you are augmenting. Looking back at Figure 3.2, you see that during the 1940s, 1950s, and into the 1960s, this was a slam-dunk as the price of electricity got cheaper and cheaper. Through the 1970s, 1980s, and 1990s, the news is not as good. Lest we forget the topic of the book, let me just point out that one of the charms of nuclear power is that, because fuel cost is such a small part of the cost of power, nuclear electricity prices are not subject to big swings in fuel cost. For fossil fuel power electrical generation plants, especially natural gas-fired plants, because the cost of fuel is a bigger fraction of their total cost, the cost of power is much more impacted by swings in fuel prices.

Electrical Power Use Today

So, how do we use electricity today? If we look beyond the United States, the shocking and sad answer is a third of us don't use electricity at all.

A third of the earth's humans, fully 2 billion people, don't have access to electricity. For those of us in the developed world, that is hard to imagine. From the middle of the last century on, our use of electricity has grown as much on life *enhancers,* such as washing machines, televisions, and electrical appliances, as with industrialization. The developing nations don't yet have the life *enablers* such as water purification and distribution, refrigeration, and electricity for cooking. Globally, this is the next big increase in electrical power production. Providing the life enablers and some of the life enhancers to the developing world is going to take a lot more electricity than we now make. Which if all done with fossil fuels is going to be a problem.

But let's get back home and look at how we in the developed world use electricity. You could name a lot of things you use electricity for around your house. Lighting, kitchen appliances, ovens, microwaves, power tools, televisions, VCRs, stereos, garage door openers, and computers might make up a list. Depending on where you live, that list might include hot water heating, and heat and air conditioning for your whole home.

But residential use is just one portion of the electricity we as a nation use. Both commercial and industrial sectors use a lot of electricity. In 1999 the distribution of electric use was about equal thirds to each sector. Over the years the fraction of industrial use has declined, especially relative to the commercial sector. That is not to say the total amount decreased, just the relative fraction. The fractional shift has been due to the shift in our economy away from the industrial sector toward the service sector.

If electricity for commercial and industrial sectors seems a little remote to you, I suggest you think a bit about it. Those sectors are working for you. First, they give you a job, and second, they produce all the goods and services you need to have a life.

As an aside into the industrial sector, electric furnaces now produce something like 40% of all our steel. Why is that important? Electric furnaces can operate using 100% scrap as opposed to basic oxygen furnaces that are limited to about one-third scrap. This makes electric furnaces much more flexible. And, by the way, basic oxygen furnaces work only because electrically-driven atmospheric cryogenic separations plants produce the massive amounts of pure oxygen needed to run them. So essentially all of our steel is dependent on electric power.

Almost anywhere in industry where you make anything move, electricity will be there. Electricity is out there pushing the assembly line forward, making motors turn, welding steel, heating, cooling, drying, mixing, smelting, and doing a million other things.

If you need one last hint of how important electricity is to us, I offer Figure 3.4 for your inspection. This is the historical per capita use of electricity in the United States. There are two important things you need to understand about this data. First, this is not a picture of how much overall electricity we have come to use over time. This is just how much each one of us uses. That is, it is the total amount of electricity divided by the number of people. If you wanted to show the total amount of electricity used, you have to multiply these numbers by the population. So if you think this curve is steep, think how steep it would be if you included the fact that the U.S. population has increased by about 3.5 times since 1920.

The second important thing is that this is not the electricity that each of us personally uses. It is rather the electricity used by *and* for each of us. That is, in 1999 at least, about a third of this is what you used in your home. About a third is used in offices and stores where you may work or use the services thereof. And about a third is used in industry where you might work and certainly where all your "stuff" comes from. So this is the total electricity used to make your life work.

So, now look at the curve. Pretty amazing, huh? After World War II, we just went nuts over electricity. It replaced so many things and gave us things we never had before. We use over 6 times as much electricity per person now as we did in 1950. I recall moving into a brand new office building in about 1978. Ten years later we were having serious problems with the wiring because it couldn't handle the electrical load of all our computers, fax machines, and

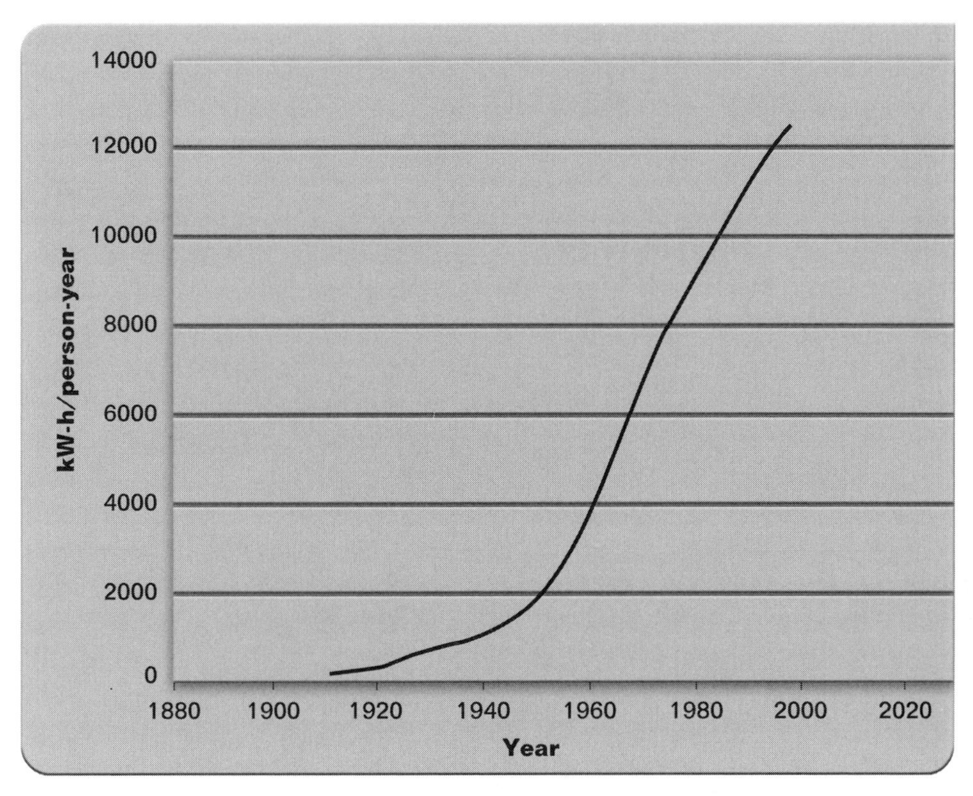

Figure 3.4. Per capita electrical use—kW-h/person-year.

copiers. The single most amazing thing about Figure 3.4 to me is the fact that the curve is still going up. There appears to be a slight curving over, indicating that recent increases might not be as strong as those of 1950s to 1980s, but the growth is still very strong and the demand continues to zip upward.

In a nutshell, my case for electricity is that it has become an integral part of life as we know it and is responsible for a great deal of what makes life both possible and largely pleasant.

4

A LITTLE BASIC NUCLEAR SCIENCE

For most people, if you are tempted to skip a chapter, this is the one you would pick. I hope you don't because it is really not all that hard, and it will help you understand many of the issues of nuclear energy much better. It is also where some of the really cool stuff is. So, release your grip on your science anxiety, and come along to find out how all this stuff really works.

Atomic Structure

Back in the days of ancient Greece, people were trying to figure out what everything was made from. About 500 B.C., two Greek philosophers by the names of Leucippus and Democritus suggested that stuff was made up of tiny, indivisible particles. They called these tiny particles atoms. Aristotle didn't like that theory, so it didn't catch on at the time. It turns out that Leucippus and Democritus were generally right, but not quite in the way they had imagined.

Stuff is made of atoms, but atoms are not indivisible. Furthermore, atoms are a lot tinier than Leucippus and Democritus could have guessed. The diameter of an atom is something like 1×10^{-8} centimeters.[1] This is very, very small. One hundred millionths of a centimeter. The largest atom appearing in nature is uranium. One single atom of uranium weighs about 4×10^{-22} grams, or 9×10^{-25} pounds.

Atoms are made up of things smaller yet. The primary building blocks are electrons, protons, and neutrons. These building blocks are made of smaller parts yet. Just what the subparts are and how they are arranged is an area of science that is still evolving. It is very complex, and we will not need that to understand basic nuclear science. We can content ourselves with understanding a little about electrons, protons, and neutrons.

Let's learn a little about these tiny particles. Electrons are very small as compared to protons and neutrons. Electrons weigh about 9.11×10^{-28} grams, where protons weigh about 1.673×10^{-24} grams and neutrons are slightly heftier at 1.675×10^{-24} grams. Therefore, electrons weigh about 1830 times less than protons or neutrons. Electrons have a negative electrical charge, while protons have a positive charge, and neutrons don't have any electrical charge. So what?

Electrical charge leads to what is called the electromagnetic force. We learned all about this back in Chapter 3, but let's go over it again here to be sure we have it. If you put two opposite charged things close to one another, they are attracted to each other. If you put two things with the same charge close to each other, you will get a force trying to push them apart. All of electricity and electronics is based on this force. It is what makes electricity flow through wires and dancing dots of light appear on your TV screen.

Normal atoms are neutral in charge. This means that the number of electrons and the number of protons any atom has needs to be the same. With the same number of negatively charged electrons as positively charged protons, they balance out with no net charge over all, or neutral charge for the whole atom.

---

[1]Let me explain the notation here. In science we get to write a lot of very big and very small numbers. Writing all those zeros is wasteful and also hard to count correctly. A method called scientific notation was developed to make very big and very small numbers easier to write and read. It works like this for big numbers: you put down the first digit, then the number of times you have to multiply that number by 10 to get the number you wish to write. So if you want to write 1,000,000 or one million, you can write 1×10^{6}. This means one multiplied by 10, six times. You only have to count the number of digits between the first digit and the decimal point to find out how many times to multiply by 10. If you want to show more than one digit, you only have to place a decimal after the first one. So, if your number is 12,345,000, then you write 1.2345×10^{7}. If you want to write a very small number, you can use the same method but use a negative number to show how many times to divide by 10. So 0.000034 can be written as 3.4×10^{-5}. Notice that the number of times you divide by 10 is one more than the number of zeros between the decimal point and the first digit of your number. This means that 1×10^{-8} centimeters is 0.00000001 centimeters. We will be using scientific notation throughout the book.

The electrons take up most of the volume of an atom. They move about the central part, which is called the nucleus. This is where the protons and neutrons are found. The nucleus has a diameter less than one ten-thousandth of the atom. It is convenient to think of electrons orbiting like little planets around the nucleus. We now recognize that their motion is a lot more complex than that. We also know that electrons have certain allowed states within their motion. Each of these allowed states has a certain energy level associated with it. If you move an electron to a higher energy state, you need to add energy to do that. If an electron moves down to a lower energy state, energy comes out. This is what drives all chemical reactions. If you burn something in air, the thing you burn combines with oxygen. This allows electrons in the burning thing and the oxygen from the air to shift to lower energy states, and energy in the form of heat and light comes out of the reaction.

The fact that electrons are on the surface of atoms makes having interactions with the electrons of other atoms fairly easy. However, since the energy shifts between the electron states are small, the relative energy released per reacting atom is very small. We will compare this to nuclear reactions in a minute.

Remember that the number of electrons and protons in an atom is the same. This is called the atomic number. It is the number of electrons and protons that sets the type of atom. A given type of atom is called an element. This is because it was discovered that all of nature is made up of the combinations of 92 elemental atoms. Just 92 different atoms can be combined in various forms to make up everything.

If the atomic number is one, one electron, one proton, the atom is the element hydrogen. If it is two, the atom is the element helium, 6 is carbon, 8 is oxygen, 26 is iron, 79 is gold, and 92 is uranium. Because the atomic number is the number of electrons and electrons drive chemical reactions, it is the atomic number that determines the chemical behavior of an element. When iron and oxygen come together, they like to form iron oxide or rust. If you bring gold and oxygen together, you get gold and oxygen since they don't react.

So far this is pretty simple, but what about neutrons? At the lighter end of the list of elements in general, the number of neutrons and protons will be about the same. As the elements get heavier, the ratio of neutrons to protons will begin to exceed one and move up to about 1.6 neutrons per proton by the time we get to the end at uranium. Figure 4.1 shows the number of neutrons compared to protons for the stable elements.

The interesting thing is that any given element can have several possible numbers of neutrons. Hydrogen can have no neutrons, one neutron, or two neutrons. Each of these are called isotopes of hydrogen. It is the same element but a different isotope. The no neutron version is called normal hydrogen. It turns out this is by far the most common kind. With one neutron, it is called

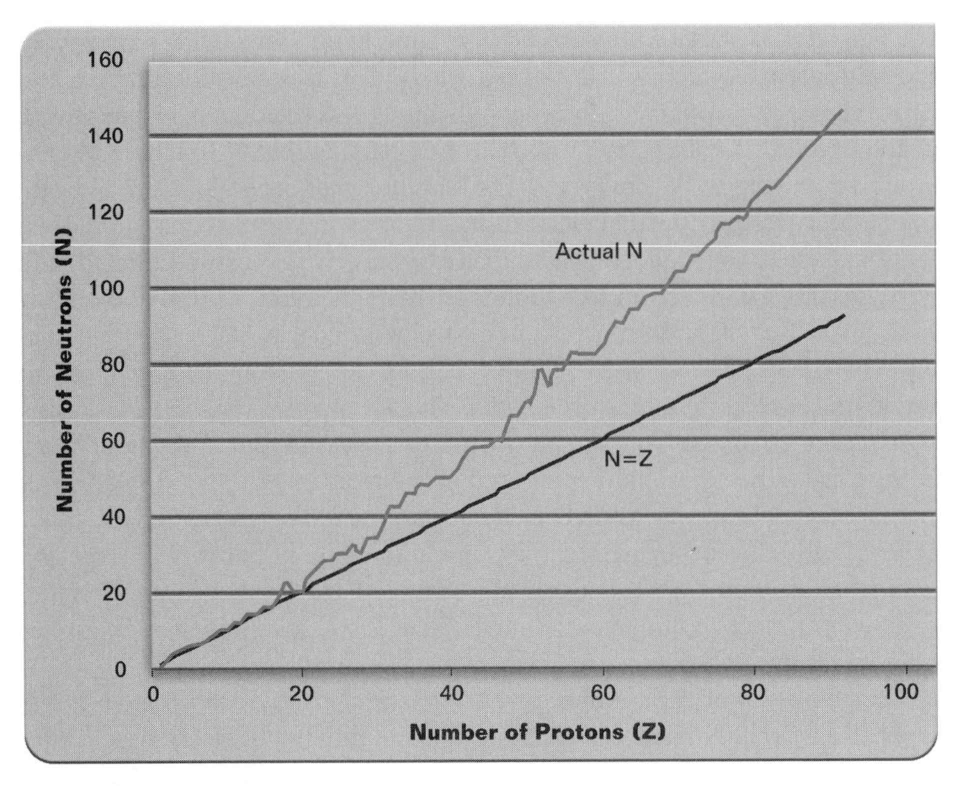

Figure 4.1. Number of neutrons and protons for stable isotopes.

deuterium, and with two neutrons, tritium. All three isotopes have the same atomic number, that is, one. Therefore, chemically they behave the same. Bring them together with oxygen and a little heat and you will get a bang that results in water being formed. You can think of water as hydrogen rust if you want. What is cool is that the nuclear behavior of these three isotopes is very different.

This makes sense because what is different about different isotopes is the number of neutrons. The difference is in the nucleus and that is where the big energy is. This is the heart of the atom, where most of the atomic mass is squished into one trillionth of its volume. As we will see, while the chemistry of different isotopes of any given element is the same, the nuclear properties can be very different. We can use these differences to release the energy of the nucleus.

Let's see how.

Binding Energy

Let's have a quick review to see if everyone is still with us.

All things are made up of atoms. Atoms in turn are made up of electrons, protons, and neutrons. Electrons are much smaller than neutrons and protons. Electrons whirl around on the outside of atoms and take up most of the space. Neutrons and protons are bunched together in the center of the atoms in what is called the nucleus. The number of electrons and protons must always match. This is called the atomic number of the atom and sets the chemical properties. Each unique type of atom with its unique atomic number is called an element. In various combinations, 92 elements make up the entire natural world. While the number of electrons and protons are fixed for any given element, the number of neutrons is not. While generally close to the same number of protons, the number of neutrons can vary, with the ratio of neutrons-to-protons increasing as the elements get heavier. Atoms with the same number of protons but a different number of neutrons are called isotopes of the same element. Different isotopes of the same element will have the same chemical behavior, because their surrounding electrons are the same, but very different nuclear behavior because the number of neutrons and therefore the nucleus is different.

Are you still with us? Good.

Above we learned that the mass of an electron was 9.11×10^{-28} grams and proton was 1.673×10^{-24} grams and neutron was 1.675×10^{-24} grams. Now, if you wanted to build an atom of aluminum, you would check the chemistry books and see its atomic number is 13. Therefore, you would need 13 electrons and 13 protons. Checking the book further you would find that all natural aluminum has 14 neutrons, so you add them.[2] If you do the math, you find out that all these parts should weigh 4.52022×10^{-23} grams. If you collect some pure aluminum, weigh it, and divide by the number of atoms[3] you would find that an aluminum atom weighs 4.48012×10^{-23} grams. This means that for each aluminum atom 4.01×10^{-25} grams is missing! Is this unique to aluminum? Nope, all atoms show this effect. You add up the weigh of its parts, and you get a number slightly bigger than what you get if you weigh the combined atom.

It turns out what is happening is that the "missing" mass is being used as energy to hold or bind the nucleus of the atom together. The missing mass is called the mass defect, and the equivalent energy is called the binding energy. So how do you figure out how much energy is in the binding energy? You use what is probably the most famous equation in all of science.

$$E = mc^2$$

---

[2]We said before that generally the number of neutrons was close to the number of protons for light elements, but as you get to higher atomic numbers, nature seems to need more neutrons to make an atom work. One way to think of it is that, with all those positively charged protons all squished together trying to force each other apart, you need some neutrons to hold them together. As you get more and more protons, just one neutron per proton is not enough. At the top of the atomic numbers, the most common form of uranium has 92 protons and 146 neutrons. You can look back to Figure 4.1 to see how the general ratio of neutrons to protons changes.

[3]There are ways to do that, just trust me. We are trying not to get too technical, remember?

We have all seen Einstein's famous equation many times. But what does it really mean? It gives us a way to calculate the amount of energy you can get if you can convert mass to energy. The "E" stands for energy in a unit called ergs. One erg is 1×10^{-10} kilowatt-seconds. That is, it would take 1×10^{9} or one billion ergs to be enough energy to light a 100-watt light for one second. But, hang on, bigger numbers are coming.

The "m" stands for mass in grams and the "c" is the speed of light in centimeters per second, which is 3×10^{10}. The superscripted "2" means you multiply the "c" by itself, that is, you multiply two of them together. If we try this for one gram, we get:

$$E = 1 * 3 \times 10^{10} * 3 \times 10^{10} = 9 \times 10^{20} \text{ ergs} = 9 \times 10^{10} \text{ kilowatt-seconds}$$

This works out to 25,000 million watt-hours. If we use the assumption we did before that 1000 million watts meets the electrical energy needs of one million people, then this is about enough energy for the electrical needs of one million people for one day. All out of the energy equivalent of one gram. Since there are 4.01×10^{-25} grams of binding energy in each atom of aluminum, it would take 2.5×10^{24} aluminum atoms to have one gram's worth of binding energy. It takes just over 111 grams of aluminum to have that many atoms. That is about the weight of 8 empty soft drink cans. The binding energy in the aluminum in 8 soft drink cans can provide enough energy for one day's electricity for one million people!

Unfortunately we can't tap into the binding energy of aluminum soft drink cans. However, we can get to the binding energy of other isotopes, that is especially light isotopes and especially heavy ones. More to come in the next section. First let's explore the binding energy of other isotopes.

If you do the same math as we did for aluminum with the most common forms of all the elements, you get a curve of binding energy shown in Figure 4.2. There is a little extra math done for the numbers in Figure 4.2. First the mass defect is given per nucleon, is the total divided by the combined number of protons and neutrons for that isotope. Also the mass defect is expressed in terms of energy using the unit million electron volts, or MeV. This is a convenient unit for describing nuclear reactions. One kilowatt-second is 6.22×10^{15} MeV, if you are trying to keep this straight.

The important thing about this plot of binding energies is that they peak about mass number 56. That is iron. This means the highest binding energy of all is around iron. If you could move up the binding energy curve from the light elements, like hydrogen, or up from the heavy elements, like uranium, toward iron, you would get a higher binding energy per nucleon. This means you would get energy out of the reaction, if you could do it. Figure 4.3 shows how much total energy is available if you could move from any given mass number to iron-56, on a per atom basis.

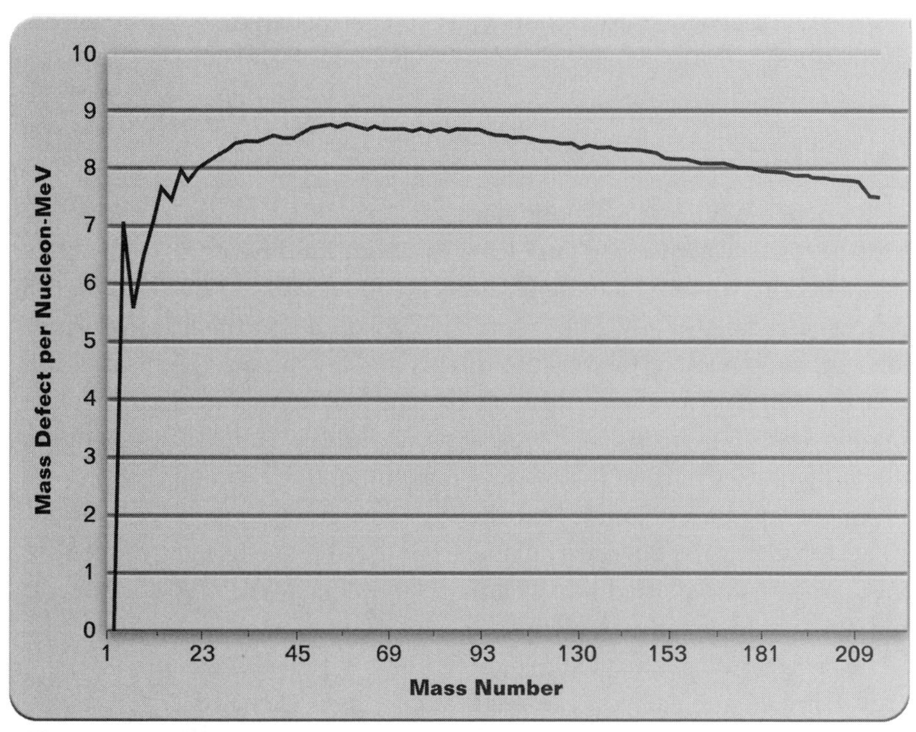

Figure 4.2. Binding energy curve.

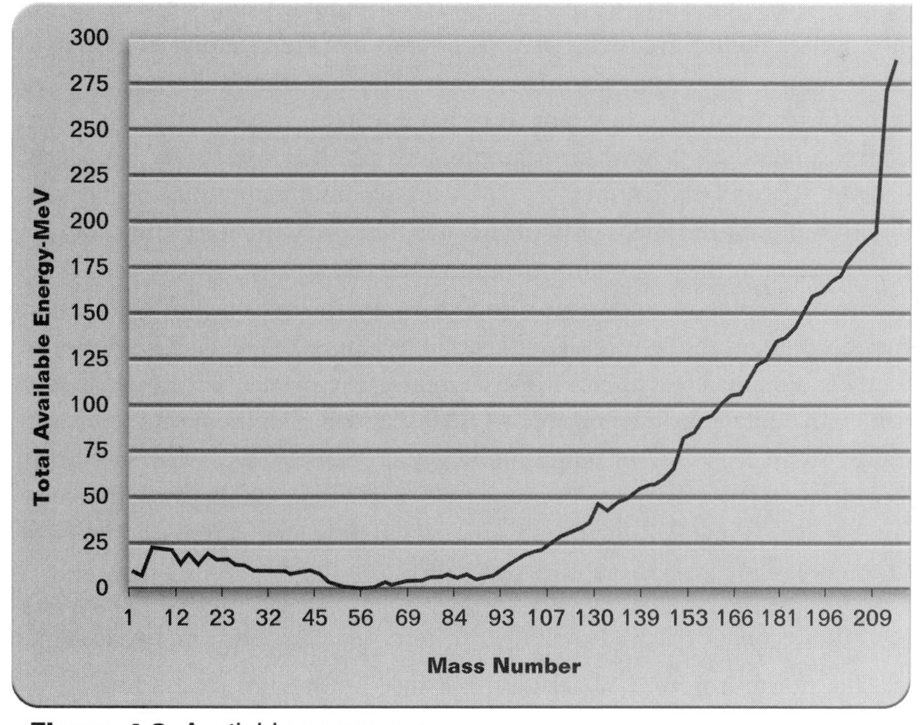

Figure 4.3. Available energy curve.

As we will see in the next section, you can't get all that energy, but you can get a lot of it.

Fusion and Fission

This is where the nuclear reactions start.

In the section above you just learned about binding energy. Again, to check to be sure you are hanging in there, if you add up the electron, proton, and neutron masses of any given atom, you will find that the mass you get is greater than the mass of the real atom. This missing mass is the mass defect, which is converted to energy to hold or bind the atom's nucleus together. Using Einstein's famous $E=mc^2$, we can convert this missing mass to equivalent energy units. Charting the missing mass for all the elements gave us the binding energy curve. Noting that the highest binding energy was around iron, we calculated the energy available if we could move from any given element to iron-56 and plotted that as an available energy curve. We then wondered if we could get at that energy.

The answer is we can get at a lot of it, but not quite all. From the binding energy curve, Figure 4.2, you can see that very light elements like hydrogen have a long way to go to get to iron. That is, there is a lot of energy on a per nucleon basis that would be released, if you could convince a bunch of hydrogen atoms to combine and form iron. It turns out that is pretty hard to do. So, what about something easier? Let's put some hydrogen atoms together to make helium. A look at our chemistry book will tell us that helium needs two neutrons to go with its two protons so we can't use ordinary hydrogen directly. To make it easy, for an example, let's use the first heavy isotope of hydrogen. Remember, isotopes are the same element (same number of protons) but different numbers of neutrons. Deuterium is the form of hydrogen that has one neutron to go with its one proton. Great! Two deuteriums have all the parts we need to make helium. Turns out this works and gives out a tremendous amount of energy. Deuterium has a mass defect of 2.25 MeV, and helium has a mass defect of 28.3 MeV. Therefore, if you take two atoms of deuterium with a total binding energy of 4.5 MeV and smush them into one atom of helium with 28.3 MeV of binding energy, $28.3 - 4.5 = 23.8$ MeV of energy will come out. Figure 4.4 gives you a picture of this reaction. Because there are about 3×10^{23} atoms in one gram of deuterium, this means that one gram of deuterium, if converted to helium, gives off 3.6×10^{24} MeV of energy. This is 1.6×10^{5} kilowatt-hours, which if we again use in our one million person city is enough energy to provide their electrical energy needs for about 9.5 minutes. That means a pound of deuterium if fused to helium would liberate as

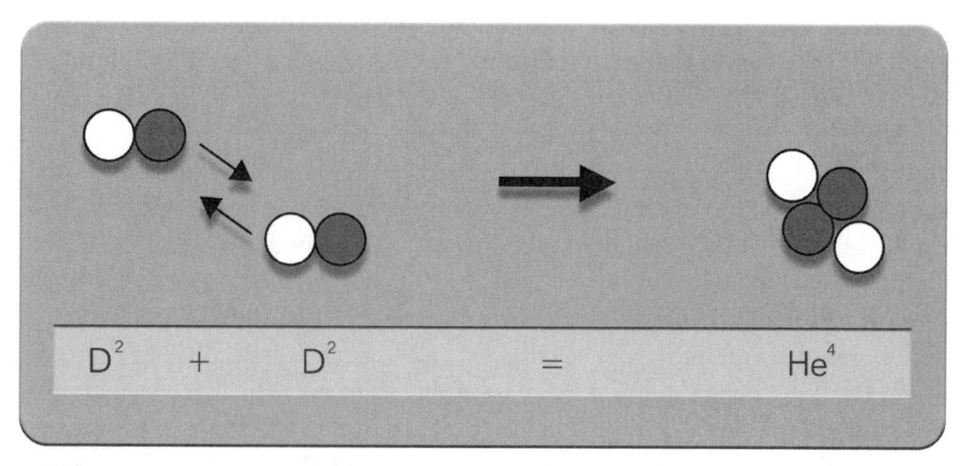

Figure 4.4. Deuterium fusion.

much energy as the electrical consumption of one million people for about 3 days.[4]

Can that be done? Is there a way to get this incredible energy? The good news is that this is the most common form of energy in the universe. The bad news is we have yet to figure out how to get that energy in a controlled fashion.

Smushing light elements to form heavier ones is called fusion, fusing together. Fusing hydrogen to get helium is the nuclear reaction that makes the sun go.[5] Fusing light elements into heavier ones and releasing their binding energy powers stars and, therefore, is the most common form of energy in the universe, out powering any other means to insignificance. The bad news for us is that it takes the gravitational force of a good fraction of the mass of the sun to cause the "easiest" fusion reaction to start. The thing is that even

[4]Please note that I said "would liberate as much energy as the electrical consumption of one million people for 3 days." In any process where you convert one type of energy into another, you can't make it all go across. Water falling in a hydroelectric plant might get as much as 90% of the water's energy into electricity. In an electrical generating plant in which we convert heat, for instance burning coal, to electricity, the efficiency is proportional to the temperature, the hotter the better. Really good systems might get as much as 55% of the heat energy to be electricity. Since we don't know how to make a fusion power plant, I can't really say how many kilowatts of fusion energy you might need to make a kilowatt of electrical power. A good guess would be 50% efficiency, which means our pound of deuterium, while liberating 3 days worth of energy, most likely would only make a day and a half worth of electricity.

[5]In our sun the fusion process is a bit more complicated than just smushing two deuterium atoms together. Our sun starts with normal hydrogen atoms, fuses two of those together to make deuterium, adds one more hydrogen nucleus to make helium-3 (two protons, one neutron), then smushes two such helium-3 atoms to make the normal helium-4 with two protons left over. In stars bigger than 1.5 times the size of our sun, it gets more complicated yet with carbon, nitrogen, and oxygen isotopes joining hydrogen in the cycle.

though we will get net energy out after the reaction is over, the reaction will not take place all by itself. Two deuterium atoms will not fuse into helium all by themselves. They are in a meta-stable state. That is, even though energy is available to be released, they are happy where they are.

Imagine this: there is a big hill with smooth sides. You have carried a bowling ball up to the top of the hill. Since you are very tired from carrying the ball up the hill, you know there is a lot of energy to be gained if you roll the bowling ball down the hill. But the thing is the hill isn't perfectly rounded. In fact there is a little depression just at the top. If you put your bowling ball in that depression, it is going to just set there, pretty much forever. To get it to roll down the hill, you are going to have to nudge it up over the lip of the depression and over the side of the hill. If you do that, then it will rush down the hill with great force, giving back the energy you invested in lugging it up in the first place.

Most reactions in nature are like that bowling ball in the depression at the top of the hill. The ingredients are in a meta-stable state and will stay that way until some energy is applied to cause the reaction to start. For each type of reaction, a different amount of energy is needed. This is the threshold energy for the reaction. Sometimes it is very small, and sometimes it is large. Let's look at burning paper. The carbon in the paper and the oxygen in the air would find a lower energy state if they could combine and make carbon dioxide. But paper doesn't just burst into flame. You have to make the paper and air mixture very hot to get the reaction going.

Fusion is like that too. The hill we come down from for fusion is very, very tall (lots of available energy), but the depression at the top (threshold energy) is very deep.

We have been able to create fusion on earth. Unfortunately the only way we can do it is in the heart of an atomic bomb. Not what you would call controlled energy. We have been working on controlled fusion for about 50 years now.

Progress is being made, but reproducing the conditions at the heart of a star is not an easy task. Estimates of the time yet needed to get controlled nuclear fusion producing practical energy range from 30 years to never, with 50 years being the accepted best guess.

So, we have seen that fusion is hard. How about getting binding energy out by working from the heavy element end back toward iron? Turns out that works and works pretty darn well. This is called fission and is the way nuclear reactors produce energy.

The first thing to look at is the binding energy curve of Figure 4.2. From that we see that if we could move from one of the very heavy isotopes back toward iron, there would be an increase in mass defect. This means there would be more missing mass which would be available to be released as energy. However, when compared to the energy release to be gained from the movement from the light isotope end, which is from fusion, it doesn't seem nearly as good. But remember Figure 4.2 is plotted on a per nucleon basis. That is, it is the binding energy for each of the total numbers of neutrons and protons. Because light isotopes have few nucleons and heavy isotopes have a lot, it is going to make the per nucleon numbers smaller for the heavy isotopes. A more revealing picture is given by Figure 4.3 which shows the available energy per atom if that isotope were converted to iron-56 atoms. This shows that more energy is available, on a per atom basis, if we come in from the heavy isotope end. While it might be hard to see on the figure, there is 290 MeV of energy available if we could convert an atom of uranium-238 to iron-56. This can be compared to 22 MeV available for lithium, which is the highest value on the light isotope side of iron.

Unfortunately you can't make uranium convert directly into iron. Iron-56 has 26 protons and 30 neutrons. Uranium-238 has 92 protons and 146 neutrons. So, if you divided uranium in two, you would get 46 protons which is palladium. Dividing by three would be between 30 and 31 protons, which would be between zinc and gallium. Actually, when uranium divides or fissions, it almost always splits into two parts. It doesn't change all the time into the same two parts but rather spreads out the possible break-up products from mass numbers 72 to 160. The spread is not even but has two peaks, one around mass 95 and one around mass 140. Since we are not moving all the way from uranium to the best case of iron, we don't get all 290 MeV of the available energy. But from Figure 4.3, you see that moving from mass 238 to masses 95 or 140 gives you a lot of energy. For uranium fission, the average energy release is just over 200 MeV. That's pretty good. If you are feeling bad that we didn't manage to get all potentially available 290 MeV, just remember that chemical reactions are measured in a few electron volts per reaction, and here we are getting 200 million electron volts per reaction!

Let's look just a minute at how the energy is released. Most of the energy released in a fission event comes out in kinetic energy of the two big break-up products (often called fission fragments or fission products). Kinetic energy is the energy of moving mass. A car parked at the top of a hill has potential energy. A car rolling down that hill has kinetic energy. In this case, the neat thing about the kinetic energy of the fission fragments is that they will bang into things pretty quickly and that energy will turn to heat, lots of heat. Heat is good because we know how to turn heat energy into electricity. These fission fragments will carry a little over 80% of the energy of the fission event. Various forms of radiation will also be released, and all of these will be converted to heat as well. They contain a little more than 10% of the fission event energy. Combined they come to just under 95% of the total energy. The other bit, about 11 MeV, comes off in particles called neutrinos. These guys are very common, and the sun makes billions of billions of them every second. The pesky thing about neutrinos is they pretty much ignore everything. They routinely pass all the way through the earth without interacting in any way. A very large number of them are passing through you as you read this. So they don't bother anyone, which is good. However, it does mean that those 11 MeV are lost to us. Still we get 95% of the energy in the nice useful form of heat.

Now, let's look at how the fission reaction happens. Recall that in looking at fusion we discovered we needed to achieve a threshold energy before the reaction would take place. This required us to heat a sufficiently dense mass of hydrogen to very high temperatures and hold that together for a long enough time to make it go. We also learned that achieving that combination was very, very hard and, so far, only stars and hydrogen bombs can do it. So how hard is it to make fission work? Fortunately for us it is just hard enough that it is hard to do by accident but pretty easy if you work at it a bit.

It is convenient that we are trying to get energy out of the heavy isotopes. Our attention was drawn there because there seemed to be a lot of potential energy available. Recall also that we learned that as elements get heavier they require more than an equal number of neutrons to protons to hold the nucleus together. You might hold Figure 4.1 handy to look back at.

You can imagine as the elements get heavier and heavier, holding the nucleus together gets harder and harder. This is indeed the case, and you can't find any elements heavier than uranium in nature. Such elements exist, that is we can make them, but they are too unstable to last long and come apart all by themselves. So, if elements heavier than uranium come apart all by themselves, it stands to reason that uranium must be close to coming apart just setting there. This implies that the threshold energy for fission in uranium is not large. Remember that threshold energy is the energy required to make a reaction happen. It is the energy it took to nudge the bowling ball out of the depression at the top of the hill in our mental picture. For uranium, the threshold energy for fission is just under 6 MeV. Pretty cool don't you think; put in 6, get 200 out. But how can we put 6 million electron volts into a uranium atom?

Well, neutrons are a good way to do that. Neutrons can give energy to an atom in two ways. First, if you just provide a neutron to an atom, it might absorb it. If it does, that isotope gets transformed into the next highest isotope for that element. If you give a neutron to ordinary hydrogen, it might absorb it and become deuterium. It is still the element hydrogen, but now is the next higher isotope which has one neutron as well as the one proton. If you give a neutron to uranium-238, it might absorb it and become uranium-239. If you give a neutron to uranium-235, it might absorb it and become uranium-236.

Fine, but why is that particularly interesting? Well, anytime you move up to a higher isotope you add binding energy to that isotope. That is, if you add the mass of a neutron to the mass of the atom before it absorbed the neutron, you would have a bigger number than the actual mass of the new isotope. This binding energy is available within the nucleus. For uranium-235, the neutron binding energy is 6.4 MeV, which is more than is needed to cause fission to happen. For uranium-238 the neutron binding energy is only 4.8 MeV which is short of the 5.9 MeV needed. Does that mean you can't make uranium-238 fission? Nope, remember we said that there were two ways to use neutrons to add energy to the nucleus. The second way is not to just provide the neutron, but to sling it in there with some considerable speed. Just like a good fastball smacking into the catcher's mitt at high velocity, a neutron can carry a good deal of energy. So, if we get a neutron moving fast enough, it might be able to provide the other 1.1 MeV of energy needed to make fission happen in uranium-238. In the scheme of things, a neutron velocity sufficient for 1.1 MeV is pretty darn fast, but perfectly doable, and, yes, we can and do make uranium-238 fission. Figure 4.5 shows a fission event.

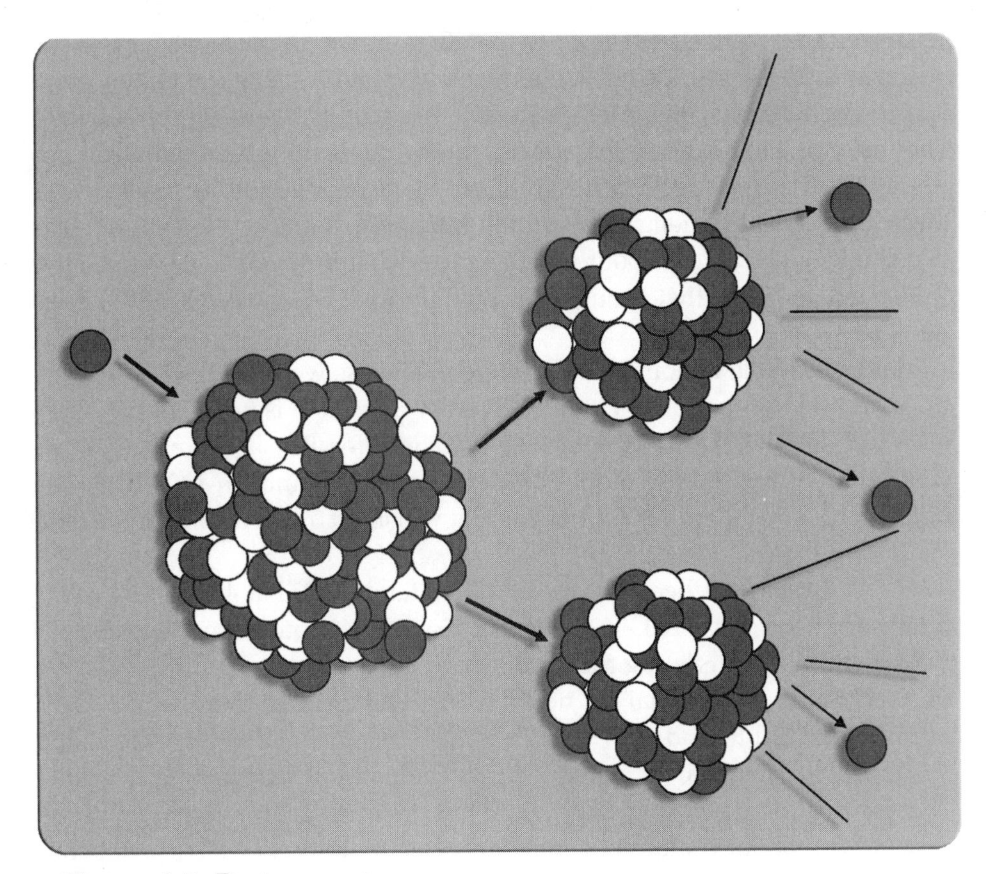

Figure 4.5. Fission event.

O.K. we can make single atoms of uranium-235 and uranium-238 fission. How helpful is that? Well, by itself, it's not very helpful. What would be really cool is if we could get the uranium to continuously fission and do that at a controlled and safe rate. Which as I will attempt to explain, we can.

If you look back at Figure 4.1, you see that for uranium-238 there are 146 neutrons and 92 protons. This is a neutron-to-proton ratio of 1.59 to 1. If we smacked a uranium-238 atom with a nice fast neutron, it might fission. If it did, it would create two break-up products, or fission fragments. As we saw above, these would be somewhere in the mass range 72 to 160. Let's assume this particular fission reaction gave us molybdenum and tin. These are mass numbers in the range of 98 and 120. Molybdenum has the atomic number of 42 and tin is 50. So, we started out with 92 protons in uranium, and we get 42 + 50 = 92 protons in molybdenum and tin. Good, that works. But now, if we check Figure 4.1, we find that molybdenum is comfortable with 56 neutrons and tin would be happy with 70. That is 56 +70 = 126, which is short of the 146 +1 =147 we started with. Don't forget the extra neutron we used to provide the threshold energy to cause the fission event to happen. Is this a problem? Well, the very good news is that this means there must be a lot of neutrons available somehow. The bad news is it means that the fission products are not going to be the normal stable isotopes of the resulting elements. They are going to be "neutron rich." The unfortunate thing about neutron rich isotopes is that they really want to be normal isotopes, so they try to move toward the "correct" neutron-to-proton ratio, which means they are radioactive. We will learn more about radiation and what it means to be radioactive shortly, so just hold that thought for now and we will get back to the good news part.

In the example fission event we were following, we started with 147 neutrons and had gotten only 126 out if we got the stable isotopes of molybdenum and tin. This leaves 21 neutrons to account for. But, as we noted above, you don't get the stable isotopes of molybdenum-98 and tin-120. What you might get is molybdenum-103 and tin-134, which accounts for 145 of the original 147 neutrons.

Let's just double-check that math. Molybdenum-103 has 42 protons, so the number of neutrons is 103 – 42 = 61. Tin-134 has 50 protons, so it has 134 – 50 = 84 neutrons. Completing the math, we can see that this accounts for 61 + 84 = 145 out of the original 147 neutrons. This means that there are 147 – 145 = 2 neutrons left to find.

These two neutrons are ejected from the fission event as free neutrons. That is, they are not bound to an atom but exist just as individual neutrons. Remember that our uranium to molybdenum and tin fission event is only one example of a lot of different possibilities for fissions and that, in fact, there is quite a range of potential outcomes and fission products. Therefore, you don't always get exactly two free flying neutrons. For uranium-235, the average number of neutrons per fission is about 2.4. This is for the case of a neutron with low initial energy. If the neutrons causing the fission have higher energy, you would be likely to get more free neutrons released. Other fissionable isotopes such as uranium-233 and plutonium-239 have different average numbers of neutrons released per fission. Uranium-233 gives 2.5 and plutonium-239 gives 2.9. In all these cases, there are more than two neutrons available after a fission event.

Now, why did we care about these free neutrons? We saw that, in a single fission event, we could get a lot of the potential binding energy out, just over 200 million electron volts per event. What we were wondering is how we might be able to get a series of fission events happening and hoping that we could control the rate of that reaction.

Since we get two or more neutrons out in each fission event, we have some neutrons available to cause the next fission event. The key to it all is to be sure that at least one of those neutrons causes another fission event. If we can arrange for that to happen, then one fission event will cause a second one, and that second one will cause a third, which will cause a fourth, and so on. This is called a chain reaction with each reaction leading to another.

So, how do you arrange for that? Let's first look at what can happen to a free flying neutron. We will study this a bit more completely later, but we will hit the high points here.

There are three things that can happen to a free flying neutron. It can bump into things, it can get absorbed into a nucleus, or it can escape our region of interest.

A neutron generated in a fission event is traveling very, very fast. But if it bumps into things, such as the nuclei of atoms, it will be slowed down. In each bump it will give up some of its energy into the nucleus it hits. We will see shortly how we design our nuclear reactor to make this happen just the way we want.

Sometimes when a neutron encounters a nucleus it won't just smack into it and bounce off. Sometimes, if the nucleus will accept an additional neutron, that free flying neutron could get absorbed by that nucleus and become part of it. The nucleus would then become the next higher isotope of that element. For every isotope there is a fixed probability of neutron absorption, which in turn is dependent on the energy, or speed if you like, of the neutron. While it gets a little complicated with fixed energy quantum levels in the nucleus, in general the lower the neutron energy, the more likely it will be absorbed in any given isotope. You can think of it as the slower a neutron is traveling, the more likely it is to be caught. The probabilities that a neutron will be absorbed vary greatly between various isotopes.

The other thing that can happen to a neutron is that it can get away from us. Now, eventually it will get absorbed into something. But remember our purpose here is to get a sustaining chain reaction of fissions going. We are going to be trying to build a machine to do that. That machine is going to be of some finite size, and therefore, it will have boundaries. If a neutron crosses a boundary outbound, then we have lost it, and we won't be able to use it to cause new fission events.

So, within our machine neutrons can bump into things or can be absorbed by the nucleus of atoms. The probability of absorption varies greatly among the various isotopes and varies with energy, but generally gets greater as the neutron energy gets smaller. Is that it? Well, no. There are a number of interesting things that can happen when a neutron is absorbed, especially if the absorbing isotope is a heavy one.

If a neutron is absorbed by uranium-235, we saw above that the binding energy of the absorbed neutron is enough to match the threshold energy for fission and cause the fission event to happen. So does fission happen every time a neutron is absorbed in uranium-235? Nope, it doesn't work that way. About 15% of the time uranium-235 will just capture the neutron and transform to uranium-236. But the other 85% of the time it will cause a fission event.

The overall neutron absorption probability for uranium-235 is relatively high, and the majority of the absorptions lead to fission. This makes uranium-235 a very good nuclear energy fuel. Most of the power reactors today run mainly on fissions in uranium-235. Natural uranium, the kind you dig out of the ground, is mostly uranium-238. Only 0.7% or one part in 140 is uranium-235. So what happens if a neutron is absorbed in the more common uranium-238? If you recall we learned above that if the neutron has very little energy from its motion, the binding energy alone is not enough to cause fission to happen in uranium-238. We come up about 1.1 MeV short of the threshold energy. It takes a darn fast neutron to have that much energy.

But absorption by uranium-238 is not a total loss. It transforms into uranium-239. This is an isotope that is seriously overburdened with neutrons, and nature just can't stand it for very long. One of the favorite things for a neutron-rich isotope to do is to convert a neutron into a proton. It does this by ejecting an electron. The neutron has no electric charge, so if it gives up a negatively charged electron, the result is a positively charged proton. Remember also that a neutron has slightly greater mass than a proton, so giving up the very small mass of an electron works out just right.

If we convert a neutron into a proton, we keep the same mass number but increase the atomic number. This means the atom that does this becomes the next highest element. But wait a minute. Didn't the list of elements stop with atomic number 92, uranium? It did for naturally occurring elements. If we let uranium-238 absorb a neutron, become uranium-239, and then internally convert a neutron into a proton, we create a brand new man-made element, which is call neptunium. Specifically we have created the isotope neptunium-239. This is also unstable and neutron rich, so it too transforms a neutron into a proton and makes yet another new element, which is plutonium. This new isotope of plutonium is plutonium-239. This is a particularly important isotope. While it is unstable and will transform itself to a more stable isotope over time, it is in no hurry about it and will exist as plutonium-239 for a long time, thousands of years. The useful thing about plutonium-239 is that if it absorbs a neutron, the binding energy is just about the same as that for uranium-235, 6.4 MeV, and its threshold energy for fission is even less than uranium-235, only 5.5 MeV. Because the binding energy of an absorbed neutron is greater than the threshold energy for fission, plutonium-239 can fission if it absorbs even very slow-moving neutrons, just like uranium-235.

Therefore, while uranium-238 will not fission with slow-moving neutrons, when it absorbs a neutron it transforms itself into plutonium-239, which will fission with slow-moving neutrons. Why is causing fissions with slow-moving neutrons so important? I will explain a bit more later but for now remember that the lower the neutron speed, the greater the probability of an absorption event. Remember also that finding very fast neutrons such that they can cause fission in uranium-238 can be hard.

What all this means is even though most of natural uranium is not the very nice nuclear fuel like uranium-235, we can turn some of the most common part of uranium, that is uranium-238, into plutonium-239, which is just as good a nuclear fuel as uranium-235.[6]

So, we have seen that because there are more than two neutrons on average released from fission, and that fission is the most likely thing to happen if uraniun-235 absorbs a neutron that, potentially, we can arrange uranium in the right way to get a sustained set of fission reactions or a chain reaction. The key will be to get at least one neutron to cause a new fission for each fission event.

If we are going to extract some energy out of the fissions, we just can't spread the uranium on the ground and hope it all works out. We probably are going to have to put the uranium in some type of container and probably have to pump some kind of fluid, that is a liquid or a gas, through it to extract the heat. This means we will have structural materials and the heat removing fluid as well as the uranium. All these "extra" materials have the potential to absorb neutrons as well as our uranium fuel. Besides that, some of the

[6]Here I am speaking only for the nuclear physics properties of plutonium-239. The fact that plutonium is a reasonably toxic material (not the most toxic on earth, which you will often hear—that is nonsense), and that it is very useful if you are trying to construct nuclear weapons, gives plutonium-239 a controversial nature. We will talk more about that in the next two chapters.

neutrons are just going to get away from us. Some are going to sneak past everything and escape the container we are using.

As you might imagine, a lot of different arrangements of uranium, structural containers, and coolants were tried. We call a machine built to cause and contain a sustaining nuclear reaction a nuclear reactor. This name comes from the chemical industry, which calls the big vessels in which various chemical reactions happen "reactors." In the next chapter, I will give you a quick tour of the common types of nuclear reactors used to produce electrical energy.

We will come back to neutrons, nuclear reactor control, and the nuclear fuel cycle, but first you need a little background on radiation and radioactive decay.

Radiation and Radioactive Decay

We need to be familiar with four basic types of radiation. These are alpha particles, beta particles, gamma rays, and neutrons. Let's look at each in turn.

An alpha particle is the nucleus of a helium atom. Alpha particles are generated when an isotope is heavier than nature wants it to be. Emitting an alpha particle is a great way to slim down because the resulting isotope skips down two in atomic number and four in mass. Plutonium-238 is a great alpha source. When it spits out an alpha particle, it jumps over the next lower element, neptunium, to go to uranium. Because it is giving up four in mass, it goes from mass 238 to 234 resulting in uranium-234. What is neat about this particular reaction is that it is a powerful and enduring heat source. That heat can be turned into electrical energy. Plutonium-238 has powered the electronics on a number of spacecraft that have continued to operate in deep space for well over 20 years.

In terms of radiation, the alpha particle is a big thing. It has a mass of two protons and two neutrons. The energy it has depends on the particular isotope that is generating it. The plutonium-238 alpha carries over 5 MeV and, therefore, is in the higher end of the range of alpha energies. In addition to its relatively large mass, an alpha particle also carriers the electrical charge of its two protons. So, when an alpha comes into any material, it interacts strongly with the atoms of that material. It has a lot of kinetic energy and a lot of charge. This means it will have a lot of impact. It can displace atoms by just smacking them out of the way. It can also strip electrons off atoms as it passes. This is because the positive charge of its two protons attracts the negative charge of electrons. As the alpha zips by, electrons are drawn off their normal positions in atoms. That alphas can have such a big impact is the bad news. The good news is that the really bad news doesn't last long. Because they interact so strongly, alpha particles do not travel very far in anything. A single sheet of paper will block the most energetic alpha particles. You can safely hold an alpha source in your hand and the layer of dead skin on the surface of your hand will stop all the alphas. However, be very sure you don't breathe in or swallow any of it. That is how alpha-emitting isotopes can be dangerous. If you get them inside you, then their short path, destructive behavior happens inside you, and that is not good.

Beta particles are just electrons. Above we saw that in a neutron-rich isotope, a neutron can be converted to a proton by ejecting an electron from the nucleus. This is a beta particle radiation. Remember that an electron is less than 1/1800 of the mass of a proton, so its mass is over 7000 times less than an alpha particle. However, it does have one-half the electric charge of the alpha particle. Just like alpha particles, betas can have a wide range of possible energies depending on the isotope they are coming from. Beta particles are not likely to be knocking atoms around like alphas do since their mass is so much smaller. However, with their charge they do disrupt the electrical nature of atoms. Betas penetrate, that is go through things, a little better than alphas, but still are stopped pretty easily.

Gamma rays can be thought of as a kind of light. In fact they represent the extreme end of the electromagnetic spectrum of which normal visible light is near the center. If any of your high school science is still with you, you will remember that the electromagnetic spectrum includes all forms of light. If we start with what we see as visible light as the center, lower on the scale is infrared, which while we can't see it, we can feel as the radiant heat from a fireplace or heat lamp. Further down the scale are the microwaves that we use to heat foods. At the bottom of the scale is a very wide range of radio waves. If we go higher on the scale from visible light, we get the ultraviolet which is that cool black light that while we can't see it, we can see its effect when it makes certain materials phosphoresce with that electric glow. Higher yet are X-rays, which we use in medicine. At the top of the scale are gamma rays.

As we move up the scale, the "light" carries more energy and becomes more penetrating. X-rays are great for seeing through soft materials like skin and muscles. Gamma rays are used to see through materials like steel to find cracks or voids.

If you can use gamma rays to see through steel, one might think they penetrate a lot better than beta or alpha particles, and that is right. Depending on their energy, gamma rays do go through things pretty well. What gamma rays do as they pass through material is interact with electrons. The most common gamma ray interaction is to knock an electron out of an atom and give that atom a bump. This leads to a heating of the material and the creation of a charged atom. If this happens in steel, the steel gets warmer and the electrons shuffle about a bit. If it happens in hamburger containing *E. coli* bacteria, and you have tuned the energy right, the bacteria are killed and the meat is healthy to eat. If it happens in you, some of your cells are likely to be disrupted. If that happens too much, you die. If it is not too much, you don't.

There is a lot of controversy about what is too much and what is not. We will talk about that more in a later chapter. For now, just recognize that gamma rays are a kind of light and just like sunlight, if you get too much it can hurt you.

The fourth kind of radiation we were going to talk about was neutrons. We have talked about them above as part of the structure of matter and as the key part to making fission reactions work. We need to think of them as radiation as well because they can come out of radioactive materials and nuclear processes and can be harmful if not protected against. Neutron interactions with materials are unlike any of the types of radiation we have talked about so far. First, they are neutral particles; that is, they have no charge. Therefore, they don't go stripping off electrons as they come zipping along. However, neutrons can indirectly cause electrical effects. Neutron absorption is often accompanied by the emission of strong gamma rays that can knock electrons off surrounding atoms.

Neutrons can affect materials in two ways. One is the same kinetic energy transfer we saw in alpha particles. That is, they can smack into atoms. With 1800 times the mass of an electron, it is no small thing to get bumped by a speeding neutron. And neutrons can really speed. A thermal neutron, one that is moving with the same vibrational energy of any material which is at normal room temperature of 20°C (68°F), is moving at 2200 meters per second. That is just about 5000 miles an hour. Neutrons given off by a fission event have a range of energies, but average about 2 MeV. This means the average neutron from a fission event starts out at 19.7 million meters per second, or 44 million miles per hour. For those keeping track, that is about 6.5% of the speed of light. That is very, very fast. So when one of those guys smacks into an atom at that speed, it gets bumped, and bumped a lot. Some of the neutron's kinetic energy gets transferred to the atom it smacked into. This can be a simple increase in the speed of the atom or it can also increase the energy state of the nucleus of the atom it hits. In either case it is likely that the atom that has been hit is going to move. In materials in which atoms have fixed locations that are important to its properties, that can be bad. In some materials a bombardment by a lot of neutrons causes the atoms to spread out which make it grow in size while reducing the density. In others it causes atoms to jump into unnatural positions in the atomic matrix making the material stronger but more brittle. Some materials, such as zirconium metal, are amazingly robust under neutron bombardment. Zirconium atoms can be repeatedly displaced from their original location and find a new stable home without degrading the properties of the metal. Zirconium also doesn't absorb neutrons very well, which makes it a very good material for use in fission reactors.

Neutron absorption is the other way neutrons interact with materials. As we talked about above, if a neutron is absorbed by the nucleus of an atom, it changes that atom to the next higher isotope of the element. Remember that often a gamma ray is given off at the time of neutron absorption as the newly formed nucleus tries to shed some of the extra energy it has picked up in the absorption.

A neutron absorbed by normal hydrogen creates deuterium. If uranium-238 absorbs a neutron it becomes uranium-239. Often the next higher isotope is not stable, and it won't stay in that form. In the uranium-238 case, the new uranium-239 converts the excess neutron into a proton by giving off a beta particle (just a normal electron) and changes to neptunium-239. That isotope is also not stable, so it repeats the beta emission and changes to plutonium-239.

While some elements have more than one stable isotope, there are also a lot of unstable isotopes. For example, tin has 10 stable isotopes and 13 unstable isotopes. That means if you get an atom to absorb a neutron, you are likely to form an unstable isotope. If that happens, then that unstable isotope is going to do something to find a stable form. It can shed a neutron by beta emission. It can shed excess weight by giving off an alpha particle. It can even fly apart in a fission event if its mass is high enough and the neutron energy is right. Neutron absorption can also leave the resulting nucleus in an excited state that is relieved through the emission of a gamma ray. So, the neutron absorption itself can lead to a number of other forms of radiation.

So, now we have described four kinds of radiation. But where does radiation come from? To do that justice we have to start at the beginning of the universe about 15 billion years ago. I promise this will be a short trip through time, so hold on.

The best guess at this time is that the universe came out of a huge explosion called the Big Bang. This is based on the observation that everything we can see seems to be expanding away from everything else. If you attempt to trace this expansion backward, you get all the mass in the universe into a single tiny dot about 15 billion years ago. I am leaving out a lot of astrophysics and the current controversies on the topic, but that is the general idea.

The important thing to us is that just after the Big Bang, the universe was about three-quarters hydrogen and one-quarter helium. However, after the Big Bang started the great spreading of this hydrogen and helium, clumps of it collected together. The gravity established by the mass drew even more hydrogen and helium to the clumps. Finally, the gravity squeezed the hydrogen and helium hard enough to get the pressure and temperature conditions needed for nuclear fusion. Do you remember fusion? Fusion is pushing some light atoms together to get some heavier atoms and, in the process, moving up the binding energy curve to give off a lot of energy. So, in these super clumps, hydrogen nuclei start fusing into heavier isotopes. That is what we call a star.

Because the majority of the universe is still hydrogen, most of the fusion that is going on in the billions and billions of stars out there is hydrogen fusing into heavier things. Just what those things are and what steps it goes through depends on the size of the star. In what are called main sequence stars, there is a cycle that involves carbon, nitrogen, and oxygen which has the net effect of converting hydrogen into helium. Our sun is just a bit smaller than the main sequence set, and it does the hydrogen into helium in a simpler cycle that has intermediate steps at deuterium and helium-3. However, even these are not the only types of fusion possible. Any reaction that moves you up the binding energy curve and releases energy is allowed. It just takes more and more energy to get the reaction going. If a star is big enough, it has enough gravity to push off other fusion reactions. Also as any star gets older and uses up a lot of its hydrogen, its energy production slows down. This upsets the balance of the outward pressure caused by the heat of the fusion and the inward force of gravity. Gravity starts to win and pulls the mass of the star inward. This increases the energy available for fusion and the higher types of fusion become possible. For a star the size of our sun, these latter stages will allow the star to move up to the creation of carbon, oxygen, and nitrogen, something bigger stars do during the main part of their life.

If a star is 8 to 10 times bigger than our sun, the end of this process is very dramatic. Fusion in such stars can go all the way to iron. Remember that if you get all the way to iron, you have reached the peak of the binding energy curve. That means you have gotten all the energy you can out of fusion. In these massive stars there is no further fusion to coast down with. When they reach the point where the fusion energy no longer balances gravity, the star starts to collapse. Bringing the star's material closer together allows gravity to pull on it harder, accelerating the collapse, bring the material closer yet, which causes greater gravity and so on. The collapse then is rapid and extremely violent. Temperatures rise up to one hundred billion degrees at the heart of the star. When these stars collapse, the crushing together is so intense it smashes the elements the star has made back into the neutrons and protons that made it up. The forces are so intense that electrons are forced into protons causing them to convert to neutrons. This creates a truly vast sea of free neutrons since most of the star's core has been converted to just neutrons. As this happens in the core of the star, the outer layers can't keep up with the collapse and hang above like crust on a pie whose filling has been sucked out. The super intense heat of the gravitational collapse sends out an intense wave of gamma rays and neutrinos. This drives some of the sea of neutrons out into the outer shell of stellar material. The neutrons pile into the carbon, oxygen, nitrogen, and other elements up to iron in the outer layers. These neutrons are then absorbed into these elements and new elements are formed. The neutron sea is so vast and so strong that the original fusion created elements up to iron are transmuted into all the elements above iron. Remember that you can get energy out if you put light pieces together up until iron. After that, it would take additional energy to make anything else. This super collapse of a star is a supernova, which in a short span will produce hundreds of times as much energy as our sun will in its 10 billion year life. The energy release is truly enormous, so there is plenty of energy available to drive over the hump of the binding energy curve and create elements above iron.

So, you can get elements up to iron built up in big stars from fusion, and you get elements above iron from the wave of transmutations that come from the neutron showers out of a supernova. Both sets are spit out into space in the terminal phases of the stars' lives. Some find their way into other later-forming stars and some find their way to planets. Which is pretty fortunate for us in that not only would a universe in which fusion never went beyond helium be boring, with only hydrogen and helium, it would not have the elements needed for the chemistry of life. Good deal, huh?

So with all this higher fusion and supernova-driven neutron captures going on, you are going to get all kinds of isotopes formed. Some will be stable. Some will not be stable. If they aren't stable, they will try to become stable. In that process, they will give off radiation. Such isotopes are called radioactive.

Let's do a check here. All elements above hydrogen and helium up to iron come about because of fusion in stars. Elements above iron come about due to massive neutron captures that happen in supernova explosions at the end of the life of bigger stars. All kinds of isotopes of the elements are formed out of these two processes; some are stable isotopes while others are unstable. The unstable ones try to become stable by giving off some form of radiation and are said to be radioactive. Are you with us? Good, we will get back to the story of the universe in a bit but let's ask another question first. How do radioactive materials behave?

Good question and one that is very important to both understanding radioactivity and to clearing up some of the common misunderstandings about radioactivity.

As we have seen, radiation is a way for a radioactive isotope to move toward a stable state. As we have also seen, it may take several steps to get there. However, it is very important to understand that each of these steps is a one-way deal. Once an atom of uranium-239 gives off a beta to become neptunium-239, that atom is done with that radiation process. That particular uranium-239 atom can't spit out another beta particle because it isn't uranium-239 any more. It has already done that and is now neptunium-239. The neptunium-239 atom will give off a beta particle, but just one, and then it is done. Then the atom is going to be plutonium-239. That atom will give off a alpha particle as its move toward stability. The point is that the original uranium-239 atom doesn't just keep on giving off beta particles forever. It does it once, and that is it.

What can be confusing is that not every uranium-239 atom gives off a beta particle at the same time. Let's say we put a whole bunch of neutrons into a whole bunch of uranium-238 and in the process made a pound of uranium-239. Some of the new uranium-239 starts changing, we call it decaying, into neptunium-239 right away, and some of it waits. You might think of it like a bathtub full of water. When you pull out the plug, all the water wants to go down the drain. But all of it doesn't go out instantly. The analogy fails at the physics of what is going on, so just look at it from the idea that even though the energy balance wants all the uranium-239 to be neptunium-239, it isn't going to happen all at once, just as all the water is not going to get out of the tub all at once.

To describe the timing of radioactive decay, scientists came up with the idea of a "half-life," the amount of time it takes half the original mass to decay, or to change to the next form. The half-life of uranium-239 is 23.5 minutes. This means if we looked in on that pound of uranium-239 we made, 23.5 minutes after we made it, half of it would be gone. So in 47 minutes it would all be gone, right?

Well, no. If you start the clock after the first 23.5 minutes, you start with half a pound of uranium-239. So after another 23.5 minutes, half of what you started with (half a pound this time) is left, or a quarter of a pound. After the third 23.5-minute span, there would be half the quarter pound or one-eighth. After four half-life spans, there would be one-sixteenth. After five, there would be one thirty-second and so on. The thing that is very important to also recognize is that the amount of radiation that is coming off is also decreasing.

In the first 23.5-minute span, we had to give off enough beta particles to transform half a pound of uranium-239 to neptunium-239, that is about 5.7×10^{23} beta particles. In the second 23.5 minutes, we only transformed a quarter pound, so the number of beta particles decreased by a factor of two as well. So, after five half-lives, the radiation has decreased to one thirty-second, or about 3%, of the original radiation level.

The neptunium-239 in our example doesn't decay at the same rate as the uranium-239. Its half-life is 2.35 days. Therefore, if you had a pound of neptunium-239 it would take 2.35 days for half of it to decay to plutonium-239. So, the neptunium-239 will stick around longer than the uranium-239. The plutonium-239 that we make out of the neptunium has an even longer half-life, 24,000 years.

Now here is the thing to zoom in on. If we look at each of these three isotopes, we see that uranium-239 is changing the fastest. That means it is going away the fastest and that at the start it is giving off the most radiation. In the first 23.5 minutes, one pound of uranium-239 is giving off enough beta particles for half a pound of neptunium to be formed. If you had a whole pound of neptunium-239 in that same 23.5 minutes, it would have only 0.7% of its half-life go by and generate only a tiny fraction of the beta particles the same mass of uranium-239 would.

If we had a pound of plutonium-239, the 23.5-minute span for the pound of uranium-239 to transform half its mass would be 1.86×10^{-9} (or about two billionths) of the plutonium-239 half-life and much, much less radiation would come out of the plutonium-239 in that same span.

Now, let's look at the long run. After one half-life of the neptunium, 2.35 days, its radiation level has been halved. But the uranium-239 has seen 144 half-lives, so it is down to 4×10^{-44} of its original rate.

So, in a nutshell, here is the deal. If the half-life is short that means the rate the radiation is given off is high, at the start. But because the rate is high, it gets used up fast, and the radiation level falls off fast. If the half-life is long, then the material is going to be around a long time, but the rate it gives off radiation is low. The next time someone tries to tell you "it is radioactive for thousands of years," you can tell yourself that this means that it must not be very radioactive.

Now back to the story of the universe.

So here we are with all different types of isotopes formed out of big stars and supernovas. Some are stable, some are not. Those that aren't stable are radioactive and are trying to become stable. These have a very wide range of half-lives. Some are fractions of seconds, some are millions of years. Those with short half-lives aren't around in amounts we can find. Those with longer half-lives are still to be found. The most common form of uranium, uranium-238, is actually unstable and slightly radioactive. Its half-life is over 4 billion years, so a lot of it has survived since the supernova that made it because, with a half-life that long, it is not decaying very fast.

So now we can get down to where radioactive stuff comes from. The answer for most radioactive stuff is that it comes from those big stars and supernova explosions. To explain a bit further. The short half-life isotopes have essentially all decayed away. However, the long-lived isotopes decay away slowly. When they decay, they often lead to other unstable isotopes. Uranium-238 decays by an alpha, which leads to thorium-234. That isotope has only a 24-day half-life, so relatively speaking it decays quickly by beta to protactinium-234, which also is unstable.[7] This leads to a long chain of unstable isotopes all with shorter half-lives than uranium-238. Table 4.1 shows all the steps in the chain of decays from uranium-238 leading eventually to the stable isotope lead-206.

[7]In this example we are going to use uranium-238, which comes out of supernova-driven neutron captures. We could have used potassium-40, which comes from the fusion process of stars bigger than our sun. Potassium-40 has a half-life of 1.28 billion years, so a lot of it has hung around since the stars that made it expelled some fraction of their mass into space in their death throes.

TABLE 4.1 Uranium-238 decay chain.

Isotope	Half-life	Radiation Given Off
Uranium-238	4.5×10^9 years	Alpha
Thorium-234	24.1 days	Beta
Protactinium-234	1.17 minutes	Beta
Uranium-234	250,000 years	Beta
Thorium-230	80,000 years	Alpha
Radium-226	1602 years	Alpha
Radon-222	3.82 days	Alpha
Polonium-218	3.05 minutes	Alpha
Lead-214	26.8 minutes	Beta
Bismuth-214	19.7 minutes	Beta
Polonium-214	0.00016 seconds	Alpha
Lead-210	21 years	Beta
Bismuth-210	5 days	Beta
Polonium-210	138 days	Beta
Lead-206	Stable	None

In this fashion, those long-lived isotopes, which survive to our time from those long ago stellar events, populate a wide range of shorter-lived radioactive isotopes which are the results of the decay of the longer-lived isotopes.

Just for fun look at the table for a minute. Look at radon and compare it to radium. What do you see?

Radon's half-life is much shorter than radium. That means radon is much more radioactive than radium, but that it goes away really fast. The interesting thing is that the slow decay of radium gives a long-term source of radon. In the same way, uranium-238 with its very long half-life slowly gives rise to all the isotopes in the table. While none of them have a half-life long enough to still be around billions of years after the supernova that created the original uranium-238, the fact that they result from the slow uranium-238 decay means you can find all of them any place you can find uranium, which it turns out is just about everywhere in very slight concentrations. If you would like some numbers, the average concentration of uranium in the earth's crust is 4×10^{-6} grams per gram of rock and between 0.3 and 2.3×10^{-6} grams per liter of seawater.

Now recall I said that stars were where most of the radioactive stuff comes from. This was the result of the nuclear reactions, specifically fusion and neutron capture driven by supernova. However, that isn't the only type of nuclear reaction that can create radiation or radioactive materials.

The most common source of radiation is the nuclear reaction of "normal" fusion going on in normal stars like our sun. The business of fusion is actually pretty messy and generates a lot of neutrons and protons that blast out into space. The "light" put out by stars also covers a wide range of the electromagnetic spectrum. Since some of this is visible light and infrared light in the form of heat, that is a good thing for us. It also includes X-rays that without the protection given to us by our atmosphere would sterilize the earth pretty fast.

Such radiation from normal stars isn't all the radiation coming from space. Not all the heavy isotopes from past stellar explosions have been collected in newer stars or planets. A lot of these are still zipping along in free space. They are big, relatively speaking, and very, very fast, so when encountered, they do a lot of damage. It is something you have to worry about in designing spacecraft and putting people in space.

Some of those isotopes that have been collected in planets are unstable and radioactive. The geo-thermal heat inside the earth is caused by the decay of naturally occurring materials like uranium and thorium. To heat the earth, which is a very big thing, it takes a lot of radioactive material especially if it is only slightly radioactive which uranium and thorium have to be if they are still around. As a small aside, the next time you admire a majestic mountain or thrill to the rush down a ski slope, thank radioactive decay. It is radioactive decay that provides the geo-thermal heat that drives plate tectonics, which drives mountain building. If the supernovas long ago and far away didn't make those radioactive isotopes, the interior of the earth would be cold and no mountains would have been made.

While radiation from space and the naturally occurring radioactive materials in the ground are the greatest sources of radiation we experience, there is also radiation that comes from human activity. Some of that is natural radiation that we just stir up. One example of that is the burning of coal in a coal-fired electrical power plant. Uranium is found in some small amounts in just about everything including coal. When coal is burned in the power plant, the uranium is put into the air. The uranium and all the shorter-lived radioactive materials that are in its decay chain are then out there in the environment instead of unobtrusively decaying away within the coal seam under the ground. For similar power outputs, a coal power plant emits much more radiation on a routine basis than a nuclear power plant.

We also are capable of producing nuclear reactions ourselves. When you go to the dentist and they X-ray your teeth, they are using a little electron accelerator to cause a beam of beta particles to in turn create a beam of X-rays. In a much less powerful version, your television causes the same reaction when the electron gun in the TV, that is the picture tube, beams electrons at the screen where in the process of stopping, those electrons give off a little bit of X-rays. That is why they tell you not to sit right up against the picture. Not to worry, the amount of radiation is very small. We will talk more about sources of radiation and at what levels one should be concerned in Chapter 6.

Other nuclear reactions we can create come from the use of neutrons. We can get neutrons from a number of radioactive materials or from a nuclear fission reaction. Remember, when we talked about neutron radiation, we said that one of the things neutrons could do is be absorbed by atoms, and if that happened, the new isotope formed would likely to be unstable or radioactive. We can use neutrons to study the nature of materials and to create artificial radioactive materials. That is, something that would not have lasted since a long ago supernova but is useful to us. There are a number of medical uses for such radioactive isotopes both to diagnose and to treat illnesses.

Also, when we talked about fission, we saw that the fission products were unstable and radioactive, so this is yet another way we can get radioactive materials from human-caused nuclear reactions.

To summarize all this, radiation comes from radioactive materials whose nuclear structure is not stable. To become stable, the radioactive isotope will give off an alpha particle, a beta particle, a gamma ray, or a neutron. The source of radioactive materials is nuclear reactions, the most common of which is fusion in normal stars and neutron capture in supernova explosions. These populate the cosmos with radioactive material that decay either quickly and are gone, or slowly with much less radiation given off but last a long time. Many radioactive materials decay to other radioactive materials leading to a decay chain on its way to a stable isotope. While nature has produced essentially all of the radiation and radioactive material in the universe, humans can also produce radiation through purposefully created nuclear reactions such as your dentist's X-ray machine or the radiation and radioactive materials created in a nuclear fission reactor.

5

NUCLEAR REACTOR PHYSICS AND NUCLEAR REACTORS

In the previous chapter we learned about some of the basic science behind nuclear energy. In this chapter we will move beyond the basic theory and start talking about how we as humans can use our understanding of this science to make energy. This moves us from theoretical to applied science and on to engineering.

To start, we need just a little more theory about how neutrons interact so that we can see what needs to be done to get neutrons to do what we want them to.

Neutron Interactions

In Chapter 4 we talked about neutrons quite a bit. We talked about them as one of the basic pieces in the structure of matter, as a key part of the fission process, and as a form of radiation. Now we are going to put all that knowledge together and see how we can use these little guys as tools to help us in getting out the binding energy of heavy isotopes and help us to make safe energy.

The first thing to do is to review and extend our understanding of neutron interactions. Let's start with a free neutron that was produced in a fission event. Neutrons from a fission event start off with energies ranging up to and above 10 MeV but averaging around 2 MeV. A neutron with 2 MeV of energy is traveling about 44 million miles per hour or 19.7 million meters per second. At the other end of the speed scale is what we call a thermal neutron. This is a neutron that is traveling at the same speed as the atoms of the surrounding material. The speed of those atoms is determined by how hot they are. The hotter the material, the faster its atoms are moving. So a thermal neutron is slowed down to the same speed as all the surrounding traffic. Recall also that our goal is to cause a neutron to create another fission event and that a slow-moving thermal neutron is hundreds of times more likely to cause a fission than a fast neutron. A thermal neutron that is in equilibrium with its surroundings will be traveling at 2200 meters per second, if those surroundings are at normal room temperature of 68°F (20°C). So we are looking for a way to reduce the neutron's speed by a factor of about 9000. Sounds pretty hard, huh? Well, it depends on what we use to slow them down. If we use the right stuff, it turns out to be really easy. Let's see how.

As you might imagine, the most likely thing to happen to a neutron zipping along at 44 million miles per hour is to smack into something. Because a neutron is so very small, those things it is going to smack into are also very small, which is to say the nuclei of atoms. While it can get complicated with the quantum energy states of individual nuclei, much of your common experience applies to how neutrons behave as they smack into nuclei.

If you throw a baseball very hard and it hits a bowling ball, you would expect the baseball to bounce off with almost the same speed it started with, and the bowling ball would roll away with just a little speed. If, on the other hand, the baseball hit another baseball, the first baseball would bounce off with much less speed and the hit baseball would go flying off pretty fast. In fact, if the hit was just right, you could imagine the first baseball stopping completely and the second going off almost as fast as the first. If you are having trouble envisioning that, then think about billiard balls. Everyone has seen a classic sharp direct hit in which the first ball stops dead and hit ball goes zipping off.

The physics of this is the ratio of the initial kinetic energy to the resulting kinetic energy of the striking "ball" is related to two things: the relative masses of the traveling and struck "ball" and the angle at which the collision takes place. I will spare you the math; however, what all that math tells us is that if you throw a "ball" at another ball the same size, which is not moving, and hit it dead on, you can make the thrown ball come to a complete stop in just one hit. This is completely independent of how fast the thrown ball is moving.

Cool, huh? Well, sure, but why is that cool? How can we use that to make neutrons useful to us? Remember, we had fission neutrons traveling at an average of 44 million miles per hour, and we need to slow them down by about 9000 times. That means if we can find something that has the same mass as an individual neutron and smack the neutrons dead on into a bunch of them, then we can not only slow the neutrons down by 9000 times, we can stop them completely. What has a mass essentially the same as a neutron? The answer is a proton. And where can we find a proton? Normal hydrogen would be a good place to look. Remember the nucleus of a hydrogen atom is a single proton. So, if we can get our fission energy neutrons to smack into a bunch of hydrogen, we can slow all the neutrons down to a complete stop in just the first hit, right?

Well, no. Remember, the object being hit must not be moving, and we are required to hit it dead on. In real life, it is very hard to make either of those things happen exactly. What we observe as temperature is really the vibration of atoms. At 68°F, hydrogen atoms, like neutrons of the same temperature, will be moving at 2200 meters per second. However, while that seems pretty fast, it is very much slower than the fission neutrons. As to hitting dead on, anyone who has ever played darts knows that you have to throw a lot of times before you get one exactly in the center of the bull's eye.

However, even though it is unlikely to get completely stopped in one hit, a neutron colliding with a hydrogen atom is the fastest way to slow it down. If you compared hydrogen to other potential materials, you could see that while hydrogen has the potential to reduce the neutron energy by 100% in a single collision, the next lightest material, deuterium, in the best case, can only reduce the incoming energy by 89%. If we move on to carbon with mass twelve, the reduction is only 28% in a perfect dead on hit.

I have used deuterium and carbon in the example on purpose. Both of these materials, along with hydrogen, are the most common types of materials used to slow neutrons down. Such materials are called moderators because they moderate neutron speeds down to where we can make better use of them.

But why would you use anything but hydrogen as a moderator when it is clearly the best at slowing neutrons down? Well if you recall, smacking into things is only one of the things neutrons can do. The other really important thing they can do is to be absorbed by nuclei. That is great if it is causing a fission event we want in order to get energy or to create a radioactive isotope we want for a medical treatment. But if that absorption isn't productive in one of those fashions and just reduces the population of neutrons, it can be a problem. Remember that if we are trying to get energy out of fission, we have to keep the reaction going. This means we get a bunch of uranium or other fissionable material together, find a spare neutron to pitch into it, cause a fission event that, along with the kinetic energy of the fission fragments, spits out two to three extra neutrons, which hopefully we can use to cause more fission events. The key to controlled energy release is to ensure that just one of the neutrons released in fission causes one more fission event.

It is hard not to lose neutrons to non-productive absorptions or have them leave your bunch of uranium and streak off into space. If your moderator absorbs very many neutrons, that is going to be a big drag on your system. Normal hydrogen does absorb neutrons. Its proclivity for that is much less than some of the really big neutron absorbers, but it is still a respectable neutron absorber. The most common form of a hydrogen moderator is just normal water. Water makes a good moderator because it provides hydrogen in a very convenient form. In terms of atoms, water, or H_2O, is two-thirds hydrogen. Water is an excellent fluid for moving heat around, and it is really cheap. But hydrogen does absorb neutrons. In a reactor using normal water, neutrons seldom get farther than one foot from where they are born. Most don't make it that far. That is a pretty impressive capture for something traveling at 44 million miles per hour to start. What happens is the hydrogen slows the fission neutrons down very quickly, and if the nuclear fuel or structural materials don't absorb the neutrons, the hydrogen in the water will.

Some nuclear reactors are designed to get around the absorption of neutrons in hydrogen. One obvious thing to try is the next lightest material, which is deuterium, the heavy isotope of hydrogen having one neutron to go along with the one proton. If you combine deuterium with oxygen, you get water that looks pretty much like regular water. Because it uses the heavy isotope of hydrogen, it is called "heavy water." Deuterium does exist in nature and makes up about 0.015% of normal hydrogen. There are ways to separate it from normal hydrogen, but as you might expect from its relative rarity, separating it isn't easy and heavy water is quite expensive. However, in a nuclear reactor, heavy water is very interesting. In normal water, which is called, in the context of nuclear reactors, "light water" to distinguish it from heavy water, neutrons are slowed down and absorbed in short order; in heavy water, neutrons are slowed less rapidly, but they just wander around until you are tired of looking at them. Well, not quite, but it seems that way to a light water reactor physicist.

Heavy water does absorb neutrons but at a rate nearly 700 times less than light water. Therefore, the choice between light water and heavy water isn't straightforward. You have to balance the neutron economy, that is how many neutrons you lose to the moderator, against the financial cost of the moderator. This gets more complicated as the choice of the moderator limits the choices you can have in the type of fuel you use.

In looking at these types of trade-offs, carbon emerges as a good candidate for a moderator. While not as good as either hydrogen or even deuterium in slowing neutrons down, carbon does have a very low probability of absorbing neutrons; while about 10 times more than heavy water, it is still 70 times less than light water. While you can't use carbon to move the heat out of a nuclear reactor, carbon does make a very nice structural material. That is, you can build your reactor out of it. It is strong, can take very high temperatures, and, if processed in the right form, can resist radiation damage very well. You will still need to put the carbon in some kind of a steel or concrete box, but the carbon can be used to support your fuel and whatever structure you might use to move out the heat. These properties made carbon-moderated reactors the first type to be built, and they may well be the future of nuclear power in years to come. I will tell you more about that later.

We have talked about neutron absorption before, but let's review that just a little more before we see just what it takes to make our sustaining fission process go.

Moderators aren't the only things in a reactor that can absorb neutrons. Remember that we are trying to ensure that one of the two or three neutrons that come out of a fission event survive until it can be absorbed in our nuclear fuel to cause another fission. Because slow-moving neutrons are more likely to cause fissions than fast ones, and neutrons born in a fission event are very fast, we are likely to include a moderator to slow the neutrons down. As we saw above, we have several choices of moderators, and we have to balance the moderator's efficiency at slowing neutrons down, against the economic cost of the material, and against its ability to absorb neutrons. Any neutrons absorbed by the moderator are lost to us. We call this a parasitic or non-productive capture.

In addition to a moderator, we are going to need something to move the heat we make out of the reactor. This is needed to keep the temperature in the reactor within the limits the fuel and other materials can safely withstand and, of course, to get the energy out so we can put it to productive use such as making electricity. We get the heat out with a coolant. This will be some fluid we pump past the fuel. This coolant will come in cool and leave hot, where cool and hot are relative terms. In a research reactor, the cool side might be 60°F and the hot might be 70°F. In a modern power reactor, the cool might be 560°F and the hot 620°F.

A very common coolant is water. Both light water and heavy water have been used. Water is convenient in that you get a coolant and a moderator all in one. As a coolant, water is really neat because it carries a lot of heat per pound, pumps easily, and is a very common substance. We can't use just water out of the tap, however. The water will have to be very clean and the trace elements in it carefully controlled. This is needed to limit the corrosive effects on the fuel, piping, and other structural materials. We also don't want a lot of potential neutron absorbers floating along in the water. The biggest problem there is that such absorbers will be changed to unstable isotopes and, therefore, make our cooling water radioactive. So, reactor coolant water isn't quite as cheap at tap water, but it is still relatively cheap. That, of course, applies to light water. We saw above that heavy water is rare and, therefore, expensive even before we make it very clean.

I said that a coolant is a fluid that is used to remove heat. Liquids like water aren't the only type of fluid. Gases are also fluids, and nuclear reactors have been designed to use high volumes of gas to take out the heat of the fission events. In one type of gas-cooled reactor, carbon dioxide is used; in another, helium is used.

There is yet another type of fluid and that is molten metal. If you heat a metal up enough, it will turn into liquid and flow like a fluid. Materials like sodium or lead are capable of transferring a lot of heat. They have some challenging properties including corrosion and the fact that if you don't keep them hot, they freeze and stick in your pipes. However, for certain types of reactors, such coolants work best. We will talk some more about that at the end of the chapter.

So we have a choice of coolants, and one of the things we have to be concerned about is parasitic absorptions by the coolant. What else can cause parasitic absorptions? Well, there are two more things: the structural materials and the nuclear fuel itself. Structural material is what we are going to use to hold the reactor together, to make the coolant flow where it needs to go, and keep all the radioactive material in a nice safe place. This starts with something to hold the fuel. Because fission produces a lot of heat, the fuel is likely to be something thin, so there is a lot of surface area relative to the volume. Some reactors have plates, but most big power reactors have some form of fuel rods. In light water cooled reactors these are long (12 feet or more) very thin cylinders. They are probably less than half an inch in diameter. The cylinder is a metal tube with a thin but very strong wall in which the nuclear fuel is held.

A number of such fuel rods will be held together as a bundle or assembly. There will be other structures used to support the assemblies and to route the coolant. All of these structural materials represent candidates to steal neutrons from us. Therefore, we try to use materials that not only can withstand the temperatures, radiation, and hot coolant, but also are not likely to absorb neutrons. We are fortunate that nature has provided us with a number of materials that do this very well.

The last type of material we have to worry about causing parasitic absorptions is the fuel itself. Most reactors will use a uranium fuel that has had the amount of uranium-235 enriched from its naturally occurring 0.7% up to 3 to 5%. The rest of the uranium is the more common isotope, uranium-238. As we saw above, it is unlikely that if uranium-238 absorbs a neutron it will fission. Since the binding energy of the neutron alone lacks sufficient energy to cause fission, only a neutron with enough kinetic energy, one moving really fast, will cause fission in uranium-238. Some do, but not many. If a slower moving neutron gets absorbed in uranium-238, it is lost to us. We aren't entirely sad about that because such an absorption is going to result in our uranium-238 atom being turned into plutonium-239, which will fission with any energy neutron, just like uranium-235. However, for just the moment, a non-fission uranium-238 neutron capture is a lost to us.

So any capture in uranium-235 is good, right? Ah, no. Recall the cool thing about uranium-235 is that the threshold energy for fission is less than the binding energy of a neutron. That means if a uranium-235 atom absorbs a neutron there is enough energy gained just in the absorption to cause a fission. However, that is not the only thing that uranium-235 can do if it absorbs a neutron. If the absorbed neutron is a slow-moving one, about 15% of the time it behaves like a normal absorption and transforms to uranium-236. The other 85% of the time it does what we want and fissions.

We begin to see that with absorptions in the moderator, the coolant, the various structural materials, and non-fission absorption in the fuel itself, getting one neutron to survive from birth until it can cause another fission is not a foregone conclusion. Did we miss anything else that can lose a neutron for us? Just one more. They can just get away. Even in a light water reactor where most neutrons never travel more than a foot before they are absorbed in something, there are still boundaries to the fueled region. At those boundaries, neutrons can escape from the system. After all, at a starting speed of 44 million miles per hour, they can be elusive.

If you made your reactor big enough, the fraction of neutrons at the edge would be very small and escape would not be a problem. In fact, the minimum size of a reactor is determined by this very factor. You may have heard and even used the expression "critical mass." What this comes from is the idea that for any given nuclear fuel type and arrangement of reactor moderator, coolants, and structural materials, there is a minimum mass of fuel needed to reduce escapes from the system to allow the fission events to be self sustaining. That is, neutron escapes can be the controlling factor as to whether or not we get a net of one new fission event out of the two to three neutrons we got out of the last fission event. The other "excess" neutrons were lost to parasitic absorptions, and we can lose only a small fraction to escape. If the fuel mass is large enough, is a critical mass, then the fraction that escapes is too small and one fission will lead to one more, which will lead to another, and so on. If the mass is below the critical mass, then too many neutrons can get away and even if you cause some number of fissions, the process will quickly die out.

Since we are trying to make the best use of neutrons, a reactor is often surrounded by a material chosen to reflect escaping neutrons back into the fuel system. Anything that scatters neutrons well—but does not absorb them too much—would make a good reflector. If it also slows fast neutrons down, that would be helpful, too. As you might guess from those criteria, water makes a good reflector. The addition of a reflector can reduce the critical mass of a given reactor configuration.

Let's summarize what we have learned about neutron interactions. In general, neutrons can smack into things, be absorbed, or get away. Smacking into things is a great way to reduce energy, or speed. Because neutrons born in fission are very speedy, and slow neutrons are more likely to cause the next fission event, and getting a chain of fission events is what we want to get energy from fission, slowing neutrons down is a good thing. We saw that the lighter the nucleus that a neutron smacks into, the better that nucleus is at slowing the neutron down. However, we have to watch out for absorptions that don't lead to fission. Such parasitic absorption can occur in the moderator, the material we use to slow the neutrons down. They can also occur in the coolant, the structural materials, and even the fuel itself.

We saw that much of the design of a nuclear reactor depends on selecting materials and arranging them in such a way as to keep the chain of fission events going. Let's look more carefully at what it takes to get this sustained reaction going and how we can control that.

Chain Reactions, Criticality, Delayed Neutrons, and Reactor Control

We have spent some time talking about neutron interactions. We saw that to get energy out of nuclear fission, we need to get at least one of the neutrons released in a fission event to cause another fission, and we looked at some of the ways we can lose neutrons to non-productive uses before they can cause a fission for us. Now let's look at this neutron balance more carefully. Our goal here will be not only to understand the neutron balance better, but also to learn how we can control fission reactions, a pretty important thing if we are going to use nuclear power safely. It is also very important to understand this if you are going to be able to judge claims that nuclear power reactors are just bombs waiting to go off. I could just tell you that such claims are wrong, but I would much rather you understand the physics and draw your own conclusions.

Recall that to get a sustaining chain of fission events, we need at least one of the neutrons released in a fission event to cause one more fission. If we want the energy release to be constant, that is not increase or decrease with time, then we must ensure that from each fission event one and only one neutron causes another fission event.

However, it is hard to look at just one neutron at a time when we have numbers like this: on the average a fission event in uranium-235 releases 2.43 neutrons. In order to think about sustained chain reactions, let's imagine we have built a nuclear reactor. This reactor will use some form of uranium fuel with enough uranium-235 in it such that most of the fission events will occur in the uranium-235. The reactor will have a moderator to slow down fission neutrons, some form of coolant to remove heat, and structural materials to hold it together and hold the nuclear reactions in. Let's also assume that in one moment of time we had one million fission events happen. That might sound like a lot, but at about 190 MeV of heat energy per fission event, one million fission events is 1.9×10^8 MeV, which works out to enough energy to light a 100 watt light bulb for 2.3×10^{-14} seconds or heat one ounce of water by 4.6×10^{-7}°F. If you might think then that we can't get much energy out, be not concerned. One ounce of uranium-235 will have 7.3×10^{22} atoms, that is seventy-three thousand trillion million atoms. So, once we get rolling along, we will have plenty of uranium-235 to have a lot more than just a million fissions.

Also, don't worry about getting the first million fission events. In a power reactor, we will have a small neutron source that will use a radioactive isotope to produce neutrons at a steady rate, and so we can be assured we will be able to start up our reactor.

So with a million fissions in uranium-235 caused by slow-moving neutrons, remember we have a moderator in this reactor, we will get 2,430,000 neutrons created out of the fissions. We can think of this as a generation of neutrons. This generation will go through their very brief lifetime, slowing down by smacking into the moderator, some getting absorbed by the moderator, the coolant, or structural materials, some escaping the reactor completely, and some getting absorbed by the fuel. Of those getting absorbed in the fuel, some will be non-productive absorptions, and some will cause fissions creating the next generation of neutrons.

We will get a sustaining chain reaction if the number of neutrons in the next generation is the same as that of the previous generation. To keep track of this nuclear physicists use:

k = number of neutrons in new generation / number of neutrons in prior generation

So, looking at that for a minute, if k = 1.00000, that is exactly one, then the chain reaction is constant. In our imagined case, if k equals exactly one, this means that of the 2,430,000 neutrons created in the first generation from 1,000,000 fissions, exactly 1,430,000 were lost to non-fission events, leaving exactly 1,000,000 to cause fissions, releasing another 2,430,000 neutrons to keep the whole thing going. This would continue until something happened to change what is happening to the neutrons during their lifetime.

One of the obvious things that would eventually happen is the neutrons would start seeing less uranium-235 because in each generation one million atoms would be fissioning and being converted to fission fragments. This would mean there would be a decreasing probability that a neutron would find a uranium-235 atom and cause a fission. Also, the number of fission fragment atoms would be building up. Some of these really like neutrons and would like to absorb them before the uranium got a chance. Now, since we have seventy-three thousand trillion million atoms in every ounce of uranium-235, at the rate of one million fissions per generation, it is going to take a long time to make much of a change to the fate of neutrons in our imaginary reactor.

Still with time the conditions will become less favorable, and the number of neutrons in some generation will become less than that of the prior generation. In that case, k will become slightly less than one. If, for instance k = 0.999, then instead of 2,430,000 neutrons, the next generation would have only 2,427,570 neutrons. Well, that is O.K.; it is still a lot, right? Sure, but after 100 generations we would only have 2,200,846 neutrons or 91% of what we started with. After 500 generations there would be 1,474,976 neutrons or only 61% of what we started with. Clearly, we are getting less and less neutrons and rather than a sustaining reaction, the reaction is coasting down to dead stop.

What happens if we change the conditions so that k is more than one, and how might we do that? One way is to do something to decrease the number of neutrons lost to non-productive absorptions. As we will see in a minute, reactor designers put absorbers in their reactors just so they can control them. That is, by sliding a little absorber in or out of the reactor, they can change the conditions so k changes just a little bit. So, let's assume our reactor is very close to k equals exactly one, and our clever reactor designer gave us a little bit of extra absorber that we could remove. We remove just a little bit, and k increases to 1.001. This means that in each new generation we will have 1.001 times as many neutrons as that in the prior generations. So in the second generation of our example, we would have 2,432,430 neutrons. In the 100th generation, we would have 2,682,748 neutrons, which is 110% of what we started with. In the 500th generation, we would have 4,001,390 neutrons, which is 165% of what we started with. In this case, the number of neutrons and the number of fission events are continuously increasing and will do so as long as k is greater than one. We already saw that k is likely to change to a lower number for purely physical reasons, such as using up the uranium fuel and creating neutron absorbing fission fragments. However, as we will see just a bit later, there are a number of other features that can be designed into a nuclear reactor that will turn such an increasing power trend back around much faster.

Before we look more at how a nuclear reactor is controlled, let's review where we are. We learned that nuclear physicists, and you are learning how to be one of them yourself, use k to define the number of neutrons in a new generation divided by the number of neutrons in the prior generation. If k equals exactly one, then the number of neutrons, the number of fission events, and the reactor power level stay exactly the same. This is called a "critical" condition. It is the set of conditions that allows one generation of neutrons to create another generation of neutrons just like themselves. If k is less than one, the number of neutrons in the next generation is less than the prior generation and the number of neutrons, the number of fission events, and the reactor power level will decrease with time. This is called a "subcritical" condition. A reactor that is subcritical is either shutting down or is shut down.

If k is greater than one, the number of neutrons in the next generation is greater than in the prior generation, and the number of neutrons, the number of fission events, and the reactor power level are all increasing with time. This is called a "supercritical" condition. This is a good thing if we ever want to get our reactor to produce more power than enough to light one 100 watt bulb for a tiny fraction of a second, but it also means we need to be sure we know how to control the increase so we can stop it when we want to.

A reactor physicist who knows what his or her reactor is going to be made of can calculate what the k will be for the reactor. That is because we know what the properties of these materials are with regard to their ability to slow neutrons down and how likely they are to absorb neutrons of various energies. We can also look at the size and configuration of the reactor to calculate how many neutrons can get out of the reactor and what the distribution of neutrons will be within the reactor. All this information and some sophisticated mathematics will allow one to calculate k.

While we saw how the number of neutrons would build up or decrease with a given value for k as a number of neutron generations went by, what we did not see was how long a given neutron generation was. For most of the neutrons in a generation, the time it takes to go from one fission event to the next is very short, a tiny fraction of a second. This is not so surprising when we remember that the fission neutrons are starting out at an average of 44 million miles per hour. In a reactor with a light water moderator, the neutron lifetime is about a ten-thousandth of a second. This is called the prompt neutron lifetime. In a heavy water or graphite moderated reactor, where neutron absorption in the moderator is much less likely, the prompt neutron lifetime might be a hundred times longer, but still only some hundredths of a second. Therefore, to go through the 500 generations in our example would take less than the blink of an eye or in the best case just a few heartbeats. Wow, that's super bad news, huh? That means, as soon as you get any condition where k is greater than one, the power level zooms up very fast, and it would be impossible to control it!

Well, here is the super good news. The cosmos was very kind to us and doesn't let just all the neutrons do that. Some of those 2,430,000 neutrons in our example, which came out of the million fission events, did not come directly from the fission event but rather from the decay of the fission fragments. Because those fragments are neutron rich, one way to move toward stability is to just pitch out a neutron. The really great thing about it is, as with all decay processes, it takes a little time. Remember that fission doesn't always give exactly the same two fission fragments from each fission event. Rather there is a range of possible isotopes that the uranium atom can split into.

These different fission fragments have a range of neutron producing decay rates. Some have half-lives (that is the time to decay to half the original amount) of a fraction of a second, some have half-lives a few seconds, and some are almost a minute. The total fraction of the neutrons produced in fission that is delayed like this (we call them delayed neutrons) is quite small. But the much longer time it takes for them to come into the neutron balance causes them to have a very big effect. It is like putting a small child on a teeter-totter with an elephant. If the board is the same length on both sides of the balance point, the kid is going to be always in the air. But if you make the board much, much longer on the child's end, then the child can out balance the elephant and lift him into the air. Delayed neutrons are like that. Their time delay is so long, up to 100,000 times longer than the "prompt" neutron lifetimes, that the effective time of one neutron generation to the next is not controlled by the very short lifetimes of most of the neutrons but rather by those of the delayed neutrons.

Now, we have to clarify that just a bit. The effective generation time does depend on the value of k relative to the fraction of delayed neutrons for the type of fuel we are using. If k is big enough, it can overpower the effect of the delayed neutrons. Let's look at that more closely.

For uranium-235, the fraction of neutrons from fission that is delayed is 0.65%. This is a very important number. It isn't a lot, only 65 out of every 10,000, but if you design your reactor right, it is enough. The key to it is to be sure that you keep k low enough that the prompt neutrons alone can't increase the power. That is, the power increase will be held to the pace of the delayed neutrons. This is pretty simple to figure. As long as k is less than 1.0065 the rate of increase will be based on the averaged neutron lifetime including the delayed neutrons. If, however, you let k get greater than 1.0065, for a uranium-235 fueled reactor that is, then the prompt neutrons are enough on their own without the delayed neutrons to cause a power increase. They don't have to wait for the delayed neutrons, so power can increase with a generation that lasts a hundredth or less of a second. That means power can increase very fast and that *might* be a bad thing. I have to stress *might*; because as we will see shortly, there can be counteracting forces that are also very fast.

Let's look at this in terms of another example reactor. Let's suppose we removed enough absorber to change k from exactly one to 1.0001 instead. This means we have a supercritical condition and power is going to increase. However, since our reactor is fueled by uranium-235, the delayed neutron fraction is 0.65%. This means that the prompt neutron fraction is 1 – 0.0065 which is 99.35%. So the number of prompt neutrons in the next generation after we changed k to 1.0001 will be equal to the number of neutrons in the prior generation times k, times the prompt neutron fraction. If you follow through the math, as shown below, you see that the number of prompt new neutrons will be less than the first generation, even through k is greater than one.

k = 1.0001

Delayed neutron fraction = 0.65%

Prompt neutron fraction = 1 – 0.65% = 99.35%

Next generation prompt neutrons = 1.0001 × 0.9935 × 1st generation fissions = 0.9936 × 1st generation fissions

These will rush right out and create more fissions resulting in another generation of prompt neutrons, which will be k times that number, times the prompt fraction. That would be 98.72% of the original number. As you can see in this condition, the prompt neutrons alone are not enough to cause the number of neutrons and, therefore, the power level to increase. In fact, the number of prompt neutrons is getting less all the time. If we wait until the delayed neutrons show up, then the net number of neutrons will increase as will the power level, but at a rate much, much slower than the lifetime of the prompt neutrons. If you bring k up to one plus the delayed neutron fraction and then repeat the math we just did, you will see that the prompt fraction and k multiplied together will give you exactly one. That means the reactor is critical on prompt neutrons alone. If you make k just the tiniest bit greater, then the reactor is supercritical on prompt neutrons alone and will increase power with a generation time equal to the generation time of the prompt neutrons, which means it will be very fast.

So, if we have a condition where the k is greater than one by the amount of the delayed neutron fraction, we call that "prompt critical." That is, the system is critical, sustaining, on prompt neutrons alone. Generally this would be a bad thing in a reactor. In a nuclear weapon, that is the idea. In fact the difficulty in getting a big boom out of a nuclear weapon is holding it together long enough to get a lot of generations of fission to happen. In a nuclear weapon, the components are smashed together to get a system with a very high k, much above the prompt critical point. Fissions build up very, very fast from generation to generation with neutron lifetimes being very short. The difficulty is that all those fissions produce a lot of heat, and when the materials in the weapon heat up to millions of degrees, they expand, expand a lot and very fast. So weapon design is all about a race to get a lot of generations in before the weapon comes apart.

For reactors, we are going to design in a lot of features to ensure we never get to a prompt critical condition, and, therefore, the reactor's neutron generation lifetimes will be controlled by the delayed neutrons, so we will have plenty of time to control the reactor power. The result of this is that a nuclear reactor behaves more like a locomotive or an ocean liner than a racecar. That is, it increases and decreases its speed (or in this analogy, power level) slowly. It takes a while to get it going, and it takes a while to stop it. Delayed neutrons are the key. Controlling the reactor with delayed neutrons is one of the reasons why nuclear reactors don't and can't behave like nuclear weapons.

You might think that getting k to be within the range of 1.0065 or less would be hard. However, in the world of reactor physics, 0.65% of k is a big range. Most parameters that effect k are measured in one-hundredths of this value.

In the next section, I will tell you about some of the things nature gives us that provide a safe control of our reactor and how we can enhance those with intelligent design. Here, I want to just close with noting some of the things we do to actively control k in a reactor. Remember, in our example, we noted that if we had k exactly at one, the reactor would stay at that power level until something changes in the conditions. We also noted that if we waited long enough, we would use up some of the uranium-235 fuel and that neutron absorbing fission fragments would build up, both of which would cause k to go down. Eventually, this would make k go to less than one, and the reactor would start coasting down to dead stop.

When you think about it, if you built your reactor so that it started out with just enough fuel and just the right configuration and shape that k equaled exactly one at the start, it would run for a second and then start shutting down. That would be pretty boring and not much good for energy production. So we build the reactor with extra fuel so that k *could* be greater than one. We also add some absorbing materials that we can withdraw later to balance the extra fuel so that k can be brought back to exactly one. Since we will want to be able to shut the reactor down at any time, we build these extra absorbers with the capability of not only balancing the extra fuel back to k of one, but also over-balancing so we can get k less than one.

There are a number of ways to have this extra absorber available. The most common is to have some form of rods made out of a strong neutron absorbing material that we can push into the reactor or pull out. Since their purpose is to control the reactor power, these are called control rods. We also can put neutron absorbers in the coolant or moderator. Boron is a very strong neutron absorber and often boric acid is added in small amounts to the water in light water reactors. The amount of boron in the water can be controlled by doing a little chemistry on a portion of the coolant flow. This is not a fast control method, but it is used to offset the slow reduction in the fuel as fissions happen. It is also more uniform than control rods and allows you to use your fuel more evenly throughout the reactor.

Reactors are always designed with extra control available so that if anything bad happens you can very rapidly insert the extra control and get k less than one and start the reactor shutting down. There are many different automatic sensors monitoring the temperatures, pressures, coolant flow rates, and reactor power levels, which are connected into the control system and will cause the shutdown without human action. And, of course, the reactor operators can always press a single button and cause the rapid shutdown.

Reactivity Feedback

Above we talked about how control of a reactor, the increase or decrease of power level, could be achieved through human actions. We also saw that as long as we designed the reactor correctly and ensured k would never get more than the delayed neutron fraction above one, the reactor would respond slowly enough that we could manage it. I also promised to tell you about things nature provided to us that make safe reactor control easier, if we chose our design intelligently.

Before I explain that to you, I need to tell you something about feedback. This is something you have seen in your everyday experience, but you might not have thought about it too much.

First of all, a dynamic system is a system that is capable of changing with time. In a dynamic system, there are factors that can cause the changes to occur in a certain way. The factors themselves may be influenced by the changes they caused in the dynamic system and thereby cause further changes. Let's look at an example. If you get in your car and press the gas pedal, the car goes faster. But it doesn't just keep going faster and faster. No matter how hard you push down at some point the car just won't go any faster. The car is clearly a dynamic system. By pushing down on the gas, you are calling on the engine to create more power and make the car go faster. However, as the speed increases, it also changes factors that impact the dynamic of the moving car. One of the most important is the resistance of the air, which increases with increasing speed. In fact, the air resistance increases at near the square of the speed. That means that increasing speed from 10 miles per hour to 20 miles per hour increases the air resistance by a factor of 4, and increasing from 10 miles per hour to 60 miles per hour increases the air resistance 36 times.

In this example, the increase in speed changes factors that impact the conditions determining how fast you can go. Because these factors act in opposition to the increase we can call them negative feedback. That is, an increase in speed causes an increase in air resistance, which in turn retards speed. Therefore, in the end, the ultimate speed is limited.

What about positive feedback? Well, if you ever heard a public address system squeal, you have experienced positive feedback. When you hear the squeal, the microphone in the public address system picks up some sound and amplifies that sound and puts it into the speakers. The microphone then "hears" the amplified sound and puts it back into the system. This is louder than the first sound because it was amplified by the public address system, so it sends it through again, but louder this time. As you might imagine, this goes around a few times, and it gets very loud, and the electronics convert the resulting rapid loop into a loud squeal. The solution is to modify the electronics in the system or simply move the microphone further from the speakers.

To see how one might employ feedback in design, let's look at another car example. Let's imagine we have a car that is so aerodynamic that air resistance doesn't impact our top speed. That is, of course, not possible, but that is the fun of "let's imagine." Now let's put a little windmill on the hood. We are going to connect the windmill to the gas pedal. This is going to be a very special gas pedal. What you are used to is a gas pedal with a spring under it that will spring back to the off position unless you continue to push down. Our imagined car will have a gas pedal that goes to whatever position you put it in and stays there until some force acts to change its position. Can you see that?

Now, if we set the gas pedal at some point between off and full speed, the car will move. If we connect a cord from the windmill to the gas pedal, as the car moves, the windmill will try to wind up the cord and pull on the gas pedal. The key now becomes which side of the gas pedal do we connect to the cord. If we connect it to the bottom of the gas pedal with the right arrangement of pulleys, as the car moves it will pull the pedal down and the car will go faster. This will cause more wind to whistle through the windmill making it pull the cord harder and pull the pedal down harder. This will cause the car to go faster and faster until something breaks or we hit something. This is a positive feedback.

If on the other hand, we had connected the cord to the top of the gas pedal with a different arrangement of pulleys, then the motion of the car would cause the windmill to pull the cord, pulling up on the gas pedal, decreasing the speed. This is negative feedback. This is what we do in nuclear reactors. We take advantage of natural factors that favor negative feedback and enhance them by design. Just like the self-limiting speed control we had built with our imaginary car when the windmill's cord was attached to the top of the gas pedal, increases tend to decrease and, therefore, inherently, by their very nature, limit increases.

Let's now look at how we do that in nuclear reactors.

There are a number of factors that cause natural feedback in a nuclear reactor. To keep things a little simpler, let's look at two of the most important ones. These are the moderator temperature and the fuel Doppler feedback effects.

As you might expect, the first one has to do with the moderator and how hot or cold it is, and the second one has something to do with the nuclear fuel. In both cases, we will be trying to see how a change in conditions in the reactor feeds back to the reactor power level.

To see how the feedback might affect the power level, we have to know something about the specific reactor. If you recall, we said the choice of a moderator was a balance between the efficiency of the material to slow neutrons down and the non-productive absorptions caused by the moderator. This not only affects what type of moderator we chose but also how much of it we use.

Using normal water for both a moderator and a coolant is the most common choice in power reactors, so let's look at that type of reactor. If the proportion of water relative to the fuel is too small, then if you added just a bit more water, the value of k would go up. Remember k? That is the number of neutrons in one generation divided by the number of neutrons in the previous generation. If the reactor is at a steady power level and k goes up, that means the reactor is going to start increasing in power and will continue to increase until something makes k go back down. Likewise, if we took water out of this reactor, k and hence reactor power would go down.

If in our reactor we had just a little too much water for the optimum mix of water and fuel, then adding a little bit of water would make k go down. Again, if we had been at a steady power level, making k go down would cause the reactor to start losing power, and it would go down until there was no fission at all unless we did something to make k go back up.

If there is just the exact optimum amount of water relative to fuel, then the moderating aspect of the water and the neutron absorbing aspect are perfectly balanced. Either adding or withdrawing water from this case would push us off this balance and cause the reactor power to go down.

For safety, we would really like to have a reactor in which any increase in power is met with a negative feedback, which tends to decrease power. This would be like our safe car in the example above that has the cord attached to the top of the gas pedal and tries to slow us down if we go faster.

The way we do that is we figure out the optimal amount of water relative to fuel for our reactor and then put in a little less than that. Why would you do that? Well, because if you have less than the optimal amount, any further decrease drives you further away from the optimal point and makes the reactor power go down. If you designed the reactor with just a little too much water, then a reduction in the amount of water would make the reactor move toward the optimal point and make power go up.

If you look at Figure 5.1, you can see how this works. Across the bottom, we have the amount of water with the proportion of water increasing as we move to the right. On the side, we have reactivity. This is just a way to say the impact of k. That is, the higher the reactivity, the higher the value of k. So if we have the amount of water just exactly right, then we get the most possible reactivity. If there is less water than the optimal point, we call that under-moderated. If there is more than optimal, that is over-moderated.

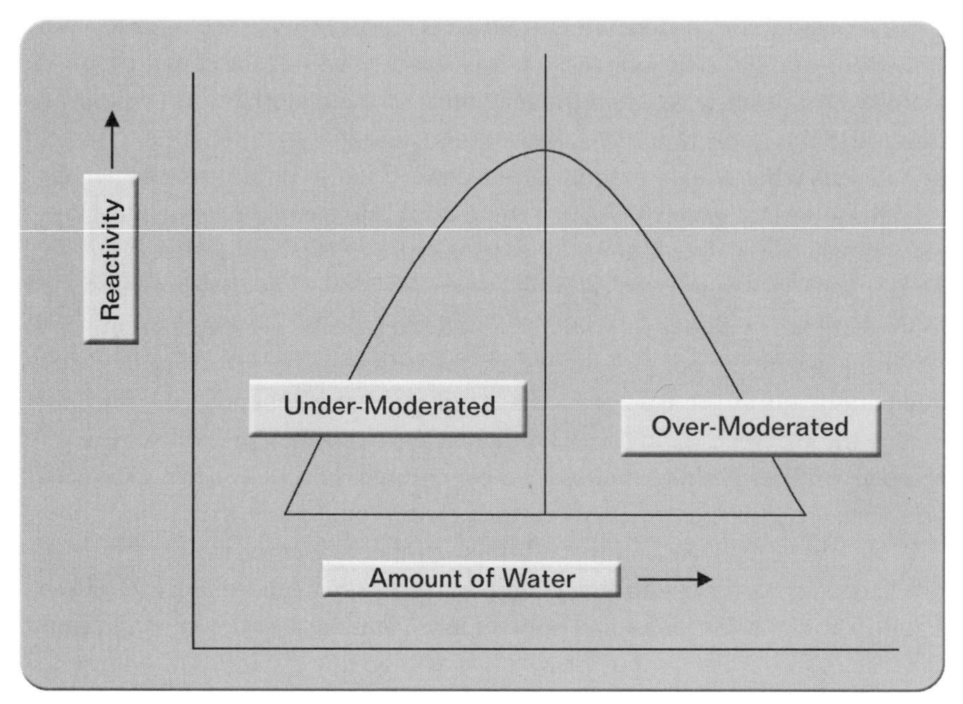

Figure 5.1. Moderation and reactivity.

So if you are on the under-moderated side and you decrease water, you slide down the curve to lower reactivity, which means lower k, which means power will decrease. If on the other hand you are on the over-moderated side and you decrease water, you move up the curve to higher reactivity, higher k, and more power.

Why is that of any help to us? When you heat water up it expands. That means per unit volume there is less water. If you designed your reactor right, you will be just a little under-moderated. Then if for some reason power goes up in your reactor, the water will get hotter, expand a bit, making less water per unit volume in the reactor, pushing you down the reactivity curve, decreasing k, and pushing against the power increase. That is the negative feedback we want.

This is really good in a water-cooled and moderated reactor because not only does that help on power increases it also works if we break a pipe and lose water or get our water really hot and make it boil. In either of those cases, the amount of water goes down a lot, and if we are under-moderated, it tends to push k down a lot.

If we look back at the power increasing case, we can't say in advance if the negative feedback is going to be strong enough to make k go back down under one. That is, will it make power go down or just decrease the rate of increase? Of the latter, we can be assured. However, we need to know what caused the power increase in the first place and how strong it was before we can say for sure if the moderator temperature effect will be enough to reverse the increase. Clearly, there will be times we want to be able to override the effect. If we couldn't, we could never get the reactor going. That would be like a spring under your gas pedal that is so strong you couldn't push it down. It would be real safe, but sort of loses the point of the car. The moderator temperature effect is like a good stiff spring under the gas pedal. It is ready to rein us back but also lets us increase if we really want to do that.

In other types of reactors, the moderator temperature effect can get a little more complicated. We will talk about the Chornobyl accident in the next chapter, but we can mention right here that the primary cause of the Chornobyl accident was a very bad reactor design. Because it used both graphite and water as moderators,[1] it had a positive moderator feedback. During a special test, they managed to operate the reactor in a very odd fashion, and this positive feedback got away from them destroying the reactor.

However, if a reactor is designed with a negative moderator temperature effect to start with, accidents like the one at Chornobyl can't happen. It is important to understand both the strength of such feedback effects and that these are inherent safety features of the design. Once you choose the amount of moderator relative to fuel, you have set the nature and size of the moderator temperature effect. It will have its effect, either as a safety feature limiting the rate of change of the power level, or forcing the power level rapidly out of control, completely without any further action. There are no pumps or valves, no levers or dials involved. The fundamental behavior of materials will play out. As long as you designed the reactor for safety in the first place, then you will get this self-limiting performance automatically. All power reactors licensed to operate in the United States are required to have such negative feedback as a basic part of their design.

As important as the moderator temperature effect is, the fuel Doppler effect is even more dramatic. Let's see why.

When we were talking about fission reactions, I told you that generally speaking the probability of neutrons interacting with a nucleus got better as the energy, or speed, of the neutron decreased. In fact, a neutron is 400 times more likely to cause a fission event in uranium-235 if that neutron is slowed down to thermal energies than a neutron traveling at the energy it had right

[1]We need to note that just because you use both graphite and water as moderators doesn't mean you have to have a positive moderator feedback. If you balance the total moderator correctly so that you are slightly under-moderated, even a graphite- and water-moderated reactor can have a safe negative moderator feedback.

out of the fission event that generated it. However, the key word in that description was *generally*. Because of the internal energy states of the nucleus, many isotopes have a complicated response to neutron energies. While the probabilities for interaction do tend to increase as the neutron energy gets lower, the transition is not smooth. In the lower to mid-range of energy, many isotopes have significant peaks of increased probability of interactions. These correspond to special energy conditions within the nucleus. You might think of it as having a neutron of just the right energy fit really well in a special slot in the nucleus. The effect is that, for neutrons of the right energy, the probability of absorption can go up hundreds, even thousands, of times from that of neutrons with just a little more or just a little less energy. These spikes of probability are called resonances. Even as they are strong, they are also pretty narrow. That is, the energy has to be very close to the "right" value to get the big increase in absorption probability.

One set of resonances that is very important to us is the neutron capture resonances of uranium-238. While in most reactors we are going to use uranium-235 as the main source of our fissions, the majority of the uranium in the reactor will be uranium-238. Recall that if you dig uranium out of the ground, 99.3% of it is uranium-238. We increase the amount of uranium-235 to get a good fuel, but this process is very expensive, so we don't do it any more than we need to. In most light water power reactors, we increase the uranium-235 up to 3 to 5%, leaving the remaining 95 to 97% uranium-238.

In getting to a sustaining chain reaction of fissions, we are trying to ensure that at least one of two or three neutrons created in a fission event can be used to cause one more fission event. Noting that the probability of causing fission is greatly increased if we slow the neutrons down, we might have a moderator in our reactor to do that. One of the challenges is getting your neutrons past the uranium-238 resonances as you slow them down. If you bounce a neutron into one of the "right" energy levels within the fuel, then one of the uranium-238 resonances will gobble it up. This is one of the reasons why normal light water is such a good moderator in that it is the best possible material to slow neutrons quickly and get them to jump over these traps along the way.

But uranium-238 resonances turn out to have a very useful property. Recall, I said, they were very narrow. That is, a neutron has to be very close to the right energy to get this big boost in absorption probability. If, however, you heat the uranium-238 up, the "width" of the resonances gets wider. This is called Doppler broadening. The magic is that if you heat uranium-238 up, it starts sucking up more neutrons. Why is that such a big deal?

If you are looking for a negative feedback effect, what could be better than finding a strong one in the fuel itself. If power in your reactor goes up for any reason, the very first thing that happens is that the fuel heats up. The uranium-238 resonances widen and suck up more neutrons and the increase is mitigated. This effect is very strong and very quick. While the moderator temperature effect provides inherent safety, it does take a bit of time for the increasing power in the fuel to heat up the surrounding moderator. The fuel Doppler effect is right in the fuel and happens almost instantly. It is also very much proportional to the size of the thing that got power going up. That is, if you get a big gain in reactivity for some reason and, therefore, get a rapid increase in power, this causes a big increase in fuel temperature and, therefore, a big counteracting fuel Doppler effect. As long as you have a lot of uranium-238 in the fuel of your reactor, this very strong negative feedback effect is going to be on your side, helping curb any power increasing in your reactor.

Types of Power Reactors

We have talked a lot about the physics of nuclear reactors, but what do they look like, and from a mechanical point of view, how do they work? There is not one answer to that because there are several different types of power reactors.[2]

In 2000, there were 437 power reactors operating in 31 countries around the world. Combined they provide about 16% of all the electricity made by humans today. The United States has 103 operating nuclear power reactors, which make about 20% of all our electricity. Nuclear energy is second only to coal as a source of electrical power in the United States, although natural gas is gaining fast and will be come the second largest source soon.

We don't have room to talk about each of the 437 individual reactors, but we really don't need to look at each one. We will look briefly at six different reactor types. While there are variations within each type as to size, manufacturer, and specific design details, looking at those six types will give you a pretty good idea of what nuclear power reactors are like across the world today. I will warn you that all of these are known by acronyms, so be ready for them.

One small detour before we start on reactors. Because we are talking only about nuclear reactors that generate electrical power, it would be good to understand how electrical power is made. Chapter 3 has more on this topic, but let's do a quick review here.

The way you make electricity is you move a material that conducts electricity through a magnetic field.[3] If you do that electricity will move through

---

[2]We will limit the discussion to nuclear reactors used for electrical power generation. There are also a lot of other types of reactors used for propulsion in ships (e.g., submarines and aircraft carriers), nuclear research, and the production of useful isotopes.

[3]At least this is the way if you are going to make a lot of electricity. You can also get electricity directly out of chemical reactions like in batteries or fuel cells. You can also get it directly from the sun using solar cells. So far these can only be used for relatively small applications. You can also get electricity directly out of nuclear power using thermionic or thermal electrical effects. Currently these are not very efficient and have been used only in a few spacecraft.

the conducting material. The normal way to do this is to have the motion be circular; that is, you rotate a loop of the conductor through the magnetic field. That way you can just keep spinning the loop rather than move something back and forth through the magnetic field. When you place a big rotating conductor in a magnetic field, you have built a generator. Now all you need is something to rotate the conductor.

If you have a good source of wind and a windmill, you could use that to rotate the generator. Falling water is a good way as well. To make use of falling water, you will need a water turbine, which is a big propeller-looking thing that will spin when water passes through it. You will also need a source of water. A large impoundment of water like a dam will do.

If we want electric power where there is no falling water and don't want to rely on the wind blowing when we need the power, it would be great if we could use other sources of energy and convert them to a rotating conductor. At the start of the Industrial Revolution, we learned that steam power was useful in many ways. We could get a big fire going and use it to boil water. The steam created was a great way to transport energy from the fire to some other place. We could use it to expand inside a cylinder and drive a piston to create rotation. That worked great for factories and locomotives. Early electrical power plants used the piston and cylinder method with a massive piston many feet across. However, a more efficient process was developed in allowing steam to expand across a set of propeller-like blades. This is a steam turbine. The blades are mounted on a shaft. The steam rushing across the blades spins the shaft. This shaft is connected to a generator, and we have electricity. I am simplifying this quite a bit. A modern steam electrical plant is very complex with several stages of turbines to use steam of different pressures and other efficiency enhancing side streams from the main steam line.

This concept of making electricity can use any heat source that will make the water boil. For that matter we don't have to use steam. Any fluid we can get hot and run through a turbine to spin it will work. Figure 5.2 shows the general layout.

One of the earliest heat sources used in producing electric power is still the most common, burning coal. We have gotten a lot more efficient, getting more productive energy out of a given amount of coal. We have also gotten better at reducing some of the pollutants that come with coal burning. We used to burn a lot of petroleum to make electricity. That has declined sharply and now oil is a relatively minor player in electricity production. The rising star as a heat source for electricity is natural gas. It has been relatively inexpensive and compared to coal or petroleum cleaner to burn. Natural gas prices do vary a good deal, however. In the winter of 2000–2001, natural gas prices rose alarmingly.

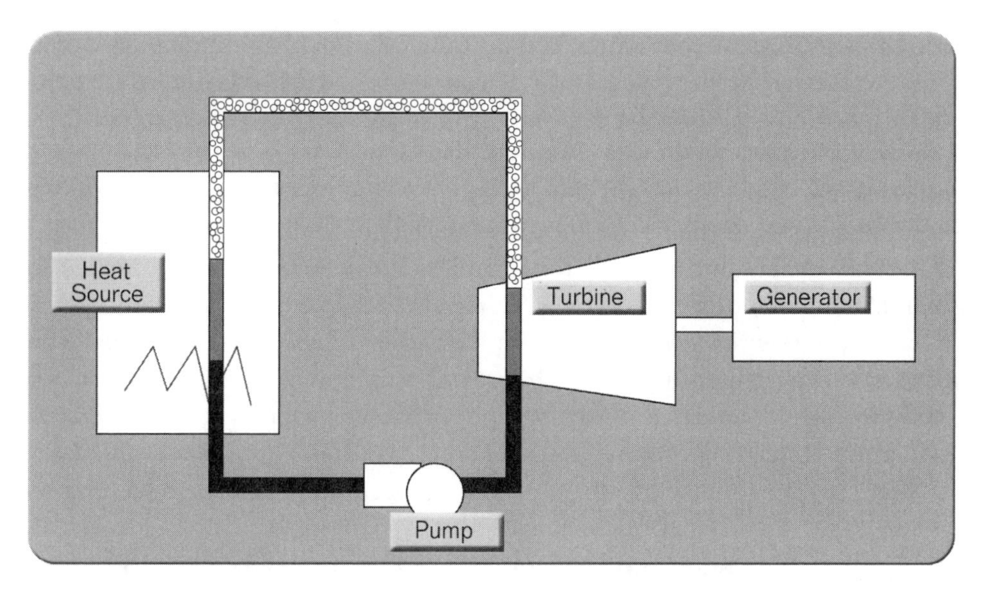

Figure 5.2. Making electricity.

Natural gas plants often use the hot gases of combustion as the working fluid and run those directly through the turbine. These look pretty much like a big jet engine. To get some additional energy out, they will put pipes carrying water into the "jet" exhaust, boil the water, and then use the steam in a separate turbine and generator.

In a nuclear electrical generating plant, we are going to use the heat produced by nuclear fission as the heat source. The working fluid for the turbine will be steam. The differences between the types of reactors are mainly in how we make the steam, what we use for a moderator, and what we use for a coolant. So let's look at some different types of nuclear power plants.

The first one we will look at is called a GCR. I did warn you about acronyms. This stands for gas-cooled reactor. These are found mainly in Great Britain although a few other countries tried them. There are two basic versions of the GCRs. The British introduced the original version in 1956. There are about 16 of these still operating. Later the British developed an advanced version, the advanced gas reactors, or AGRs. The first of these was completed in 1976. While there are only 12 of them, they tend to be larger than the older version. Both types use graphite as a moderator. The coolant is a gas, carbon dioxide. The good thing about carbon dioxide as a coolant is that it is not going to strongly react chemically with anything in the coolant path. It is also pretty cheap; after all, burn anything with carbon in it, and the primary product is carbon dioxide. The bad thing is that, compared to water, it can't carry as much heat, so you have to pump a lot of it to get the heat out of the reactor.

In the AGRs, the fuel is uranium dioxide. This is a ceramic material. If you don't know what a ceramic is, look at your coffee cup. If it is not metal, plastic, or glass, it is probably a ceramic using silicon dioxide (sand) rather than uranium dioxide. We use a ceramic for nuclear fuel because it is a very stable material. It will not melt until you get it to 2880°C (5220°F). It is tough and will hold on to most of the fission products well. The uranium dioxide is formed into little pellets and placed in a stainless steel tube about 0.6 inches in diameter and about 39 inches long. Thirty-six such "fuel pins" are arranged in a fuel element with sufficient space for the carbon dioxide gas to pass between the pins. This element is about 9-1/3 inches in diameter. Figure 5.3 shows an AGR fuel element being assembled. Eight of these elements are loaded end to end into a single fuel channel. The standard AGR has 332 such fuel channels.

The fuel channels run through a large graphite block. Because carbon doesn't slow down neutrons as well as water, there needs to be a greater volume of the moderator, and therefore, these reactors tend to be larger than water-cooled and moderated reactors. Large fans pump the carbon dioxide through the pipes, which hold the fuel. When the reactor is operating, the fission-generated heat is passed to the flowing gas. The hot carbon dioxide leaving the reactor is sent to a set of steam generators. These are big heat

Figure 5.3. Advanced gas-cooled reactor fuel element (courtesy British Nuclear Fuels, Ltd.).

exchangers. They are a bunch of tubes inside a bigger tank. The hot gas is on one side of the tubes, and water is on the other. As the hot gas flows through the tank, the water on the other side of the tubes picks up the heat from the gas. The pressure of the water is carefully controlled to allow the water to boil creating steam. This steam is then sent to a turbine causing it to turn, which then turns a generator to make electricity. Figure 5.4 shows the layout of an AGR power plant. While the British have had reasonable success with this type of reactor, they never found much following in other countries. The British themselves have built their first water-cooled reactor, moving away from their former sole reliance on the GCR technology.

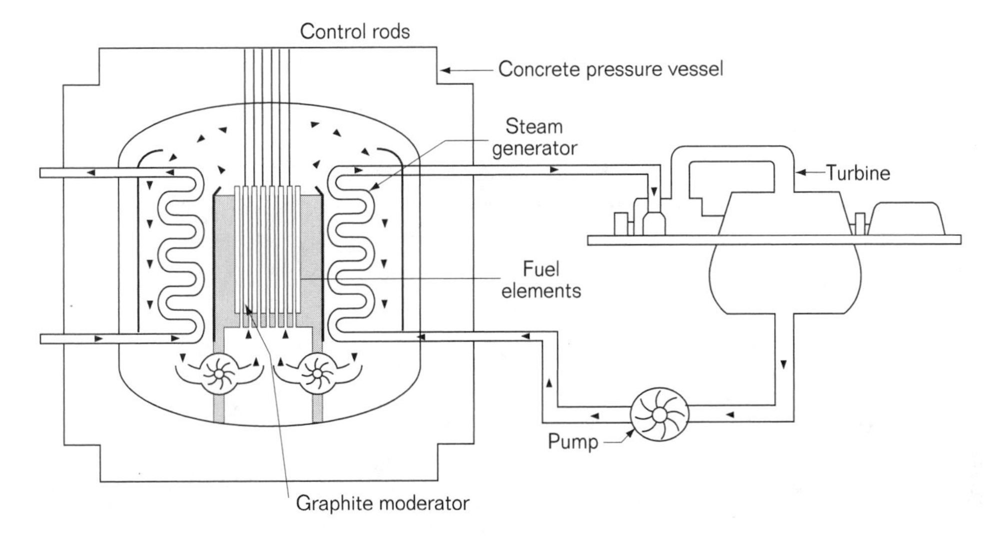

Figure 5.4. AGR plant layout (courtesy World Nuclear Association).

The next type of nuclear power plant we will look at is a PWR, a pressurized water reactor. This type of reactor used normal "light" water as both a moderator and a coolant. This is the most common reactor type used today, comprising just about 60% of the total number of electrical generating reactors in the world. There are several manufacturers of this type of reactor, and you can find PWRs in the United States, France, Japan, Korea, Ukraine, Russia, Germany, Spain, Belgium, and 15 other countries.

This type of reactor also uses a uranium dioxide fuel. There are a number of variations in the exact dimensions of the fuel, but typically the fuel pellets are stacked in long thin tubes about a third of an inch in diameter and about 12 feet long. The tubes are made of an alloy of zirconium, which is a very good material for nuclear applications.[4] These "fuel rods" are combined in an assembly of up to 17 rods on a side. Figure 5.5 shows a typical PWR assembly.

Some of the positions within the assembly will be left open for control rods to move in and out of or for instrumentation, so there will probably be less fuel rods than the 289 potential in each assembly. There will be a lattice framework holding the rods apart at about a half-inch center-to-center spacing to allow the water to pass through the assemblies. A big PWR producing 1100 million watts of electricity will have about 200 such assemblies. The arrangement of assemblies will produce an active region about 12 feet in diameter and 12 feet tall.

The assemblies are mounted in a large steel tank, called the primary vessel, which is typically 6-inches thick. Water is pumped through this vessel to carry off the fission heat. This water is kept pressurized to about 2000 pounds per square inch (psi) to prevent it from boiling. That is why it is called a

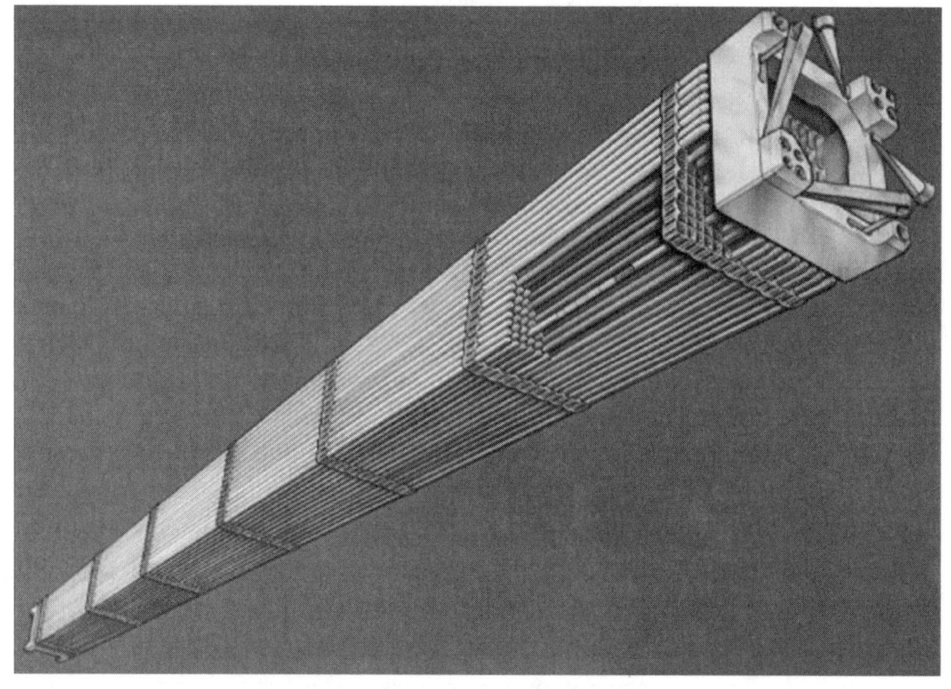

Figure 5.5. Pressurized water reactor fuel assembly (courtesy British Nuclear Fuels, Ltd.).

[4] Zirconium alloys are strong, do not absorb neutrons hardly at all, and are very resistant to irradiation damage. As neutrons smack into the zirconium atoms and push them out of alignment in their metal matrix, zirconium atoms are especially good at finding their way back, keeping the metal strong.

pressurized water reactor. The flow loop through the reactor is called the primary circuit. This pressurized water is run through a steam generator just like the carbon dioxide of the AGR. The primary water goes through a lot of tubes, which are in a tank of water, which is kept at a lower pressure, around 1000 psi. This water picks up the heat of the primary water and it boils. This steam is sent to a turbine, which then turns the generator. Figure 5.6 shows how a pressurized water reactor is laid out.

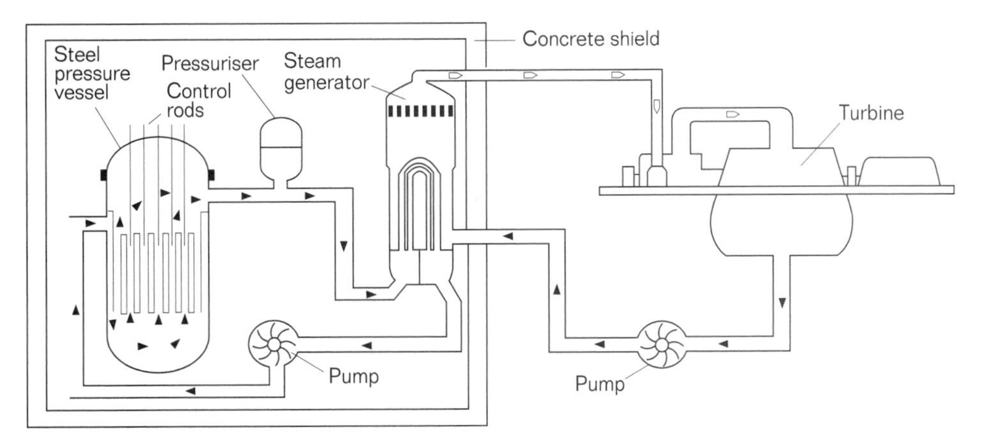

Figure 5.6. Pressurized water reactor plant (courtesy World Nuclear Association).

A BWR, boiling water reactor, is similar to a PWR in some ways and different in a few. Like the PWR, it uses normal light water as a coolant and a moderator. Sometimes you will see BWRs and PWRs referred to as LWRs for light water reactors. BWRs are the second most common reactor type and are found in the United States, Japan, Europe, and Taiwan. In addition, Mexico and India have two BWR plants each.

The fuel is similar to that of a PWR, using uranium dioxide clad in zirconium alloys. The fuel rods tend to be a little larger, but the assemblies are smaller with the assemblies ranging from 7 × 7 to 9 × 9 arrays of fuel rods. The rods are about the same length as PWRs at 12 feet. Because the assemblies are smaller, more of them are used. The result once again is a fueled region about 12 feet in diameter and 12 feet high in a 1000 million watt sized electrical power reactor.

The big difference between PWRs and BWRs is that in BWRs, the water is allowed to boil within the primary vessel. Instead of the 2000 psi of the PWR primary circuit water, the BWR primary pressure is only about 1000 psi. Water is constantly fed into the bottom of the primary vessel. As the water rises through the fueled region, the heat of the fission reaction causes it to boil in the upper end of that region. The steam then is routed directly to the turbine. Figure 5.7 gives you a view of a BWR plant layout.

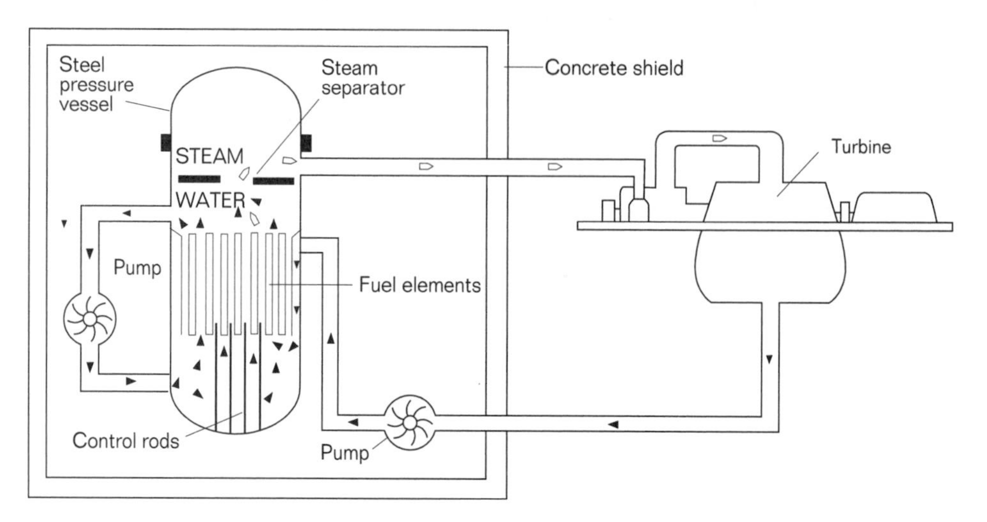

Figure 5.7. Boiling water reactor plant (courtesy World Nuclear Association).

The obvious advantage of the BWR is it doesn't require separate steam generators. This is especially good, as steam generators have turned out to be difficult to maintain. This should not be a big surprise. You are pumping millions of gallons of very hot water under high pressure through thousands of individual tubes. That is tough on the plumbing and provides a lot of opportunities to get a leak. Most PWR plants can expect to replace their steam generators at least once during the life of the plant (40 to 60 years).

Another advantage of a BWR is, with the lower primary pressure, the primary vessel need not be as thick (typically 4 inches instead of 6 inches). That saves a lot of steel and a good deal of money. It also means some of the reactor safety features such as the containment building can be less robust. The containment building is the massive steel-reinforced concrete structure built around the reactor to ensure that in the worst-case accident, the radioactive material stays inside. That is, it will be contained. The strength of this structure is set to contain the potential energy release if a major pipe breaks and that energy is set by the pressure of the primary coolant. Therefore, the BWR gets credit for its lower pressure.

However, the PWR does have the advantage of a steam cycle that uses water that has never seen the reactor. While the BWR circulating water is kept very clean, there is always a little bit of minerals or corrosion products in the water that are activated by the neutrons in the reactor. This means the primary circuit contains some radioactive materials. In a PWR, this is only a problem if there is a primary pipe leak. In a BWR, this primary water circuit includes the steam that goes to the turbine. Therefore, all that piping and the turbine have some exposure to radioactive material. This complicates maintenance of the BWR plants.

The next type of reactor we will look at is the RBMK. This stands for Reactor Bolshoi Moschnosti Kanalyni. These are commonly known for their most infamous plant and are called the Chornobyl-type reactors. These were designed in the Soviet Union. With the closure of Chornobyl Unit 3, which is in the Ukraine, on December 15, 2000, only Russia and Lithuania continue to operate RBMK reactors. The Russians have 11 reactors and the Lithuanians have two. The Lithuanian plants are unique in that they are a special jumbo-sized version at 150% the nominal size of the standard 1000 MW electric Russian reactor making them the two largest nuclear power plants in the world. Originally built when Lithuania was a Soviet satellite state, the intention was to send a good portion of the power back to Russia. Therefore, Lithuanian reactors provide most of the electrical power to the now independent country.

The RBMK reactor in some ways looks like a cross between a GCR and a BWR. Developed from the Soviet military production reactors that were used to make plutonium for nuclear weapons, the RBMKs are moderated by graphite and cooled by light water. Like the GCR, the fuel is held in channels that run through a large graphite stack. The fuel is again a metal-clad uranium dioxide. Fuel pins are held in small clusters, which are mounted in the fuel channel. Light water is pumped into the bottom of the fuel channel. Because water can carry heat better than carbon dioxide gas, these channels, at a little over 3 inches in diameter, are much smaller than the AGR channels. As the water rises in a channel, it picks up the heat from the fission reactions. The pressure and flow are controlled, so the water boils near the top of the fuel channel. This steam is routed from each of about 1600 channels into 4 large steam drums and from there to the steam turbine. Figure 5.8 shows a RBMK plant layout.

The RBMK well deserves the very negative opinion it gained as the reactor type that caused the worst nuclear reactor accident of all time. We will look at that accident a bit closer in the next chapter. However, it is instructive to recognize that there are a few good features of the RBMK design, which explains why the Soviet designers would choose to build a reactor like this.

The large graphite mass of the RBMK represents a large, slow-acting heat sink. This means you can soak up a lot of energy in just the graphite if anything goes wrong. Well, if a limited set of things goes wrong, that is. Using over 1600 flow channels instead of one big pressure vessel means that you don't have to worry about a big vessel failure that causes all the cooling water to be lost. There have been several individual flow channel pipe failures over the years, and the results have been pretty minor. The four massive steam

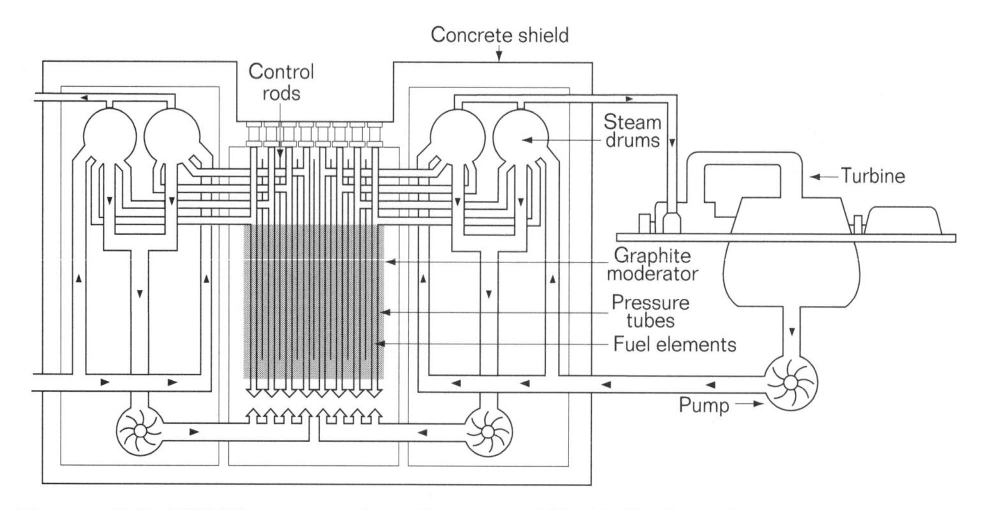

Figure 5.8. RBMK reactor plant (courtesy World Nuclear Association).

drums with hundreds of individual pipe connections each significantly mitigate this advantage. A major break in one of the steam drums would be very much like a primary vessel failure in a PWR or BWR in that all your cooling water would get away pretty fast.

The major advantage of the RBMK to the Soviets was that, with the graphite moderator, they could operate with a low enrichment of uranium-235 in their fuel. If you recall, most reactors cannot sustain a fission chain reaction on the uranium that comes out of the ground. That uranium is 99.3% uranium-238 and 0.7% uranium-235. Because uranium-235 fissions better, we increase the amount of uranium-235 relative to uranium-238 to make most nuclear fuel. This is called enrichment. I will explain how we do that in just a bit. PWRs and BWRs work best with enrichments of 3 to 5% uranium-235. With a graphite moderator rather than a light water moderator, the RBMKs could run with only 2% uranium-235 enrichment. With their centralized economy, the Soviets could use uranium coming out of military reactors and their pressurized water civilian power reactors as a feed source for the RBMKs. That is, when the fuel in those reactors was "used up," it still had some uranium-235 left. You might think of it as the last bit of peanut butter you can't get out of the jar. That fraction was still enough to run in the RBMKs. This was a significant savings to them and made efficient use of all their uranium.

As we discussed above, the primary design fault that caused the Chornobyl accident was the positive void coefficient,[5] which caused the

[5]We talked earlier about the moderator feedback. The term void coefficient focuses on the specific effect on reactivity if the coolant, which in this case is also a moderator, disappears. Because in boiling the density changes so much, it is essentially as if the coolant disappears or turns into void. Therefore, we can "void" the coolant by either losing it through a big leak or by rapidly boiling it. A positive void coefficient means reactivity increases if the coolant voids.

power to increase when water was lost from the reactor coolant loop. No reactor licensed to operate in the United States is allowed to have a positive void coefficient. As we saw in that earlier discussion, having a positive void coefficient is a pretty dangerous thing. The Soviets mistakenly thought that the effect would turn back negative once enough water was lost. It took some years and some very special conditions before this mistaken belief was put to the test, but the operators at Chornobyl Unit 4 put themselves in that state April 26, 1986. The resulting power excursion was so violent that it blew blocks of graphite out of the building. We will look more at that accident in the next chapter.

Another interesting variation of reactor design is the CANDU. This stands for Canadian Deuterium Uranium. As we noted for the RBMKs, if you shift from light water to a moderator that absorbs fewer neutrons than light water, you can operate the reactor with a lower uranium-235 enrichment. The RBMKs use graphite and were able to use 2% enriched fuel rather than the 3 to 5% used in light water reactors. The Canadians decided to go one better and use heavy water as their moderator. This allowed them to use natural uranium, that is uranium as it comes out of the ground with no enrichment at all. Because enrichment is expensive, this is a pretty important advantage. However, it is offset by the fact that instead of enriching the fuel, you have to enrich the moderator. Normal water is 0.015% heavy water and 99.985% light water. That means that for every 6600 molecules of light water you can find one of heavy water. This can be compared to uranium-235 which is 0.7% of natural uranium or one part in 140. But when you are trying to enrich an element in one isotope over another, the relative difference between the isotopes becomes important. In uranium, the difference between uranium-238 and -235 is 3 atom mass units, that is only 1.3%. The light water molecule will weigh in with 18 atomic mass units while heavy water will have 20. That is a proportional difference of 11%. So while heavy water makes up a smaller portion of "normal" water than uranium-235 makes up of natural uranium, the heavy water is easier to distinguish and separate from normal water. While enriching normal water to heavy water is easier than enriching uranium, it isn't very easy and heavy water is very expensive.

The CANDU reactor is somewhat like a GCR in that it has metal-clad uranium dioxide fuel pins arranged in small clusters lined up in fuel channels. These channels, rather than going through a block of graphite, are suspended in a tank of heavy water. Inside the fuel channels, the fuel is cooled with heavy water.

The cooling water works just like a PWR flow circuit in that after being heated by the fission energy as it flows through the reactor, it is routed to steam generators. There it is used to boil water in the "secondary" circuit, which in turn runs the turbine and generator. Figure 5.9 shows a CANDU plant.

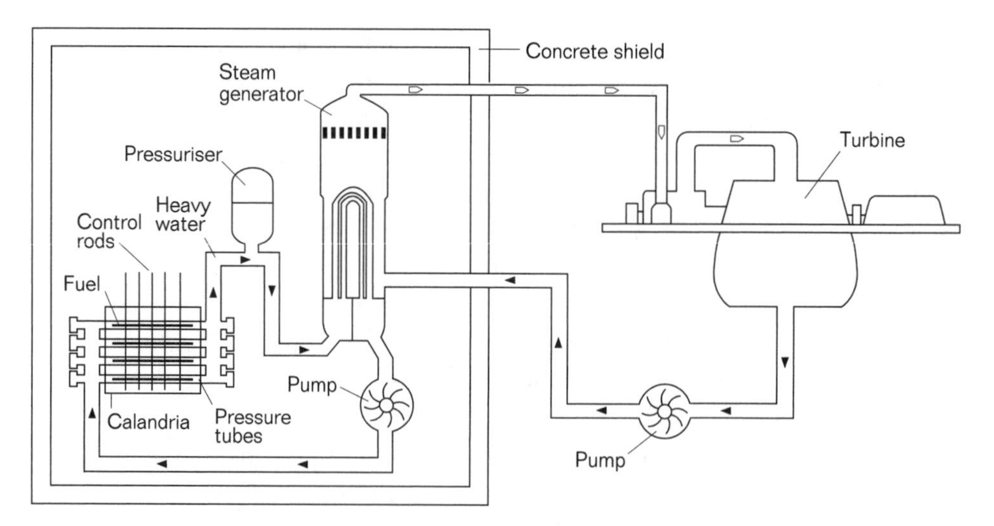

Figure 5.9. CANDU reactor plant (courtesy World Nuclear Association).

Because the fuel is not enriched, it runs out of sufficient uranium-235 pretty quickly. If they had to shut down to refuel the reactor as soon as the uranium-235 got too low, it would be shut down a lot of the time. The Canadians beat this problem by developing an automated system to refuel the reactor while it is operating.

The Canadians built quite a few of these reactors and sold a number of them to other countries, specifically South Korea, Romania, Argentina, and Pakistan. However, the ability to operate on natural uranium, the use of heavy water, and the ability to refuel when operating made these reactors a great way to make plutonium for nuclear weapons. The United States had five somewhat similar heavy water reactors at Savannah River in South Carolina whose primary mission was making materials for nuclear weapons.

When it became clear that nations might be interested in using the Canadian technology for these purposes, the Canadian government imposed very strict controls. This hurt them in the international marketplace, and it has been hard for them to make new sales. They had hoped for additional sales to South Korea and China. It now appears the South Korean sale, at least, will not happen.

Both India and Germany have built heavy water reactors. The German design is unique, but the Indian design is largely taken from the Canadian concept.

The last reactor type we will look at is a FBR, or fast breeder reactor. I have included this type in the list, not because there are a lot of these; in fact there are very few. I have included this type because they are very different, and they represent a different future for nuclear energy.

The reason these are call "fast" reactors is that unlike all the other types we looked at, this type of reactor doesn't depend on slowing neutrons down and, therefore, uses "fast" neutrons to cause a fission events. If you recall, we saw that in general the probability of getting a neutron to interact with something increased as the neutron was slowed down. Because the interaction we were trying to get was having the neutron cause a fission event in our nuclear fuel, we saw that slowing the neutrons down was a good thing. However, that isn't the only way to get a continuous chain reaction of fission events going. The key to a continuous chain reaction is to be sure you have at least one neutron causing a new fission event for each previous fission event. For fission events from uranium-235 using slow neutrons, we got between two and three neutrons, averaging 2.43, per fission. We needed to get one neutron past the non-fission absorptions in the fuel, absorptions in the structural materials, and absorptions in the coolant and moderator. Essentially all power reactors use moderators and operate off "slow" neutrons. So, clearly we have made this work. But just as the probability of causing fission gets less as neutron speed gets greater, so do the probabilities of neutron absorption in all the other stuff. It is possible to get a balance of fuel and reactor configuration that will be critical, that is, support a continuous chain reaction, without slowing neutrons down. Generally, it will mean we have to increase the enrichment quite a bit, perhaps to 20% uranium-235, or use fuels other than uranium.

Another really important thing happens if we get a reactor to operate off fast neutrons. One of the challenges in making a reactor critical is that not all the neutron absorptions in the fuel lead to fission. Remember that uranium-238 can't fission with slow neutrons. With fast neutrons, some of the absorptions in uranium-238 will lead to fission and the rest will lead to the production of plutonium-239. The production of plutonium-239 in the reactor is a good thing in that it is a good fuel, fissions with any energy neutron, has a high probability of doing so, and gives off on the average 2.9 neutrons per fission. In a reactor with a moderator, you might get as many as 7 atoms of plutonium made for every 10 atoms of uranium-235 you use up. That is pretty cool since it helps you keep the reactor running longer before you have to stop and put in fresh fuel. But what would be really neat is if you could get more plutonium than uranium-235 used up. And why not? If you have over two neutrons per fission event, why can't you get one to cause the next fission, one to be absorbed by uranium-238 and make a plutonium atom, and let the other fraction get sucked up elsewhere? Well, in a reactor using slow neutrons, we just can't get that to work. The higher probabilities of neutron reactions with the coolant, moderator, and structural materials as well as the non-fission absorptions in the uranium-235 just won't let us make it balance out.

However, in a fast reactor, all those reactions in the non-fuel parts of the reactor are smaller. The ratio of fission to non-fission events in the fuel is also better, plus the higher the energy of a neutron causing a fission event, the more neutrons are released in the fission. So, in the end, it comes out that you can configure a fast reactor that will make more plutonium-239 than the uranium-235 it uses. To be cooler yet, you can take the plutonium you make, mix it with normal non-enriched uranium, put it back in the reactor, and still make more new plutonium than you use up! You can do this with a fast reactor and that is where the term breeder comes in. You breed more new fuel than you use up.

If this sounds like an impossible "energy for free" machine, it isn't. All you are really doing is using the reactor to convert uranium-238 to plutonium-239. However, this is a really good thing. Recall that uranium-238 is 99.3% of natural uranium. Current power reactors that operate with slow neutrons run off uranium-235, along with the amount of plutonium they make as they run. We aren't getting very much value from the uranium-238. In fact something like 95% of the potential energy content of the fuel is not used.

If you need a little feel for the significance of that, try this on. In the United States we have several large enrichment plants. Because uranium is enriched by taking some uranium-235 from natural uranium and giving it to the uranium you want to use as fuel, you get a lot of uranium that has less than the natural amount of uranium-235. This is known as depleted uranium. We have been making this for a lot of years and have been storing it. We currently have something like 600,000 metric tons of the stuff. If you could put this into fast breeder reactors and convert it to plutonium, you could make a lot of fuel. How much? Well, because you also get non-fission captures in plutonium-239 you can't just keep going around the cycle and use all the uranium up. Let's guess you could get half of it. You probably could get more than that, but half will show the point. If you did that you would be working with 300,000 metric tons of fissionable material. That is 3×10^{11} grams, which is 7.6×10^{32} atoms we can fission. At 180 MeV of useful energy per fission and 2.24×10^{19} MeV per kilowatt-hour, that gives us 6.1×10^{15} kilowatt-hours of energy. But that is thermal energy, not electrical energy. The thermal efficiency of a nuclear power plant is about 33%, so that means we would only get 2×10^{15} kilowatt-hours of electricity. How much is that? Well, worldwide we used 12,260 billion kilowatt-hours (that is, 1.226×10^{13}) in 1997. At that rate, the depleted uranium we already have sitting around would power all the electrical needs of the planet for 164 years at the 1997 rate. This is just using uranium we have already gotten out of the ground and just the stockpiles in the United States. Using the depleted uranium already mined and sitting around from the former Soviet Union and European enrichment programs would extend it much longer. Then you could start mining uranium ore again. That is why breeder reactors are important. They greatly increase the available energy from the uranium resources we have.

So what do fast breeder reactors look like?

First of all, they are not going to have a moderator. This probably means they can be more compact; that is, they don't need space to allow for the neutrons to interact with the moderator and slow down. However, they can be more compact only if we can still get the heat out. As we have seen, a gas coolant needs a pretty big flow channel because gases can't carry very much heat, at least not relative to liquids. Convenient liquid coolants like water would not be a good choice for a fast reactor because water is such a good moderator. O.K., what is left? Well, a really great way to transport a lot of heat is to use a molten metal. So, we are looking for a metal that doesn't absorb neutrons very well, transports heat well, flows easily so it isn't hard to pump, and doesn't corrode the piping. It is hard to get all those properties at the same time. So far the best compromise has been sodium. It has very good heat transfer properties, melts at a conveniently low temperature of 208°F, and is just about as viscous[6] as water. Stainless steel does very well in liquid sodium, so as long as you make everything inside the reactor out of stainless steel, corrosion will not be a big problem.

However, if you remember what was probably the single most exciting experiment from your high school chemistry class, sodium and water don't get along very well. In fact, they react violently. If we are going to use a conventional steam turbine to make electricity, we are going to have a challenge keeping the water and the sodium apart. There is also the small problem of sodium-24. Normal sodium is sodium-23. While it doesn't absorb neutrons well, it does absorb some. When it does it becomes sodium-24. This isotope has a 15-hour half-life. Since you are all now experienced reactor physicists, you will recognize that is a fairly short half-life, which means, while it goes away pretty fast, it is also pretty intensely radioactive. It decays with a strong gamma ray, which means you need some serious shielding between people and the sodium that goes through the reactor.

One way to deal with both these problems is to have three nested coolant loops. The first loop is liquid sodium that goes through the reactor. The second loop is also liquid sodium, but it gets its heat from the first loop in heat exchangers just like a pressurized water reactor primary loop heated water in a secondary loop to make steam. The difference is that in this type of FBR, the secondary loop is liquid sodium and it doesn't boil. In this arrangement, the secondary loop of sodium will not be radioactive because it never sees the neutrons from the reactor. The third loop is water and uses the secondary loop to heat it. This looks more like the primary to secondary loop interface for a pressurized water reactor in that the water loop is allowed to boil, making

[6]Viscosity is a measure of the resistance to flow. Think of it as the runniness. The lower the viscosity, the easier it is to pump; therefore, being nearly as good as water is good.

steam to run the turbine. A lot of care is taken in the design and manufacture of the steam generators that bring the second and third loops together. Clearly, a leak that allows water and sodium to mix would be very bad.

Four countries have worked on FBR designs in a big way. These are the United States, Soviet Union, France, and Japan. In the 1970s breeder reactors appeared to be the future of nuclear power. The ability to make more fuel than you used, that is, convert non-fuel uranium-238 into plutonium-239 fuel, was very attractive. The United States built a large 400 MW test reactor, the Fast Flux Test Facility (FFTF), which operated successfully for many years. This reactor did not have a power generation capability because its mission was to test fuels and materials for the power reactors to follow it. We designed and partially built a power-generating FBR, the Clinch River Breeder Reactor, but we never finished it.

Two things got in the way of the U.S. breeder program. The first was politics. One of the things that constantly troubles nuclear power is the linkage to nuclear weapons. In a power system that uses breeder reactors, you are going to be making plutonium and then separating that plutonium to make it into fuel to go back in the reactor. Many people got concerned that having a lot of plutonium moving around in that fuel cycle would give terrorists or other malefactors the opportunity to steal the plutonium to make nuclear weapons. The thing is you can also make nuclear weapons out of uranium-235. However, most nuclear power reactor fuel has no more that 5% uranium-235 and you can't make that into a weapon. But if you had pure plutonium, you would have fuel for the weapon. It takes a lot more than that to make a weapon, and we will visit the topic in the next chapter and learn a little more. However, the point is people got worried and put on political pressure.

Added to the political concerns were economic concerns. While breeder reactors make much more effective use of uranium than normal power reactors, they are not currently cost competitive. The power plant itself is more expensive to build and operate. Using three loops instead of two adds cost, as does the special care needed to use the chemically reactive sodium.

You also have to build a plant to take the fuel from the breeder reactor and separate out the new plutonium you made and then you have to have a plant to make new plutonium-based fuels. The separations plant has to handle very radioactive material. Because plutonium is more radioactive than uranium, although it is still only weakly radioactive, and more chemically toxic than uranium, it is more expensive to make plutonium-based fuels.

The costs of the Clinch River Breeder Reactor and its supporting separations and fuels plant were climbing. At the same time, uranium seemed pretty plentiful, so the economic case in the United States just didn't support the breeder. Because there was no compelling economic case to counter the political opposition, the U.S. breeder development program was stopped.

The experience in the Soviet Union was a little different. They built two power-producing breeder reactors. They built the BN-350 in Kazakhstan in 1973, which had a thermal power of 350 MW and an electrical generation of 135 MW electric. The Kazakh government inherited this reactor upon the breakup of the Soviet Union and has since shut it down. The Soviets built a bigger breeder in Beloyarski in 1981, the BN-600. This has an electrical generation capacity of 560 MW electric. Because this is in Russia, the Russian government now operates it. The Soviets had separation capabilities and had partially completed a large fuel fabrication plant. The Russian government inherited these facilities along with the overall plan for moving their nuclear power system into breeder reactors.

They maintain this view; however, the economic difficulties in Russia have significantly slowed down the process. Their current thinking includes shifting to a liquid lead system as opposed to sodium, and they believe this will improve the economics of the concept. Lead does not have the chemical reactivity of sodium, but it is harder to pump. The Soviet navy built and operated a number of lead-cooled reactors for its submarines, but these appear not to have been completely successful.

The French had the most advanced breeder program. They built the 233 MW electric Phenix reactor at Marcoule in 1974. They have the separations and fuel fabrication facilities needed in full operation. The reactor was something less than trouble free, but provided a great deal of operational experience and served the function of a test bed very well. The French followed this with a bigger reactor, the 1200 MW electric Super Phenix, but just like the U.S. breeder program, the French program ran into cost and political challenges, eventually being canceled.

The fourth nation to work on breeder reactors is Japan. The Japanese government traditionally takes a long view toward planning. Recognizing their total dependence on foreign suppliers for the energy that drives their economy, the concept of a breeder power system appealed greatly. They have consistently included breeders in their long-range planning and have moved slowly toward that goal. They built a 280 MW electric breeder reactor in 1996, the Monju. They have small-scale separations and fuel fabrication facilities and bigger plants in progress. However, the Japanese program has not been a happy one. There have been a series of embarrassing accidents including a fire at the reactor. They have maintained that they have not diverted from their long-term goals, but current progress has slowed significantly.

O.K., here is a quick review. There are 437 nuclear power reactors worldwide. Over half of these are pressurized water reactors that use normal "light" water for the coolant and moderator. About a fifth of the remainder are boiling water reactors, which are like pressurized water reactors except they let water boil directly in the reactor vessel. These are the mainstream reactors today. The CANDU design uses heavy water as a coolant and moderator and can operate with natural uranium, but, while there have been some foreign sales, it is largely restricted to the homeland of its birth, Canada. Likewise, the graphite-moderated, gas-cooled reactors are limited to Great Britain. The RBMKs, a graphite-moderated, boiling light water-cooled reactor, are an unfortunate product of the former Soviet Union now operated only by Lithuania and Russia. The few fast breeder reactors are an interesting concept for the vastly greater utilization of the uranium resource but currently are not economical.

One last thing and your education in nuclear science and engineering will be complete. I want to tell you a little about how the nuclear fuel cycle works. That is, how we make fuel from the uranium ore and what happens to the nuclear fuel after we get the energy out. It will be valuable to understand this before we get into Chapter 6 in which we will explore the "bad" things about nuclear energy.

Nuclear Fuel Cycle[7]

To make nuclear energy work, we need nuclear fuel. The process of making and using nuclear fuel is called the nuclear fuel cycle. Figure 5.10 gives an overview of the process, and we will look at each step.

If you recall, we learned that uranium is found just about everywhere but in very small amounts. It makes up about 4 parts out of every one million in common rocks and is dissolved in seawater to a concentration between 0.3 and 2.3×10^{-6} grams per liter. If you were to dig up all the earth's crust down to a depth of 12 miles and sift it for uranium, you might get as much as 10^{14} tons or 9×10^{16} kilograms. Clearly that would be a pretty messy process, but the point is, while rare, uranium is not uncommon. In fact, it is more abundant than silver.

Because digging up the entire planet would be wasteful, we instead look for ores that have a concentration higher than the 0.0004% average. There are some very high-grade ores in Canada and central Africa that range from 1 to 4%. Mid-grade ores are found in the United States, Canada, and Australia. These might have uranium concentrations as high as 0.5%. Uranium can also be obtained as a by-product in the extraction of other useful minerals. Gold ore in South Africa and Florida phosphates are examples.

After the ore is mined out of the ground, we start the process of separating out the uranium. (Figure 5.10 takes us through the nuclear fuel cycle. Key terms from the figure are italicized in the following text). First, we go through

[7]We will be looking at the uranium nuclear fuel cycle. While it is possible to have a nuclear fuel cycle built around thorium, essentially all nuclear reactors use uranium.

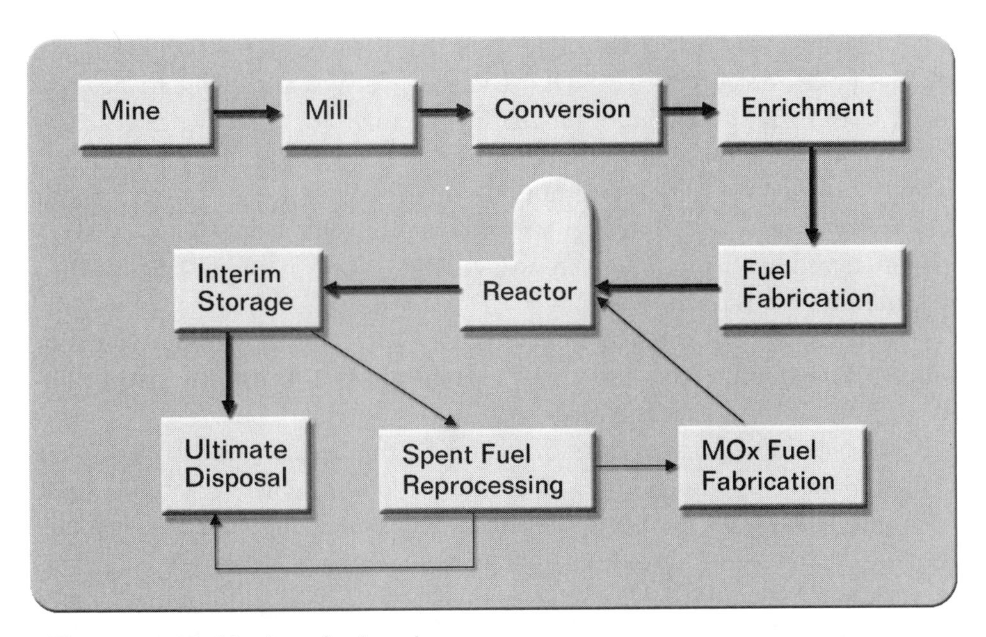

Figure 5.10. Nuclear fuel cycle.

a series of typically metallurgical processes to crush, screen, and wash the ore letting the heavy uranium settle out as we wash away the lighter debris. This is usually done right at the mine.

The next step is the *mill*. This is likely to be sited near the mine, so we don't have to move a lot of material. There is still a lot of non-uranium stuff in the ore at this stage. In the mill, we will use either an acid or alkali bath to leach the uranium out of the processed ore. Going through this chemistry gives us a yellow powder that is about 75% uranium oxide. This is not the uranium dioxide we will use for nuclear fuel; we have quite a way to go yet to get to that. The chemical form of this yellow powder is U_3O_8, and it is called yellowcake.

The next step is to purify the yellowcake and prepare it for the following steps in the process. This is done in a *conversion facility* in which a series of chemical processes are conducted. These result in removing the remaining impurities. For most nuclear reactor fuel cycles, the next step will be enrichment. Only the CANDU reactors can use natural (that is non-enriched) uranium. To prepare the uranium for enrichment, the conversion plant puts the uranium in a convenient chemical form for enrichment. As we will see in a minute, a gas would be a very convenient form. If you combine uranium with fluorine, a very reactive element, you can get uranium hexafluoride, UF_6. This chemical form is quite handy in that it is a solid at room temperature but becomes a gas if heated above 134°F.

Following along on Figure 5.10, we see that the next step is *enrichment*. For all fuel cycles except the CANDU reactor cycle, we are going to need to increase the fraction of uranium-235 in our fuel. For light water reactors, PWRs, and BWRs, this will be in the range of 3 to 5%. Recall we are starting from 0.7% uranium-235 in natural uranium. The goal then is to process our UF_6 in some way to create two streams, one in which the uranium-235 is increased and the other, where we will get the "extra" uranium-235, in which the uranium-235 is decreased.

There are several ways to create these two streams; however, two major processes have come out to dominate current practice. These are gaseous diffusion and centrifuge technologies.

Gaseous diffusion was developed during the Second World War in the United States as part of the Manhattan Project, which created atomic weapons. Nuclear weapons require a core that is essentially pure fissile material. For a bomb made of uranium, that means essentially all uranium-235. This was a tremendous challenge in 1944, to start with 99.3% uranium-238 and 0.7% uranium-235 and get out just the uranium-235. Remember that uranium-235 and uranium-238 are the same element and will behave chemically the same. That means you can't use any of the normal chemical processes to get the uranium-235 out. So, what can we do?

Well, clearly the difference between uranium-238 and uranium-235 is that the 238 is about 1% heavier. What can you do with that? Again, a number of things were tried. Many worked but were very slow or required a huge amount of energy and/or equipment. What was chosen was gaseous diffusion. This uses the principle that the lighter a gas is the better it diffuses, or drifts, through a medium. What they did was to pump UF_6 gas through a diffusion barrier similar to cloth. The lighter uranium-235 hexafluoride moved through just a tad better than uranium-238 hexafluoride. If you took a stream off the down stream side of the cloth, you got a stream with a little bit more uranium-235 in it. The up-stream side had a little less uranium-235. If you kept this up, pumping the enriched side through more and more sets of cloths, pulling the depleted stream off, eventually you got a significant change in the uranium-235 fraction.

While the principle that light gas diffuses better than heavy gas is simple, the application is not simple. The fractional enrichment at each stage is quite small so it takes a lot of stages to make a significant enrichment. This means there is going to be a lot of gas being pumped which means it takes a lot of energy to run the pumps. As a result, gaseous diffusion enrichment plants are very big and take a very large amount of energy.

Enriching uranium for weapons required an effort that was nothing less than heroic. For nuclear power reactors, since we need only to get to something at or less than 5% uranium-235, it is much easier, but gaseous diffusion is still a pretty energy- and equipment-intensive process. Which is why better methods were sought.

The "better" method that was found was centrifuge technology. So, how does that work? Once again we are going to take advantage of the fact that uranium-238 weighs just a bit more than uranium-235. If you could put a bunch of natural uranium atoms in a pile and shake them up long enough, gravity would pull the heavier atoms to the bottom and let the lighter ones float to the top. That would work, but it would take a very long time and a lot of shaking. So, how can we speed up that process? Well, we can make an artificial gravity with a spinning cylinder.

Just like the force you feel when you make a tight turn at high speed in your car, by moving in a curve you get an acceleration, not unlike gravity, directed outward. This is simply because you and the car were going forward and the process of turning requires a force to bend your direction into the curved path. You could continuously get that force if you just kept the car going in a big circle. That is what a centrifuge does. It puts uranium into a big spinning cylinder and uses that spin to create a force like an artificial gravity pulling the heavier atoms to the outside and causing the lighter atoms to float toward the inside of the cylinder. Then all you have to do is suck the enriched stream out of the center and bleed the depleted stream from the periphery of the cylinder. Again, you will use many stages to build up to the enrichment you want.

While centrifuges were challenging to make, it is not easy to get big cylinders to spin very fast and not break, they were made to work. The real advantage to centrifuges over gaseous diffusion is that it requires significantly less energy. This makes enrichment a lot more economical.

The Europeans and the Soviet Union shifted to centrifuge technology for enrichment. In the United States, we had built truly huge gaseous diffusion plants for our nuclear weapons program. Because these plants were already there and operating, the cost of enriching fuel for civilian power reactors was just the energy cost to run them. That is, you didn't have to build a big facility, it was already there, built by the weapons program. These plants then, even with their relatively inefficient process, could make enriched uranium for civilian power reactors pretty cheaply. A fully commercial process that had to start from a vacant lot, even if it were using the more efficient centrifuge technology, would be hard pressed to compete. There have been a few companies exploring the idea of commercial enrichment in the U.S, but none have been completed. The U.S. government has since turned over the operation of the civilian portion of the big enrichment facilities to a private firm.[8]

[8]That is, they turned over that part of the process that is capable of only a little enrichment, enough for commercial power reactors but far from what is needed for weapons.

When we leave the enrichment plant, we have UF_6 enriched to the amount our particular reactor needs. The next step is the *fuel fabrication* plant. In this plant we will convert the UF_6 to the final form of UO_2 which is uranium dioxide. This ceramic material will be very carefully prepared to get just the right density and chemical properties. The powered UO_2 will be formed into little pellets of exact dimensions. These will be loaded into the long, thin tubes that make up the fuel cladding. These tubes will have end plugs put on and welded shut. A series of tests and checks are performed to ensure everything is O.K. inside these fuel rods and that the rods are leak tight. The individual rods are then combined into the type of assembly needed for the type of reactor we are going to use them in. These finished assemblies are then ready to be shipped to the *reactor*.

At the reactor the new fuel will be stored for a bit waiting for the next refueling outage. That is the time at which the reactor is shut down to take out spent fuel and add new fuel. Both CANDU and RBMK reactors refuel while the reactor is operating, but that is the exception, not the rule. Most reactors now refuel every 18 months with some working toward a 2-year refueling schedule. In a refueling outage, you are probably going to take advantage of the down time to do a bunch of other maintenance as well. There are places in the reactor that are hard to get to when the reactor is hot and operating, so this is a good time to replace parts there, do inspections, or conduct upgrades to equipment. While the reactor is shut down it isn't making any power and, therefore, not making any money, so there is great incentive to keep the refueling outage as short as practical. However, shortchanging on maintenance to shorten an outage is not only unsafe, it is economically dumb. If something breaks during operation, the unplanned outage to fix it is always more costly than a planned outage. Because commercial nuclear power plants are in business to make money, "run until you break" is never the rule. It is "keep it running well." The key to shorter outages is careful planning of all the things you want to do and being sure all the people and parts are ready. It might look like a stock car racing pit crew, where everyone knows just what to do and does it in the least amount of time but is darn sure they do it right.

A lot of work has been done in the last few years to make the fuel last longer to extend the time between outages, to decrease the length of outages, and to reduce the number of unplanned outages. All three have been quite successful. We will look more at that in Chapter 8.

Back to our fuel cycle. For light water reactors, typically you will replace about one-third of the fuel each outage. You will also shift the other two-thirds of the fuel around to get the fuel to produce power evenly. Reloading patterns are a fine art and a lot of work is put into them to get the best use of the fuel.

So, if you follow a given fuel assembly in a light water reactor, it will see three operating periods, totaling 4.5 years or more. As it is used, a lot of its original uranium-235 will be used up, but some plutonium-239 will have be made from neutron captures in the uranium-238. This plutonium and the remaining uranium-235 will provide power up to that third refueling. At that time we have gotten just about all you can get from that assembly. Therefore, we will finally, after three operating periods, take that assembly out, along with its brothers and sisters of the same age, and put them in storage. At this point, these assemblies are considered spent fuel.

Spent nuclear fuel is highly radioactive. Mostly it is still uranium dioxide and most of that is uranium-238. We still have some uranium-235, more than natural uranium, and we have some plutonium. All the uranium-235 and plutonium we have fissioned has been turned into fission products. If you recall these fission products are going to be neutron rich and really want to get back to stable isotopes. They do this through radioactive decay. Because there are a lot of potential fission products and each of these has a different decay rate, the radiation from spent fuel is a mixed bag. However, recall that if something is very radioactive that means its half-life is short and it is going to go away pretty fast. Conversely if something radioactive lasts a long time, then it isn't very radioactive. While the combination of isotopes gives a pretty complicated picture, Table 5.1 gives you an approximation of the energy produced over time in the spent fuel removed in a refueling outage. The fuel removed in the refueling is going to be about one-third of the total fuel load. The values given in the table assume we have a normal large power plant producing 1100 MW of electrical power. Since nuclear reactors are about 33% thermal efficient, that means such a plant would produce about 3300 MW of thermal power. So, the

TABLE 5.1 Decay heat in spent nuclear fuel.

Time After Shutdown	Power (MWt)	Percent of Operating Power
Operating	1100	100%
1 minute	27	2.5%
10 minutes	17	1.5%
1 hour	11	1.0%
1 day	5.2	0.5%
1 month	1.9	0.2%
1 year	0.6	0.05%
10 years	0.09	0.008%

third removed for refueling would have then had an operating thermal power of about 1100 MWt.[9]

After the first minute of shutdown, essentially all of the power is from radioactive decay. During that first minute we are still getting some fissions from the delayed neutrons. What you see in looking at the table is that the radioactivity of the spent fuel decreases quite rapidly in the beginning. This is when the short-lived radioactive fission products are going quickly through their decay and being strongly radioactive in the process. As these short-lived, highly radioactive isotopes decay, the radiation level falls off. However, because there are long-lived, less radioactive isotopes in the spent fuel, even 10 years after the fuel has been removed from the reactor, there is still some radiation.

All this about the radiation of spent nuclear fuel is to help us understand about the next step in the fuel cycle, which is *interim storage*. After we discharge the spent fuel from the reactor, we have to keep it somewhere. Because it is radioactive and it is still producing heat, we have to both shield the radiation and keep the spent fuel cool. The best way to do both of these is to put the spent fuel in a pool of water. The water is circulated through a set of heat exchangers to remove the heat. The pool is also deep enough to leave sufficient water above the spent fuel to shield the radiation. The major forms of radiation are beta particles and gamma rays. Beta particles are easily stopped by just a little bit of water. The gamma rays take more water to stop them, but water does a good job of it. There is a great line in a nuclear radiation shielding manual: "Water has no cracks."

When commercial nuclear power was first planned, it was intended that the interim storage period would be a few years, just enough to allow the radiation and thermal energy to decrease to where the spent fuel was easier to handle. Therefore, interim storage was provided for right at the reactor. Reactors typically have a spent fuel pool that can hold several years' worth of spent fuel discharges. After a period of interim storage, the original plan was to ship the fuel to a *reprocessing plant* where the remaining unused uranium-238 and uranium-235 along with the useful plutonium would be separated from the radioactive fission products. The useful materials would be made into new fuel and the waste stabilized and sent to final storage.

However, it didn't work out that way. On the surface, this concept of a fuel cycle was pretty clever. Because most of what made up the fuel, that is uranium-238, had not been used and you had good fuel material in the form of unused uranium-235 and new plutonium, getting that out of the spent fuel

[9]Often we need to distinguish between thermal power and electrical power. To do this, we write megawatts thermal as MWt and megawatts electric as MWe.

and reusing it seemed like a really good idea. This was not a breeder cycle in that you were using light water reactors. While light water reactors make some plutonium, they can't make more plutonium than the uranium-235 they use up. Since the total amount of good fuel is not increasing, these aren't breeders. But still you would get the remaining value from the unused part of the fuel. You would have to make new fuel because you would need new cladding tubes, and you would have to add more uranium-235 to the mix but not as much as if you started from natural uranium.

What happened was the same political and economic forces that made breeder reactors not viable got in the way of light water reactor fuel recycling. Within the United States the concern about having a fuel cycle in which plutonium was separated was politically strong. Several large commercial reprocessing plants were proposed. One was built and operated for several years. Another bigger one was finished but never operated. At the same time as the political forces were lining up against light water reactor recycle, the economics were becoming more questionable. Uranium was relatively inexpensive, and the reprocessing and plutonium fuel fabrication was getting more expensive. It became harder to show that the apparently more efficient recycling was in fact the less expensive way to go.

You might compare this to the decision between paper and china plates. Clearly using paper plates and throwing them away after one use is much more wasteful than using china plates and washing them after each use. But if you got to a point where there was a lot of concern about dishwasher soap phosphates polluting water ways and the price of soap became greater than paper plates, well you might just go with paper.

In Europe they didn't see it this way. Both the British and the French built and operated large reprocessing plants. They also built *MOx fuel fabrication* plants that take the recycled materials from spent nuclear fuel and create new fuel. Because this is a mixture of uranium and plutonium dioxides, the fuel is called mixed oxide or MOx for short. They are producing MOx fuel now and using it in light water reactors. They also have reprocessed fuel for other nations and made MOx fuel for them. Japan has a large program in this area and hopes to soon complete their own commercial-scale reprocessing and fuel fabrication plants. They have built small test-scale facilities, and the larger plants are in construction.

The Russians also have large reprocessing plants. Their commercial and military programs are sometimes hard to separate. While they have reprocessed a lot of "commercial" fuel, they have not recycled this into their power reactors and have amassed a good store of separated "civilian" plutonium.

In the United States, however, the government chose to use what is called the "once-through" fuel cycle. On Figure 5.10 that is what is shown in the darker arrows. After interim storage the spent fuel is to go to final storage or disposal. After imposing the once-through fuel cycle on U.S. nuclear utility companies, the government promised to take the spent fuel from them and provide the *ultimate disposal*. But this was not a gift. The government started charging nuclear utilities one dollar for every megawatt-hour of electrical energy they made to go into a nuclear waste fund. For a large reactor, this could be more than $8 million per year.

The U.S. government program went through a series of advances and retreats. A number of disposal methods were examined, and a number of sites were considered. The government promised to take the fuel by 1998. However, the government programs have been running behind. I will explain more in the next chapter. The bottom line is the final storage step is still not available. The problems are far more political than technical, but again, more of that later.

If we look at Figure 5.10 and trace the lighter lines, we can see the original plan for a nuclear fuel cycle and the fuel cycle the French and other countries are using. That is, after interim storage, the spent fuel is sent to the reprocessing plant where the useful uranium and plutonium are separated. The radioactive fission products are removed and put in a stabilized form so they can be sent to a final storage facility. The recycled uranium and plutonium are sent to a MOx fuel fabrication plant, where they are mixed with fresh enriched uranium to make new fuel. That fuel is then sent back to the reactors to make more power. The French have concluded that recycling and the once-through fuel cycle are essentially equivalent on cost for them.

So there, now you know some basic nuclear science and some basic nuclear engineering. You know how fission works and how we can control fission in nuclear reactors. You know what the major reactor types are and how the nuclear fuel cycle works. The next step will be for us to look at some of the scary things we have all been told about nuclear power. We will shine a bright light into some of these dark corners and see if there really are monsters under the bed.

6

ALL THE BAD STUFF ABOUT THINGS NUCLEAR

Nuclear Is Scary

Let's just face it. Nuclear things are scary. Why? Because they are unknown, little understood, out of our common experience, and worst of all invisible.

As a 5-year-old kid quaking in the front row of the Roxy theater in Rock Springs, Wyoming, the invisible monster in "Forbidden Planet" scared the willies out of me.

I recall the time some 15 years later when I was first working in a radiation laboratory all on my own. Now the university didn't let undergraduates do much of anything dangerous unsupervised, but still, in order to do my assigned experiment, I had to arrange the radiation source in its lead brick cave, line up the detector, and take my measurements. I remember clearly the little thrill of fear as I recognized that if I screwed up somehow, I would have no way of knowing. The radiation was totally undetectable by my senses. I could be doing harm to myself, and even to the folks across the hall, and none of us would know it until they processed our radiation film badges the following week.[1]

[1]For truly hazardous radiation work, there are real-time radiation detectors that let each worker know how much radiation is around them at all times.

For the reader's assurance, let me tell you that during the next 30 years as a nuclear engineer, the largest sources of human-caused radiation that I was exposed to came from dental X-rays and from "excess" cosmic radiation exposure during the time I have spent above much of our protective atmosphere riding in jet airliners. This is despite visiting a wide range of nuclear facilities and working for 3 years at a nuclear power plant.

The fact remains, however, that for the average person, anything "nuclear" or having to do with radiation is scary. They see it as something that can *get* them, but they will never see it coming.

To help you deal appropriately with this phobia, I want to show you that the hazards of things nuclear are less than often portrayed. The first thing we are going to look at is human health effects of radiation. After that, we are going to look at several other aspects of things nuclear that cause people concern. Specifically, I am going to try to help you understand nuclear power plant accidents, the issue of nuclear weapons, and nuclear waste. There are legitimate concerns on each of these issues, but I hope to show you that, in general, these have been significantly overstated and taken out of context over the years.

Let's have a look at each of these concerns.

Human Health Effects of Radiation

There are several ways to measure radiation, each helpful in its own way.

As we saw in Chapter 4, radiation comes from things that are radioactive, that is, materials and processes that give off radiation. Radioactive decay happens when a material is in an unstable state and gives off radiation as part of the process of moving to a more stable state. One way to measure radiation is how fast the material is giving off radiation. The units for this are the Curie and the Becquerel. A Curie is 3.7×10^{10} disintegrations per second. A Becquerel is one disintegration per second, which means one Curie is 3.7×10^{10} Becquerels. While these units are helpful in telling us how active a radioactive material is, they don't tell us much about how dangerous it might be. To do that, we need information about what type of radiation is being given off, how much energy that radiation might have, and what is the nature of the transmission of the radiation.

To get a better understanding of radiation, we use another set of units that take into account these things. It measures the absorbed energy per unit mass of the thing (e.g., human tissue) absorbing the radiation. These units are rads, which is 100 ergs per gram and Grays, which are one Joule per kilogram. If these units don't help you much, there are 10 million ergs in a Joule, and it takes 36 billion ergs to make a watt-hour worth of energy. Which means

100 rads of radiation being absorbed in a kilogram of tissue would be enough energy to light a 100-watt light bulb for 1/100 of a second.

However, we are still a bit short of understanding how dangerous the radiation might be. That is because the amount of damage from various forms of radiation is dependent on both the energy and the type of radiation. If we take that into account, we measure radiation dose in either rem or Sieverts. One rem equals 0.01 Sievert or 10 mSv (milli-Sieverts). The Sieverts is an international standard unit, and the rem is an older unit but still found in much of the literature. I will stick with rem, but if you run across Sieverts, remember, the conversion is pretty easy, 100 rem per Sievert.

If the radiation is gamma rays or beta particles (see Chapter 4 now if you skipped that part), then rems are the same as rads. If the radiation is neutrons or alpha particles, which are relatively speaking much, much bigger, the rem value is 10 or more times that of the rad. That is, you calculate the absorbed energy in rads and then multiply by the appropriate factor to get the effective damage done as measured in rem.

Now that we have some terms to measure radiation, let's look at what radiation can do.

The types of radiation we are talking about here, gamma rays, beta particles, alpha particles, and neutrons, are called ionizing radiation.[2] This means these types of radiation can knock electrons out of atoms leaving them with a positive charge, which is to say we have created a positively charged ion. This is to distinguish these types of radiation from visible light and radio waves, which do not have sufficient energy to peel electrons away from atoms and hence are called non-ionizing radiation. That doesn't mean they are harmless. Laying in the sun's ultra-violet radiation for a few hours or leaving a potato in your microwave oven too long should be a sufficient demonstration of that.

But at the heart of it, what nuclear radiation does is it ionizes atoms. In biological systems, nuclear radiation impacts organic molecules. This can be either direct effects such as breaking up a DNA molecule or indirect effects such as ionizing water in a cell. Since on a volume bases, a lot more of cells are water than DNA molecules, the indirect effects are more common. When water is ionized, the hydrogen and oxygen atoms are broken apart. If they recombine into water, nothing bad happens. However, if they reform as H_2O_2, hydrogen peroxide, that isn't good; hydrogen peroxide is toxic to cells.

If the DNA is damaged, it may cause the cell to die. It may also be a non-fatal damage and nothing bad will happen. It might weaken the cell such that when it divides the daughter cells die, or they may carry a weakness on as mutations in future generations.

[2]Just to be clear, neutrons do not directly cause ionization. However, since neutron capture often results in the emission of gammas that do cause ionization, neutrons do cause indirect ionization.

It is important to note at this point that radiation doesn't create new forms of mutations. All it does is speed up the natural mutation rate. Some have pointed out that without natural radiation the mutation rate would be much slower and evolution would, therefore, be much slower. This begs the question of mutation rate versus changes in the environment which allow the mutations to be successful, but it does point out the fact that not all mutations are bad. At least, I enjoy my opposable thumbs. However, the point is radiation is not going to create giant atomic mutant ants or fire-breathing Japanese dinosaurs, not unless evolution was going that way anyway. So to answer the issue raised in the introduction to the book, no, there are no giant atomic mutant ants. Radiation is a natural process, and all life on the planet is adapted to a base level of radiation. More radiation just changes the pace.

Different cells have varying degrees of susceptibility to radiation damage. Cells that are reproducing quickly are the most susceptible. Blood-forming cells, especially white blood cells, are the most susceptible. Reproductive and gastrointestinal are the next most susceptible with slowly reproducing nerve and muscle cells less susceptible. It is the sensitivity of quickly reproducing cells to radiation that gives us one of our strongest weapons against cancer. Because a cancer is a set of cells reproducing out of control, radiation is particularly effective at killing cancer cells. The challenge is being able to totally eradicate the cancer without also inflicting unsustainable damage to surrounding healthy tissue.

We do need to stop for a minute and recognize how tough a human is. We are able to take a lot of damage and repair ourselves essentially as good as new. We get seriously hurt or killed only when the damage crosses a threshold of non-repairability. This is true of radiation as well.

When people look at the whole person effects of radiation, they generally divide it into the effects of short-term, high doses and long-term, low doses. We can speak more authoritatively on short-term, high doses because we have actual case histories. These come from the survivors of the two atomic bombs dropped in Japan at the close of the Second World War, and from industrial accidents.

In rough terms the effects of massive short-term radiation exposures are as follows:

1000 or more rem = death
300 to 500 rem = death in 50% of those exposed depending on individual and level of medical care
100 rem = no fatalities, illness with flu-like symptoms
10 rem = limit of any evidence of latent cancer effects[3]

[3] I have seen several different values cited for the "limit of discernable effects." These were 50 rem, 30 rem, and 10 rem. To allow the greater margin, I will use the more restrictive number.

For long-term, low-level effects, the evidence is much more difficult to determine, and there is a lot of controversy on just what the effects are. At one extreme are those who say any damage is harmful and the harm is cumulative. At the other extreme are those who say life has dealt with natural radiation for 3.5 billion years and knows how to repair minor damage, as demonstrated by our very existence.

The difficulty in finding the "correct" place between these extremes is that you are looking for a small effect hiding in many other competing factors. Cancer is going to claim about a quarter of all of us, so trying to pick out those cancers that are due to this cause or that is really, really hard, especially if the thing you are looking for doesn't have much impact. For example, making the link to smoking wasn't all that hard. You needed only to look at the cancer death rates of smokers and non-smokers, and it was pretty obvious. Linking cancer to radiation is much more difficult.

Let's step aside for a minute and look at how much radiation exposure we get and where it comes from.

Data for the average U.S. citizen, produced by the National Council on Radiation Protection, is shown in Table 6.1. Recognize that this is a composite average. For example, most of us don't have major nuclear medicine performed on us, but the value in the table is the dose of those who have experienced nuclear medicine divided among all of us.

TABLE 6.1 Average annual radiation dose to U.S. public.

Source	Dose (mrem)[4]	Percent of Total Dose
Natural Sources		
Radon	200	55%
Cosmic	27	8%
Terrestrial	28	8%
Internal	39	11%
Natural (Subtotal)		82%
Artificial Sources		
Medical X-Rays	39	11%
Nuclear Medicine	14	4%
Consumer Products	10	3%
Occupational Sources	0.9	Less than 0.3%
Fallout	Less than 1	Less than 0.03%
Nuclear Fuel Cycle	Less than 1	Less than 0.03%
Artificial (Subtotal)		18%
Total	**~360**	**100%**

[4]A mrem is 1/1000 of a rem, so 1000 mrem is 1 rem.

In *Radiation and Life,* Eric Hall gives very similar numbers, but he offers us the additional specificity of 1.4 mrem for fallout and 0.14 mrem from nuclear fuel cycle.

Let me clarify some of the radiation sources in Table 6.1 for you.

The single largest source of radiation for the average human is radon. Back in Chapter 4 we talked about radioactive materials. We learned that thanks to ancient supernova stellar explosions, the earth has got a lot of uranium, which is a naturally radioactive material. In the process of going from the unstable state of uranium toward the stable state of lead, uranium transforms itself into a number of other radioactive materials. Table 4.1 shows you the decay chain for uranium. One of the radioactive isotopes is radon-222. From Table 4.1 you will notice that at 3.82 days, its half-life is pretty short. That means it goes away pretty fast, but it gives off a lot of radiation in that short time. However, even if it goes away quickly, it is coming from things that have much longer half-lives, so there is a long-term supply of "new" radon. The next thing of interest about radon is that it is a gas. This is particularly bad news since it means it doesn't stay quietly in the ground like the rest of the uranium decay chain, but floats out into the air where we can breathe it. The last thing to notice about radon-222 is that it decays with an alpha particle. If it stayed in the ground, that would be a good thing, as it takes almost nothing to block an alpha particle, a sheet of paper would do. However, because it is a gas and we get it inside us by breathing, it is a bad thing because alpha particles do a lot of damage in the short distance they travel. Because uranium is essentially everywhere on the planet so is radon.

The next item in Table 6.1 is cosmic radiation. This is radiation coming from the sun and from other stars including distant old supernovas.

Terrestrial radiation comes from all the other radioactive materials in the earth. This includes all the other decay chain isotopes in the uranium chain and a similar set from the thorium isotopes. It also includes radiation from naturally radioactive potassium-40. This isotope has a half-life of 1.3 billion years, which means that it is still around after its long distant creation in stars and it is not very radioactive. It does make up for not being very radioactive by the fact that relatively speaking there is quite a bit of it.

That brings up the next category, internal sources. With all the naturally radioactive substances around us, it is to be expected that we would pick up some and get it inside of us. Potassium-40 is a big contributor here as are the uranium series and the thorium series. The average 70-kilogram person (154 pounds) has about 17 mg (0.0006 ounces) of potassium-40, which works out to a bit over 5 tons in the U.S. population. The amount is actually very small, but it does add up in our total radiation exposure.

All together naturally occurring sources of radiation give us about 80% of our total radiation dose.

Of the "artificial" sources almost all of it is from medical X-rays. A full set of dental X-rays will give you 40 mrem.

The next highest artificial contributor is nuclear medicine. While few of us undergo nuclear medicine treatments, those that do may take very large doses. Diagnostic procedures might give moderate doses, but aggressive cancer treatments can give extremely high doses.

The consumer products category represents a wide range of things. Often it is a product made out of things that are high in naturally occurring radioactive materials.

Occupational sources include workers within the nuclear power business, but a lot of this is doses to medical workers and those using industrial radioactive sources for things like X-raying welds.

Fallout is that radiation we still get from the aboveground nuclear weapons testing that was done before the ban on such tests. The United States, the Soviet Union, the United Kingdom, France, and China set off hundreds of nuclear weapons releasing tons of fission products to the atmosphere. The good news is we stopped doing that, and essentially all of the highly radioactive short-lived fission product isotopes from these tests have decayed away. It is only the less radioactive, longer-lived fission products that still concern us.

And the last category is the nuclear fuel cycle. This includes all the fuel cycle activities including reactor operation. Compared to the other numbers it is so small as to be indistinguishable. But what about accidents you say? Well, we will look at that in the next section.

Let's go back to the difficulty in determining the effects of low-level, long-term radiation effects.

We saw that on average a member of the U.S. population got about 360 mrem per year. Just how average is the average? The answer is that it varies quite a bit. While uranium and, therefore, radon are everywhere, the local concentration varies widely. Also, the closer you are to sea level the less cosmic radiation you get because the greater thickness of air helps protect you. Terrestrial sources vary a lot as well. The amount of terrestrial sources we are exposed to is also impacted by our choice of building materials with granite being a particularly strong natural radiation source. Other variables include things like airline flights. A coast-to-coast trip is worth about 5 mrem, which means a 100,000-mile-per-year frequent flyer is getting 1.5 times the average dose from normal background radiation.

But recall Table 6.1 was for the average United States citizen. What about elsewhere in the world? It turns out the natural exposure does vary a good deal across the planet. While we get about 360 mrem, the folks in Finland get about 760 mrem and those in Sweden get 600 mrem and the Australians get only 160 mrem. For most, radon is the primary driver.

Well, with this kind of variation surely we could see some trends in radiation linked health effects, right? Well, the story gets better and worse.

Beyond the variation in national averages there are some locations with strong terrestrial sources or very high radon concentrations. There are places in India with natural doses of 1500 mrem per year. In Brazil, Iran, and Sudan local factors produce natural doses of up to 3800 mrem per year. In a few isolated places in Europe, radon pockets have given annual doses of 5000 mrem per year. In studying the populations in these areas, it is hard to find any clear radiation linked health effects at low doses. The General Accounting Office (June 2000) report on radiation standards[5] referring to studies on the variation in natural background radiation and the attempt to link this to increased human health risk put it this way:

> ". . . we examined 82 studies, which generally found little or no evidence of elevated cancer risk from high natural background radiation levels. A large number of studies reported a lack of evidence of cancer risks; some others reported evidence of slightly elevated risk, and some others reported evidence of slightly reduced risks. Overall, the studies' results are inconclusive, but they suggest that at exposure levels of a few hundred millirem a year and below, the cancer risks from radiation may either be very small or nonexistent."

Other studies have looked at workers within the nuclear weapons production complexes and alternately found no linkage or a strong linkage, depending on the study.

Some examined this data and came to the conclusion that some "extra" radiation is actually good for you, citing data that indicates that those with somewhat elevated radiation doses were healthier and lived longer than those with average or lower radiation doses.

Even at higher than long-term, low-level doses there is some mystery in how humans respond to radiation. Animal studies indicate that radiation can damage reproductive cells that result in genetic effects in the offspring of the exposed animal. However, exhaustive studies of the atom bomb survivors show no genetic effects at all, none.

So what are you going to do? Lab tests and first principles physics tell you harm is being done, but, when you look in the real world, the effects are sufficiently small that they are hard to find. In 1994, the United Nations Scientific Committee on the Effects of Atomic Radiation summed up the scientific dilemma this way:

> ". . . there are theoretical reasons based solely on the nature of DNA damage and repair to expect that cancer can occur at the lowest doses

[5]General Accounting Office, Radiation Standards, GAO/RCED-00-152 (Washington, D.C., June 2000).

without a threshold in the response, although this effect would perhaps not be statistically demonstrable."

Which translated from the careful words to plain speak is, "well, any level of harm is harm, but it is unlikely that you will even see it at the low dose levels."

The current treatment of low-level radiation effects is called the "linear no-threshold" model. What this comes down to is that we can't see clear effects below 10 to 50 rem (10,000 to 50,000 mrem), but it still bothers us. So we draw a line from the last clear impact we can see downward to lower doses making the assumption that even the tiniest dose does something bad. This means that in estimating the impact of some radiation release, we would treat 1 mrem given to 100,000 people the same as 100 rem given to a single person. Those opposed to this view say this is silly because with 100,000 people, you have 100,000 repair mechanisms working to fix the minor harm. The proponents say that might be, but since we can't see the actual effects, the linear model assures us that we are bounded because it can't be any worse.

Recently there has been quite an outcry against the linear no-threshold model attacking both its scientific bases and the very large negative impacts on medical and nuclear fuel cycle operations with its extreme and unwarranted conservatism.

I am not a doctor or a nuclear health physicist, so I can't really offer an opinion on a scientific basis. However, logically it would seem that there is a level at which we can repair ourselves. The fact that globally there is a wide variation in natural radiation without clear health impacts seems to support that. It would appear that we are straining at gnats when we should be a lot more worried about other things.

Nuclear Power Plant Accidents

O.K., fine, normal nuclear fuel cycle operations don't have much of a contribution to the radiation doses we all get naturally. But what about accidents? Aren't nuclear power plants poised to kill millions?

Well, no. Let me explain.

Let's start with the most important point to get across. NUCLEAR POWER PLANTS CAN'T EXPLODE LIKE A NUCLEAR WEAPON! Sorry to shout, but if you still believe that, then this book isn't working for you. To explain a bit more.

Nuclear weapons are not easy to make. The really tough part is getting the nuclear material to hold together long enough to get a big boom.

I am going to assume you read Chapters 4 and 5. If you didn't, some of this might not make sense (a not-so-subtle hint that all that nuclear science stuff would be helpful if you really want to understand this subject and a big reward for those of you who stuck with me through all that).

Let's do a quick review of the physics of nuclear energy. The way we get energy out of nuclear fission is we cause heavy elements, like uranium or plutonium, to split apart, e.g., fission, giving off a lot of energy. We get atoms of these heavy elements to fission by whacking them with a neutron. The really cool thing is that in the fission process two or three "new" neutrons are given off. If we can arrange for one of these "new" neutrons to cause another fission, we can keep a steady chain of fission reactions going. That is what we do in a nuclear power plant.

In a nuclear weapon, the goal is not a long-term steady chain of reactions; the goal is to get as many reactions as possible, as fast as possible. Rate of reaction is the difference between fire and explosion. Earlier we found out that nuclear physicists use the term "k" to represent the number of neutrons in one generation to the next. In a nuclear power plant, we want that to be exactly 1.0 so the reactions go on at the same level over time. In a nuclear weapon, that number needs to be as big as it can be made.

We also learned that in order to increase power from the starting point of shut down, nuclear power plants must be able to raise k to somewhat greater than 1.0. However, if we keep the excess to less than the delayed neutron fraction (see, you really have got to read the basic science stuff), the rate of increase is quite slow. In a nuclear weapon, the key is to assemble the nuclear materials in such a way as to get k much greater than the delayed neutron fraction. Not only does this mean you get a new generation of fission events in a fraction of a second, it also means the multiplication of each generation (that is, how many more fissions each generation has than the preceding) is large. So that is it, right? Is that all it takes to make a weapon?

Well, no. Even if you can get a large k and get a big multiplication over very short times, the really, really hard part is to get a lot of generations. You see, even if the time between generations is short, after you get a few generations into the process you are going to be creating a very large amount of energy. This energy is going to melt, then vaporize the bomb, causing it to turn into a very rapidly expanding cloud of vapor. At that point the nuclear material will be too far apart to sustain the chain reaction and the nuclear processes will quit.[6] Which means the key to a big boom is to keep the components together as long as possible to get as many generations as possible, to get as many fissions as possible, and to get as much energy released as possible. And trust me, that is hard.

---

[6]In case this isn't obvious, what is happening is that when the cloud of material gets bigger it is more likely that a given neutron will just zip out of the cloud without encountering a heavy atom. When the material is in solid form, the atoms are so close together it is almost impossible for a neutron not to hit a heavy atom.

Nuclear power plants and nuclear weapons are arranged very, very differently. In a nuclear weapon all the fissionable material is concentrated in as small a space as possible. That makes sense if you are going to try to get as much of it as possible to react as fast as possible. In a nuclear power plant, you are trying to get the energy out over a long period of time (several years instead of a fraction of a second), and you are trying to extract the energy in a constructive fashion. This means the fissionable material is going to be distributed over a fairly large volume with a lot of other stuff like fuel rod cladding, fuel spacer grids, control rods, and coolant, mixed in to do the job of getting the energy out of the sustained fission process.

As I said, in nuclear power plants, we intend to operate in such a way that k never exceeds the delayed neutron fraction. However, that can happen in certain accident cases. So then, it is a bomb, right?

NO, NO, NO! If you do get a circumstance where k exceeds the delayed neutron fraction, you will get power increasing on prompt neutrons alone and power will increase very quickly, too quickly for most control systems to detect the increase and take corrective action. However, what normally happens is that the inherent safety features, primarily the Doppler effect, will provide the negative feedback needed to terminate the increase and bring the power level back under control. In a nuclear weapon, the nature of the fissionable material is such that there is essentially no Doppler effect.[7] But even in the extreme case, in which Doppler and other negative power coefficient effects (e.g., fuel expansion, coolant expansion) fail to control the increase, the worst that can happen is the fuel comes apart. To be sure, this is what we call a bad day at the plant. You are going to have quite a mess, but the nuclear explosion—well it just won't be an explosion. The heat released is going to push things apart too fast to get very many generations going and the total nuclear energy released is just not going to be much. A nuclear power plant just lacks the needed design features to make a bomb.

Designing a nuclear power reactor to provide a continuing steady chain reaction of fissions is difficult. Compared to designing a device to get a near instantaneous burst of nuclear energy even as it destroys itself, power plants are easy.

With the question of behaving like nuclear weapons out of the way, I want to briefly describe types of accidents we worry about when we design nuclear power plants and what we do in the design to try to address them.

---

[7]This is achieved by making the core of the weapon out of either uranium-235 or plutonium-239. Both chose to fission more often than to absorb neutrons without fissioning. Nuclear power plants usually are fueled with a mix of uranium-238 and uranium-235, about 96% uranium-238, and 4% or so uranium-235. It is the uranium-238 that has the large Doppler effect in which heating the fuel causes a strong increase in non-fission absorptions in the uranium-238, hence very quickly turning the increasing power back around.

To simplify quite a bit, there are two kinds of accidents that might happen at a nuclear power plant. One is the loss of the ability to cool the fuel at normal power levels. The other potential is an increase in the reactor power level beyond the ability of the cooling system to remove heat.[8] Now, a modern nuclear power plant is a very complex machine, and there are a lot of different routes to get to either of those two end points. It would take many volumes to discuss the potential scenarios and all the mitigation features nuclear power plants have to deal with them, but let's touch the high points.

To keep things simple, we will focus on pressurized water reactors (PWRs) as these are the most common form of commercial nuclear power plant. There are close parallels for most of the other types of power reactors, however. O.K., how might you lose the ability to cool the reactor fuel? Well, the means to circulate the coolant might fail. In a PWR, coolant circulation is done by large electrically driven pumps. There are usually multiple loops or paths in the cooling system, each with its own big pump. To operate at full power, you would like to have all loops running, but if you can get the power down usually a single loop will do. Also you will have two independent sources of off-site power to drive the cooling pumps plus you can use the electricity the power plant itself makes to power the pumps.

So, it is hard to lose the pumping of the main or primary cooling system, but that just circulates water to and from the reactor core. You need another system to cool that water. This is the secondary cooling system, the one that makes the steam to drive the turbines, to make the electricity, which is the whole point. The secondary cooling system is also arranged in loops, matching the primary system, and each can be isolated from the others in case anything goes wrong in one. The secondary system has large makeup water sources, or feedwater systems. In addition they have backup auxiliary feedwater pumps and sources, all with redundant sources of electrical power.

If for any reason the connection of the secondary system to the steam turbine and condenser, the pathway to the so-called ultimate heat sink, is interrupted, you can vent the steam and provide makeup water and thereby cool the system.

Another way to lose the ability to cool the fuel without losing the ability to pump the primary water is to lose the water itself. This can be either from small pipe breaks that drain the water slowly, or large pipe breaks, which drain the water quickly. In the case of small pipe breaks, the normal charging system, the high-pressure injection system, can provide enough water going in to match the water going out. For larger pipe breaks the plant has an Emergency Core Cooling System (ECCS), which provides an independent

[8]We will talk more about these "famous" accidents in a bit, but the Three Mile Island accident was a failure to cool type and the Chornobyl accident was a reactor power type.

source of water to cool the fuel. These systems are very conservatively designed to be able to safely cool the fuel in the event of the worst possible pipe break at the highest allowed operating power level of the plant. There are multiple sources of water for the ECCS including the ability over the long term to recirculate the water spilling out of the broken pipe.

Independent of all of this is a residual heat removal system (RHRS) designed to cool the reactor while it is shut down. Why do you need that? Remember that, while the reactor operates, it creates fission products. These are radioactive and give off heat. There is quite a mixed bag of these guys. Some are short-lived and, therefore, very radioactive. Some are long-lived and, therefore, not very radioactive. So just after you shut the reactor down, the fuel is very, very radioactive and gives off a fraction of its full power heat. After 10 minutes the heat of the radioactive fission products will be about 1.5% of the full reactor power; after an hour it is down to 1%. This heat must be removed, so you have a RHRS to do that. The RHRS also provides you another source of cooling if you need it after an accident, provided you get the reactor shut down.

So now let's look at accidents in which your cooling systems are working fine, but you are making too much power for the normal capability of the cooling system.

First of all, it is very hard to get in that state. The reactor will have a lot of instrumentation and controls designed to prevent it. The instrumentation monitors an array of conditions such as power level, coolant volume, coolant temperature, coolant pressure, and so on. If any of these parameters exceeds a set point, the reactor will automatically start a decrease in power. If they exceed a slightly higher threshold, the reactor will automatically be shut down.

The instrumentation is independent and redundant. That is, there will be multiple sensors monitoring the same parameter so that the failure of any one sensor will not cause the system to fail to see a bad condition in the reactor. They will be powered by independent electrical systems and the critical instruments and controls will be backed up by constantly recharged batteries to ensure their operation even if all other electrical sources are lost.

If a sensor detects a parameter out of range, it calls for a shut down or, as it is called, a reactor trip (as tripped off line). The individual trips back each other up as well. For instance, in addition to a high power trip, there will be a rate of rise trip. So if the reactor power starts to rise, if the rate of increase exceeds a slow value, you will get a rate of rise trip. But if for any reason the redundant rate of rise detectors fail to do their job, the high power trip will shut the plant down. Then there are the high coolant temperature and high coolant pressure trips waiting to save the bacon if the power-related trips fail and allow the reactor coolant to get too hot, which would also increase the coolant pressure.

So what do these trips actually cause to happen? In a PWR, the "control" is provided by a set of control rods that are inserted into the fuel bundles. Recall that, in a PWR, the fuel is a set of about 1/3 inch diameter rods, about 12 feet long collected together in "bundles" of 15 to 17 rods on a side. Within this array, some of the positions are left open. It is into these open places in some of the fuel bundles that the control rods are dropped.

The control rods contain strong neutron absorbing material. Once they get into the fuel too many neutrons are absorbed to allow a sustained chain reaction and the reactor coasts down in power. However, it is important to recognize there is no real "off" switch in a nuclear power plant, only a "down." Because we are running off delayed neutrons, it will take a minute or so for the neutrons coming from the previous fissions to show up and for them to be absorbed by the control rods. So when you drop in the control rods, the power will take a sharp drop and then coast down. After all the fission power goes away, we will still have the fission product radioactive heat, the decay heat, to deal with. Remember that after 10 minutes this is only about 1.5% of the operating power. However, in a 3200 MW reactor that would be 48 MW of heat that, while a small part of the operating power, is still something that must not be ignored.

The control rods are pulled up out of the fuel during normal operation. If we want to decrease the reactor power, we can insert a fraction of them to push k a little under 1.0, then pull them back once we have gotten to the lower power level we were seeking. If there is an emergency, we can drop in all the control rods. If we are inserting the rods in a slow fashion, we use a screw or a jack arrangement, like an old car jack. If we are in an emergency, we just release the rods and let them fall. In many plants, the catch that holds on to the rods is an electro-magnet. So if you lose power, the electro-magnet stops working and lets go of the control rods. The rods are appropriately weighted and shaped to allow them to fall quickly even against the flow of the cooling water.

Well, fine, so it is hard to get a condition where the reactor power increases above the normal ability of the cooling systems, but why bring it up if you can never get there? In reactor safety, we very seldom say never.

One of the reactor overpower accidents we worry about is called a cold water injection accident. Does that seem odd? Why is cold water a bad thing? It would seem like cold water would be a good thing, a better way to cool the fuel. It is a bit complicated, but it goes like this. Because we have designed the reactor with inherent safety, if you increase the temperature of the fuel or the coolant, the natural response of the system, just the physics now, no control systems at all, is to decrease the power. Good, right? Sure, that is good. However, it also means that, if for any reason you decrease the temperature of the fuel or coolant, the natural response is going to be to increase the power level. That is a convenient thing if you want the plant to automatically seek a power level. Heat it up, and it powers itself back down. Cool it off, and it powers itself back up. But the down side is that, if you get a sudden shot of cold water, the reactor power level goes up a lot, and that ain't good.

How would you get such a shot of cold water? One way might be a big break in your secondary cooling system. Whoa now, a break in the secondary system means you can't remove heat from the primary so things are going to heat up not cool down. Well, yes, that is the long-term effect, but in the short term, the water escaping from the secondary system is going to be going from 1000 pounds per square inch to atmospheric pressure. This means it is going to boil and boil very fast. When water boils it extracts heat from its surroundings,[9] and in this case, it will be a lot of heat and the surroundings will include the water in the primary cooling loops. Therefore, for the first bit of such an event, the primary water is going to take a real big dip in temperature. This cold water is going to come zipping into the reactor, which is going to respond with a big power increase.

So this is a bad thing, a big deal? Well, the secondary break would be a real mess, but the reactor wouldn't mind all that much. The Doppler effect is an amazingly strong actor. As soon as the power increase causes the fuel to heat up, the uranium-238 will start increasing its non-fission absorptions of neutrons and act like an almost instant set of control rods and shut the power back down.

Another overpower accident to worry about is a control rod ejection accident. Let's say you had some of the control rods in the reactor. If somehow the upper housing of the control rod mechanism were to break, you would have 2000 pounds per square inch reactor coolant pressure trying to force the control rods out with nothing pushing back. In that event the rods would leave, rapidly. This would cause a sudden spike in reactivity and the reactor power would increase rapidly. Is this a bad thing, a big deal? Again, there would be real mess with reactor coolant squirting out the control rod housing hole, and you would call on the other systems to shut the reactor down and ensure fuel cooling, but the overpower once again would be handled by the Doppler effect. The power would go through a wild ride way up and way down, but the ride would only be a few seconds long. The fuel and control rod systems are designed with just this accident in mind and assured to survive the worst-case conditions.

Let me also add that on top of the inherent safety features built into the design of a nuclear power plant and all the redundant safety and backup systems, all nuclear power plants in the United States have a "containment" building as the final barrier between the radioactive stuff in the reactor and the public. The containment building is a very large pressure vessel in which all the primary reactor systems are found. This includes the reactor, the

[9]This little bit of physics is why evaporating sweat cools you off.

primary coolant pumps, and the steam generating equipment (steam generators in a PWR, the reactor itself in a BWR). This containment building is designed and constructed to withstand the highest possible pressure pulse from any large pipe break or the worst-case natural external impact such as a telephone pole driven by a tornado. The idea is that, if all other safety systems fail, the containment will maintain a pressure-tight seal and keep all the radioactive material inside, hence protecting the public. These are the domes you are used to seeing at reactor sites. They are very impressive structures and capable of withstanding truly amazing forces.

That ought to give you some feel for the types of nuclear safety things we think about in designing nuclear power plants and some sense of how robust we make them. But that doesn't answer all the questions. Let me give you a little more background by introducing you to two studies on reactor safety done by the U.S. Department of Energy. This will give you some history on how our thinking about nuclear power plant accidents has evolved. We will follow that with an examination of the two most infamous reactor accidents: Three Mile Island and Chornobyl. A realistic look at what did go wrong in each of these accidents is required for an honest balance.

The first major reactor safety study goes by the name of WASH-740. The full title is *Theoretical Possibilities and Consequences of Major Accidents in the Large Nuclear Power Plants*. This work was done in 1957 by Brookhaven National Laboratory at the behest of the Atomic Energy Commission, the predecessor of today's Department of Energy. At that time, people were just getting serious about putting nuclear energy to use to make productive electrical energy. Folks recognized that the core of a nuclear reactor was the site of a great deal of concentrated energy and that there were quite a lot of radioactive materials produced in the process of making nuclear energy. So, the question was asked, how bad could an accident be if the worst were to happen? This genuinely honest question ran the good scientists at Brookhaven down a dark path of "but what if?" questions. What resulted is a pretty spectacular example of bad public relations.

If one might summarize the results of WASH-740 into three findings, they would be that:

- If the containment were to remain intact, the worst accident inside the reactor would not harm the public.
- A major accident would be likely to happen once in about a million years. That is reactor years, so if you had 100 reactors you would expect a major accident every 10,000 years.
- If the containment did fail, a major reactor accident would kill thousands and cause millions to billions of dollars of property damage.

Well, of course, the third of these findings makes great negative press. For all the years after this study whenever a nuclear safety person were to say we have this or that safety feature, the critic's response would be well, what if that were to fail? The nuclear safety person would say well, we have this or that system to back it up. The critic would say but what if that failed? Eventually the nuclear safety person would try to explain about the containment building, but when given the final "what if" would have to give the WASH-740 answer, that is only the third bullet from the list above, ignoring the other two. The problem is that the context of how unlikely the ultimate failure was would be lost.

I once read somewhere a description of an accident at a neighborhood gas station done in the WASH-740 style worst case, "what if" fashion. The scenario went along the lines of a fully laden cement truck loses its brakes and its steering and careens down a steep hill at high speed into a gas station. At that moment, a tanker truck is filling up the underground tanks. The cement truck rips out the individual pumps and rams the tanker truck. The tanker and the underground tanks are in the optimum state of partial fill to provide the most explosive mixture of liquid gasoline, gasoline vapor, and air. The resulting explosion takes out most of the block. The adjacent underground tanks either explode or burn as needed for maximum havoc. There is, of course, also a large propane tank at the gas station that, after the fires burn a bit, explodes creating a second massive blast. The lead-acid auto batteries, tires, and contaminated oil on site from oil changes give rise to massive clouds of toxic vapors as they are consumed in the fire. Hundreds are killed in the explosions and thousands poisoned by the toxic cloud. Would that be enough to make you give up on cars and just ride your bike? The point is, if you play the "what if" game and never ask how likely each aspect you add to the scenario is, you get a very skewed view of the actual hazard.

In 1975 the Nuclear Regulatory Commission (NRC) took another look at the problem. At that time the science of probabilistic risk analysis was becoming fairly powerful. This was originally developed as part of the aerospace business. When you are putting people on top of a rocket filled with many tons of high explosives you want to be sure you have looked at all the ways things can go wrong and done all you can think of to make sure that they don't go wrong. That is what probabilistic risk analysis is all about. In very simple terms you look at what can go wrong and break that into individual failures.

What the NRC's 1975 WASH-1400, *Reactor Safety Study*, did was to look at a large number of pathways leading to reactor accidents. They looked at the various combinations of individual component (e.g., individual valves or sensors) failures required to cause a system failure and what combinations of system failures with all the redundancies and backups were required to cause an accident sufficient to hurt anyone outside the plant. By looking at the probabilities of the individual component failures, they could combine those to create an overall estimate of causing the harm. They also did extensive work in estimating the consequences of reactor accidents. In WASH-740, they just assumed a large part of the radioactive material in the reactor got outdoors, and then blew away in the wind. In WASH-1400, they tried to develop good science on how the radioactive material would really behave in a major accident and how it might move into the environment.

One of the more important results from WASH-1400 was the use of relative risk. That is, you could break down the various aspects contributing to risk and see where you could profitably improve your safety. For example, if you discover in the analysis of your specific power plant that risk coming from the control rod drive system is a larger fraction of the total risk than that for analyses done for some other plant, you might want to look at what aspect of the other plant's control rod drive system accounts for a lower fraction of risk and upgrade yours. You might also just recognize some simple things you can do to improve your safety.

In one plant I was involved with, we found we could make the plant about twice as safe by changing two valves. There were four valves in this essential system and at least two of them were required to work to keep the plant safe. As originally specified all four valves were the same type powered by the same system. Therefore, a single failure in the design of those valves or in the system providing the power to make them operate would fail all four valves at the same time. By changing two of the valves to a different type and providing an independent and redundant power system, we eliminated the common failure path.

After WASH-1400, a significant number of other reactor safety studies were done. The science of risk analysis and the understanding of accident consequences were greatly improved. While the risk analysts caution us against it, people naturally look for the bottom line numbers. In terms of probabilities, those are usually summed up as the probability of core damage, the probability of containment failure, and the probability of large release. The probability of core damage is the probability that all normal and emergency cooling systems fail to keep the nuclear fuel from significant melting. This runs from about 10^{-4} to 10^{-6}. That means that in any given year the chances are 1/10,000 to 1/1,000,000 that such an accident would happen. Another way to look at that is if you stood and watched, you would expect to see that event at least once in 10,000 to 1,000,000 years. The probability of containment failure runs around 10^{-2}. If that seems a little high, once in a hundred years, you need to understand that the containment is more than the big pressure vessel. There are a lot of pipes and instrument lines that go through the containment wall. The pipes have valves that have to shut and all the penetrations have seals that have to work.

The probability of large release is the combined probabilities of core damage and containment failure. So the probability of large release is on the order of 10^{-6} to 10^{-8}.

If those numbers don't mean much to you, I have a little example I like to give. If you accept that written human records go back 6000 to 8000 years, then we have a recorded history of about 20,000 to 30,000 springs that have followed winter. So the probability that spring will not come, based on record observed evidence, is on the order of 10^{-4}. You get on the order of 10^{-6} for the sun not coming up tomorrow, using the same logic.

The point is that 10^{-4} and 10^{-6} are small numbers.

Yes, but accidents do happen. What about that? O.K., let's talk about the two most infamous accidents: Three Mile Island and Chornobyl.

The accident at Three Mile Island occurred on March 28, 1979. The Three Mile Island plant had two reactors. The accident took place in Unit 2, which had at that time been operating for about one year. It was the worst accident to happen at a U.S. nuclear power plant and to those of us in the nuclear business who watched it play across the evening news for nights in a row—a major disaster. Before we go on let's state right here that exactly **zero** people were killed and arguably no one was hurt. That is not to say it didn't frighten a lot of people.

O.K., what actually happened? It went something like this.

Three Mile Island Unit 2 was a pressurized water reactor with an output of 880 MW of electrical power. Remember that a PWR has a primary coolant loop that circulates water from the reactor core to the steam generators where it boils the water in the secondary loop, which in turn drives the turbine, which drives the generator making the electricity. The spent steam from the turbine is cooled in the condenser and then pumped back to the steam generators as cooled liquid water, ready to start the circuit through the steam generators and turbine again.

At 4 a.m. on the day of the accident, a valve shut in the line from the condenser to the pump in the secondary system. Without feed to the pump, the feedwater pumps automatically shut themselves down and the turbine also tripped off. At this point, there is no way to remove heat from the secondary system and, as soon as the secondary system absorbs all the heat it can from the primary coolant, the primary coolant and the reactor fuel will start to heat up. But the reactor tried to do what it was supposed to do if feedwater fails. After all, we have backup systems, right? Yes, there was an auxiliary feedwater system and the emergency feedwater pumps started up automatically just as they should. However, in a previous maintenance procedure, they had closed valves that isolated the auxiliary and main feedwater systems and by error failed to reopen them. So the vital cooling water was blocked from entering the steam generators.

The job of the steam generators is to boil water. Denied new feedwater, the steam generators quickly boiled off all the water they had. At this point the secondary system could remove no more heat from the primary coolant and the primary coolant started to heat up. This caused the pressure to rise in the primary coolant, which, correctly, caused an automatic trip of the reactor. Again, remember, getting a reactor trip doesn't get us out of the woods just yet. We still have to get the decay heat out of the reactor. The primary coolant pumps are still running, but without any water in the secondary system there is no place for the heat in the primary to go. The coolant pressure continues to rise.

The primary coolant system is designed to take pressure beyond the normal operating pressure, but there is a limit to what it can do. To protect against overpressure there is a pressure relief valve. This is a valve with a big spring pushing the valve shut. If the pressure gets large enough, it pushes back on the spring and lets some water out. The neat thing about such a valve is that, as soon as a little bit of water comes out allowing the pressure to decrease, the spring can push back and close the valve. At least that was what was supposed to happen. What did happen was the valve hung up and did not close.

This is the first operator error. They didn't understand that the relief valve had hung open. If they had, they could have prevented a lot of damage. But the plant itself helped to fool them. The outlet of the pressure relief valve traveled through a pipe into a quench tank. There was a temperature sensor in that pipe, which should have alerted the operators that the relief valve had not closed. However, the relief valve was known to have a small leak and that sensor always read high. Therefore, the high temperature that morning didn't alert them to the problem.

Now, the errors begin to compound. The reactor is continuing to heat up, water is being slowly lost through the stuck open relief valve, and the pressure continued to decrease. With increasing temperature and decreasing pressure, water beings to boil inside the primary coolant system. This caused a false reading of a high coolant level. The operators were carefully trained to ensure that the reactor was not overfilled, because that could cause uncontrollable pressure increases that would break the primary piping. With falling pressure, the emergency water injection pumps had come on to fill the primary system. The operators, seeing what appeared to be an already too full system, shut these pumps off. This was big mistake number two.

Another thing operators were trained to worry about was cavitation of the primary coolant pumps. The way a pump works is to create a localized low pressure that causes water to flow into it. If the pressure on the suction side of the pump is too low, the localized lower pressure in the pump could allow boiling inside the pump. This creation of many tiny little pockets of rapidly expanding steam is called cavitation. The effect is like a lot of tiny explosions which, when combined, will shake a pump to pieces. With the pressure still falling, the operators saw the cavitation point coming at them and shut off the primary coolant pumps. This was mistake number three.

Now they had shut off the addition of water to the reactor and stopped the circulation of what water they had left and still had not stopped the leak through the stuck open relief valve. The reactor got hotter and hotter, and as later inspections showed, over half the fuel melted.

The basic problem was the operators didn't understand what was going on. Unless you have been in a reactor control room, you really can't appreciate their fix. With all their instruments, and those have gotten a lot better since 1979, you really can't see what is going on inside the machine. What was happening made no sense to them. How can you have too high a liquid level, but pressure going down? It would be like turning on the element on your range under a teakettle and having the water freeze inside. It was just backward.

Eventually someone suggested they close the block valve in the line beyond the pressure relief valve. They saw the pressure stabilize and got water back into the system. But by that point they had a real mess in the reactor. Most of the fuel cladding had failed, releasing the more volatile of the fission products and over half the fuel had melted into a big slag of debris.

Now it is absolutely correct to say this was a very significant accident. Tremendous damage was done to the reactor and errors of design and operation were revealed. It is also absolutely true that no one was killed. The pressure vessel held in the melted fuel. The containment building held in essentially all the radioactive materials. There was a small release, when the quench tank that was taking the output of the stuck open relief valve overflowed before the containment was isolated, that is, closed down all the penetrations to the outside. The maximum radiation exposure was a fraction of natural background and the average radiation exposure given to the general public was trivial.

Not everyone accepted this assessment of the impact of the accident. There were those who were convinced that the releases were much greater and that significant harm had been done to them. This ended up in the courts and over 15 years bounced around in the legal system. After a trip to the Supreme Court, the final ruling came in June 1996 from Middle District of Pennsylvania Court Judge Sylvia Rambo. She dismissed the lawsuits granting in favor of the utility and others being sued. The conclusion of her judgment reads in part:

> "The parties to the instant action have had nearly two decades to muster evidence in support of their respective cases. As is clear from the preceding discussion, the discrepancies between Defendants' proffer of evidence and that put forth by Plaintiffs in both volume and complexity are vast. The paucity of proof alleged in support of Plaintiffs' case is manifest. The court has searched the record for any and all

> evidence which construed in a light most favorable to Plaintiffs creates a genuine issue of material fact warranting submission of their claims to a jury. This effort has been in vain."

I translate that to say, "We have been working this for 20 years. It is clear there is a really big difference in the view of what harm has been done. However, it is clear that the case that harm was done is poor. I tried very hard to find a reason to support those claiming harm, but there just wasn't any valid reason to believe that view."

The accident at Three Mile Island generated a vast number of changes within the nuclear power business. Most of these were positive but, like any massive response to a big scare, some of the changes were wasteful without really helping, which unfortunately increased the cost of nuclear power to an undue extent. A whole set of equipment and procedural upgrades was made across the industry. The volume of the requirements became massive. Everyone with any idea on nuclear safety got on the bandwagon. At one point, I joined an effort to help the Nuclear Regulatory Commission evaluate the real safety effectiveness of these. We were successful in helping them focus their resources on the most effective changes. Good things that happened were we improved our understanding of things that could go wrong and greatly improved the operator training to handle those. All plants now have full scope simulators. These are exact copies of the control rooms in which computer-driven controls and displays allow operating crews to practice a wide range of potential upset conditions.

The results can be seen in the excellent safety and operational performance of the current fleet of U.S. power reactors. That is not to say there haven't been equipment failures and human mistakes in operating these plants. With over a hundred complex machines operating for many years, things are going to go wrong from time to time. However, with the improvements to the equipment and the training of the operations staff, no problem has been allowed to escalate to serious damage or significant radioactive releases in any U.S. reactor in the intervening 24 years, not once, none.

So, now let's talk about Chornobyl. The first and most important thing to say is that the accident at Chornobyl was a direct result of design and operational features that are not allowed in power reactors operating in the U.S. I will return to this point a bit later, but we need to understand this fact before we get into the accident itself.

The April 26, 1986, accident at the Chornobyl plant was the worst reactor accident ever and represents just about as bad as it can get. The "good" things that happened at Three Mile Island, the integrity of the primary system and the success of the containment, failed at Chornobyl.

The Chornobyl complex consisted of four big RBMK reactors, each at 1000 MW electrical output. The accident occurred in Unit 4, the newest and most modern of the reactors. The RBMK design is an outgrowth of the Soviet military production reactors. The original Soviet military production reactors were very similar to the U.S. designs developed during World War II. These were large graphite blocks penetrated by pipes carrying cooling water and holding uranium metal fuel rods. The mission of the military production reactors of both nations was to fission uranium and to produce plutonium by capture of the excess neutrons in the uranium.

By modern standards the original production reactors were pretty crude, lacking instrumentation and many safety features. The single most important safety feature that was lacking was the absence of a negative void coefficient. If you made it through the nuclear science and engineering chapters, you know how important that is. As a quick refresher, prudent reactor design calls for the moderator, the material used to slow neutrons down so they are more effective at causing fission, to be balanced against the fuel such that if you lose some of the cooling water, which is also a moderator, you drive the reactor into a less favorable condition, which causes the reactor to shut down. This is the void coefficient. If you have a safe reactor, the void coefficient is negative. That means voiding the coolant either by boiling or just plain spilling it out has a negative effect on power. To get a negative void coefficient, you have to set the moderator-to-fuel balance a bit off the optimum point. In the military production reactors, the push was on national defense and creating the biggest possible stockpile of plutonium for weapons as fast as possible was a national priority for both the U.S. and the Soviet Union. In both countries the original graphite moderated military production reactors had positive void coefficients. That means, if you were to have lost coolant or boiled coolant in the reactor, the power would have increased. This is particularly scary because if you were in some event which caused you to void coolant, the resulting power increase would probably cause you to lose more coolant, which would increase power more, which would cause more coolant loss, which, well, you get the picture. You would go rapidly into a very bad day.

In the U.S., our commercial power reactors were mainly based on an outgrowth of the naval propulsion reactors. These light water reactors did not use graphite as a moderator, and it was easier to design with a negative void coefficient. The Soviets, on the other hand, developed both a light water reactor family of civilian power reactors and a graphite-moderated family.

While the RBMKs were substantially advanced from the original military production reactors, they still shared a number of the unhappy features, the worst of which was the positive void coefficient. The amazing thing to Western eyes was that the RBMKs were boiling water reactors. That is, in each of their 1600 individual pressure tubes, water was allowed to boil. This had the charm of removing the need for steam generators and the entire secondary coolant loop. However, operating a boiling water reactor with a positive void coefficient is like doing a high wire act without a net in a strong wind. Every time you started up the reactor, you had to start boiling water. When you did that, you got an uncommanded power increase that you had to dampen back out with active controls. Unlike Western plants, which were inherently stable, the RBMKs were inherently unstable. Now to be sure they had a strong Doppler effect, which helped them out, but any positive coefficients are basically scary to a designer or operator of a nuclear reactor.

To give the Soviet designers some justice, they had done studies and experiments on the void coefficient and had convinced themselves that while the initial result of coolant voiding would be a positive effect on reactor power, with sufficient voiding the effect would turn around and become negative. Therefore, in a really serious voiding event, it would all be just fine and behave much like a Western designed reactor. After the accident it became very, very clear they were wrong. The effect was positive right down to the last ounce of water in the plant. So instead of the premise that a little voiding is bad but a lot is good, it was actually a little voiding is bad and a lot is very, very bad.

What happened in the accident was that as the most modern of the RBMKs Chornobyl 4 was chosen to conduct a special test. Like other reactors, the RBMKs depend on electrical power to drive their main coolant pumps. Like other reactors, the RBMKs were designed with redundant sources of electric power. However, unlike Western reactors, the offsite electrical grids were not exceptionally reliable. The Soviets wanted to know just how long they could use the reactor's own electrical generation to power their pumps. To that end they were running the reactor at a very low power trying to still generate electricity and make the pumps go.

They were having a lot of trouble with the test and had to force the reactor into a very unnatural state to keep trying to complete the test. In the process they violated a long set of their own safety procedures.

Just before it all came apart, they had the reactor at just a fraction of its full power, but in that, they had all the fissioning going on in just a small part of the reactor near the coolant inlet. All the neutrons were down there and when that last bit of liquid water started to boil, it became clear they were in a very unstable condition. And here is the really heartbreaking thing. At that point they realized they were in trouble, so they pressed the emergency shutdown button, which turned out to be the worst possible thing to do.

The RBMKs use control rods somewhat like a PWR does. The difference is, in a RBMK, the control rods slide through the graphite block not into the fuel bundles. In order to cool the control rods, the tubes that the rods move in are filled with water. To make matters worse yet, the tips of the control rods, which can be in the top of the reactor during normal operation, are not neutron absorbing materials but graphite instead. Now with all the reactor power at the bottom of the core, when the control rod tips came zipping into that region, the first thing that happened was the water in those control rod channels was displaced with graphite. Instead of shutting down the reactor, the neutron absorbing water was replaced with moderator-enhancing graphite, and that set the positive void coefficient off with a vengeance. The reactivity effect was greater than the delayed neutron fraction and the reactor went prompt critical, that is, critical on prompt neutrons only. The power increased by over a thousand fold in a heartbeat. The reactor structure could not take this magnitude of energy release. The pressure tubes burst. This over pressurized the reactor cavity, and the big lid that all the pressure tubes go through was ripped off, breaking any pressure tubes that were left. The fuel and cladding materials were then so hot that they stripped oxygen from the remaining water as they corroded. This released hydrogen from the stripped water and the hydrogen burnt and exploded. This propelled pieces of the core out of the reactor and set the graphite burning.

Like the original military production reactors, the RBMKs do not have a containment building. Instead they have an accident localization system that strengthens the structure around the reactor core but provides no ultimate barrier to the environment. The forces of the massive over-power, steam explosions, and hydrogen fires smashed through the accident localization defenses and the heart of the reactor became a burning pyre of graphite exposed to the environment.

There has been a lot of widely varying claims as to the human heath impacts of the Chornobyl accident. In trying to find a credible resource, I have captured the following findings from the April 1996 International Atomic Energy Agency symposium—One Decade After Chernobyl:[10]

- 240 people were hospitalized at the time of the accident suffering from acute radiation. Of these, 31 died shortly after. In the intervening 10 years, 14 more had died, however, not all from radiation effects.
- A large number of people were conscripted to be "liquidators." These people were used to conduct the most hazardous parts of the clean up and took large radiation doses. It is estimated that 2500 additional cancers would occur in that population.
- In the population at large, it is expected that 2500 additional cancers would also occur. However, because in that exposed population, 500,000 cancers would be expected in the same time period, it will be difficult to see the impact of the Chornobyl-related cancers.
- After the accident, a large number of leukemia deaths were predicted. These have not been observed.
- Thyroids in developing children are particularly sensitive to the release of iodine-131, which is a short-lived but strongly radioactive material.

[10]You might note that Chernobyl is the Russian spelling. I have generally used the Ukrainian spelling which is Chornobyl.

> Of children under 3 years old at the time of the accident, 4000 to 8000 additional thyroid cancers are expected. However, 90% of these are expected to be curable. As of the date of the conference, 500 cases had been confirmed.

Regardless of the precision of the numbers, it is clear the Chornobyl accident was a major tragedy. But what does it mean to us? I think it means we have both an opportunity and an obligation to help the rest of the world operate nuclear power safely. The United States has risen to that challenge and has done a lot of great work in helping improve the safety of Soviet-designed reactors in terms of design, equipment, procedures, training, and regulation. I have been proud to be associated with the folks doing this great work.

But what does it mean in terms of commercial power reactor safety in the United States? Well, it does mean we have to stay sharp, but it is important to also recognize the fundamental differences between the Chornobyl design and operation and what we have and do in this country. I think the three big factors are:

- No commercial power reactor in the United States is allowed to have a positive void coefficient, none. They have to prove this to the licensing authority and all have. You just can't get the Chornobyl accident in a U.S. plant.
- All U.S. reactors have a containment structure. It is not an absolute guarantee, but it is a very robust line of defense.
- No U.S. reactor operations staff would put a plant in that extreme a condition and knowingly violate so many safety procedures. I have had the opportunity to conduct a safety audit at a U.S. reactor. One question I liked to ask was, "Who is responsible for nuclear safety?" I asked the operators, the shift supervisors, and the plant manager. I got the right answer each and every time. "I am." I also worked for years alongside of folks who gave the qualifying exams for commercial nuclear power plant operators. These were very dedicated people. If they could not be convinced the candidate really knew their stuff and had the intelligence, commitment, and cool to operate a plant safely, they would not pass them.

O.K., I want to make one final point about nuclear accidents. Despite all I have said and all I have tried to explain, you are still going to hear people tell you that nuclear reactors are dangerous, very hazardous. I want you to look at Table 6.2. I want you to look at it for two reasons. One, I spent a lot of time digging up all these numbers, and two, I think it makes a really important point.

The point is that if nuclear power is so dangerous, why aren't we killing any of the public with it. In the 31 years of data I could collect, we lost over 1.5 million folks to highway fatalities. We lost 457,389 to falls; 186,354 to poisons; 175,074 to fires of which about two-thirds are in private dwellings; and 2954 to lightning. In the last 18 years, over 6000 people have drowned in bathtubs. Along the way, I learned about 20 people a year are killed by dog bites, and in the last 15 years, over 30 people have been killed by tipping vending machines over on themselves.

One might wonder why I chose 1966 as a starting point. I will confess that this is as far back as I could locate National Safety Council information, which was my primary resource. Were there commercial nuclear power plants before 1966? Yes, but there were only a few and they never hurt any of the public either.

If nuclear power is so dangerous, why aren't they killing some members of the public? After all, safe, reliable natural gas kills on average something like 20 people a year through pipeline accidents. If nuclear power is more dangerous than bathtubs, why do roughly 300 people a year die in bathtubs and exactly none in nuclear power accidents? I think the answer is that nuclear power isn't all that dangerous. But I will let you study Table 6.2 and see what you think. Remember that my job is to supply you with the facts and numbers and your job is to decide their significance.

Before closing this section on accidents, I have to say something about terrorism. I was working on this section during the month of September 2001. Having been to the World Trade Center and worked with folks there, I have some small appreciation for the devastation the events of September 11 represent.

Shortly after the attacks people started thinking of other vulnerabilities and things the terrorists might go after next. Nuclear power plants appeared on some such lists. There has been a good deal of discussion on this, and some of it is pushed to be pretty scary. Here are a few facts and observations.

The containment on a modern nuclear power plant is pretty darn strong. However, they were never designed to withstand a direct impact of a purposely aimed jet airliner. There are those, including a very impressive list of experts in an article in the September 2002 issue of *Science*, who say an airliner will just crumple up and slide around a reactor containment building. I haven't seen the definitive engineering analysis, but this makes some sense to me. The *Science* article refers to tests done years ago where a jet fighter plane was slammed into a section of containment building wall. Only a small fraction of the thick wall was harmed. Some dismiss this, saying a jetliner is a lot bigger than a jet fighter. The *Science* article authors say, no it is valid, because the jetliner just spreads its greater mass over a larger area so the force on any portion of the containment wall would be roughly the same, and therefore, the jet fighter test shows containments can take the jetliner crash. One additional thought is that the hardest things on a jetliner are the jet engines. One of the engines on the plane that hit the Pentagon struck a sloping side of the building

TABLE 6.2 Selected U.S. accidental fatalities.

Year	Highway [a]	U.S. Airline Passengers [b]	Rail [c]	Natural Gas Pipelines —Trans & Dist [d]	Falls [c]	Poisons [c]	Fires [c]	Electrocution —domestic [c]	Drowning in bathtub [c]	Lightning [c]	Venomous Plants and Animals [c]	U.S. Public Fatalities Nuclear Power Plant Accidents
1966	50,894	no data	1027	no data	20,066	3931	8084	no data	no data	110	48	**0**
1967	51,559	no data	967	no data	20,120	4081	7423	no data	no data	88	44	**0**
1968	53,763	no data	849	no data	18,651	4109	7335	211	no data	129	69	**0**
1969	55,043	no data	884	no data	17,600	4300	6900	222	no data	131	48	**0**
1970	53,816	no data	852	no data	17,500	4600	6700	211	no data	122	51	**0**
1971	53,907	no data	750	no data	16,755	5356	6776	216	no data	122	48	**0**
1972	55,600	no data	692	no data	16,744	5418	6714	206	no data	94	42	**0**
1973	55,096	no data	789	no data	16,506	5335	6503	232	no data	124	49	**0**
1974	46,049	no data	716	no data	16,339	5534	6236	203	no data	112	53	**0**
1975	45,500	no data	608	no data	14,896	6271	6071	258	no data	124	50	**0**
1976	45,523	no data	552	no data	14,136	5730	6338	203	no data	81	53	**0**
1977	47,878	no data	576	no data	13,773	4970	6357	212	no data	116	55	**0**
1978	50,331	no data	602	no data	13,690	4772	6163	140	no data	98	62	**0**
1979	51,093	no data	551	no data	12,700	4500	5700	156	373	87	50	**0**
1980	51,091	no data	632	no data	12,300	4300	5500	177	331	94	47	**0**
1981	49,301	no data	580	no data	11,700	4300	4900	150	294	87	54	**0**
1982	43,945	210	545	no data	12,077	4733	5991	151	356	100	76	**0**
1983	42,589	8	544	no data	12,024	4633	5028	160	362	93	64	**0**
1984	44,257	1	570	no data	11,937	4911	5010	148	353	91	62	**0**

continued

Year	Highway [a]	U.S. Airline Passengers [b]	Rail [c]	Natural Gas Pipelines —Trans & Dist [d]	Falls [c]	Poisons [c]	Fires [c]	Electrocution —domestic [c]	Drowning in bathtub [c]	Lightning [c]	Venomous Plants and Animals [c]	U.S. Public Fatalities Nuclear Power Plant Accidents
1985	43,825	486	551	no data	12,001	5170	4938	146	353	85	55	**0**
1986	46,087	4	556	35	11,444	5740	4835	150	343	78	63	**0**
1987	46,390	213	624	11	11,733	5315	4710	121	348	85	68	**0**
1988	47,087	255	571	25	12,096	6226	4965	122	364	73	48	**0**
1989	45,582	259	608	42	12,151	6524	4716	143	345	75	76	**0**
1990	44,599	8	663	6	12,132	5803	4175	100	318	89	55	**0**
1991	41,508	40	651	14	12,662	6434	4120	82	312	75	90	**0**
1992	39,230	25	542	10	12,646	7082	3958	66	345	53	68	**0**
1993	40,134	0	670	17	13,141	8537	3900	82	306	57	57	**0**
1994	40,718	228	635	21	13,450	8994	3986	84	301	84	63	**0**
1995	41,770	152	569	18	13,986	9072	3761	88	281	76	81	**0**
1996	43,649	319	565	48	14,986	9510	3791	66	330	63	68	**0**
1997	43,458	2	527	10	15,447	10,163	3490	53	329	58	68	0
Total 1966–1997	**1,511,272**	**2,210**	**21,018**	**257**	**457,389**	**186,354**	**175,074**	**4,559**	**6,344**	**2,954**	**1,885**	**0**

[a] 1966-1995 Federal Highway Administration, 1996–1999 National Safety Council
[b] National Transportation Safety Board
[c] National Safety Council—Accident Facts 1971–1998 Editions; Injury Facts 1999–2000 Editions
[d] Office of Pipeline Safety—U.S. Department of Transportation

and skimmed off, coming to rest in a parking lot some distance away. Reactor containment buildings are cylindrical so there is no flat surface to hit.[11]

Beyond that, we can also observe that the reactor core itself is usually below grade, that is, below ground level. Therefore, if an aircraft were to crash into the side of a containment building, and even if it did manage to breach the containment, it would probably not actually hit the reactor core itself. It very likely would destroy a lot of the cooling systems and things would get bad quickly, but it would be unlikely to blow all the radioactive material in the reactor into the environment in the initial impact.

If an airliner crashed into the less well protected supporting equipment of the plant rather than the containment building, it might render the cooling systems inoperable. In that case, you might well get a Three Mile Island like accident, which makes a mess of the reactor, but no significant radiation is released.

In the worst case, in which the terrorists are successful in breaching the containment and failing all the cooling systems, the resulting accident would be bounded by the Chornobyl event. Recall in that case the reactor cooling was defeated and the "confinement" was breached. In addition there was a massive graphite fire that burned for *days* giving a strong driving force to put radioactive materials high into the air. An airliner into a light water reactor would not have that driving force because there would be little to burn after the initial fire of the aircraft fuel.

The Chornobyl event had 31 deaths at the time of the accident and potentially several thousand deaths from cancer in the decades to come. These would appear in a population that will see half a million cancer deaths from other non-related causes. While a terrorist might gain a lot of terror from hitting a nuclear plant, the truth of it is he will kill very few people at the time of the event, and those that will later succumb to cancers will be lost in the larger population.

But, of course, airline crashes are not the only way for a terrorist to attack a nuclear power plant. Truck bombs and armed attacks are certainly something to consider. It turns out that nuclear power plants are one of the few facilities in our national infrastructure that does consider these things. Every U.S. nuclear power plant has a trained armed security force who is authorized to use deadly force to protect the plant. Not wanting to give any terrorists alternative ideas, but if I had a choice of going after a facility either totally unprotected or protected with only a night watchman versus a facility with a team of military capable troopers armed with automatic weapons, it would

---

[11]If you see a commercial reactor that looks square, you are looking at a structure that goes around the containment. Some boiling water reactors, in particular, are built this way. The actual containment is inside that structure and is cylindrical.

not be a tough choice. That is not to say these wackos are afraid to die. Clearly, they have demonstrated that they are not. However, one would assume that they do want to have a reasonable chance of successfully completing their vile mission. In that regard, a nuclear power plant would be a tough nut to crack.

There has been some talk about how the tests of the armed guard force have had a number of failures. These tests are what they call force on force. That is, a bunch of very smart and tough guys—with full access to plans of the plant—figure out how to attack the plant and then they do that while the plant forces try to keep them out. They use military grade "laser-tag" type equipment that records hits on all parties recording them out of the action if "killed."

I had the experience of working late one night at a plant when we knew an exercise was going to happen. We had been warned to stay inside the buildings and out of the way of either the attackers or defenders, but no one knew just when or how the attack would come. It was sufficiently dark that I could not see what was going on, but I certainly could hear it. The automatic weapons fire went on in spurts for over an hour. It was pretty fierce. As I recall the replay, the "bad guys" got inside the perimeter, but they didn't get to any of the vital buildings.

But in some exercises, the "bad guys" are going to win. When that happens, the security force learns how they were defeated and they fix it. The test is repeated until either they pass or the Nuclear Regulatory Commission pulls the operating license of the plant. The fact that sometimes a security force fails the test doesn't worry me. What would worry me is if every security force always passed. That would suggest the test was not very tough and that it was just a show. If some fail and get retested until they can make the grade, that suggests to me that everyone in the system, testers and tested alike, takes this appropriately seriously.

Nuclear Weapons

One of the truly unfortunate things about peaceful civilian nuclear power is the association with nuclear weapons. While there are places in the fuel cycle of a civilian power program and a nuclear weapons program where there will be similarities, there is no essential connection. That is, you can have a civilian nuclear power program without any nuclear weapons, and you can have nuclear weapons completely without civilian nuclear power. One does not require, lead to, or even imply the other. I will explain more in a bit.

First, I do have to at least get you to consider the question of whether nuclear weapons are evil or not. How can you even ask that? Of course they are evil. What else could you call a weapon designed to kills millions of civilians in an instant? Well, one potential thing you could call them is guarantors of the peace. Here I am talking about nuclear weapons held by superpowers, not rogue states or terrorists. We will discuss them in a bit. The theory for superpowers, which of course you all understand, is that nuclear weapons have made global warfare so unthinkable, so unwinnable, that no one will ever attempt to instigate a global war again.

If one studies the horrific impacts of the Second World War, you get an in-the-heart feeling of why it is so very, very critical that we never allow that level of global insanity loose again. While we in America certainly did our part in the conflict, you have to see the War from the viewpoint of those upon whose soil the battles were fought to truly appreciate this. The destruction in Europe, especially in Germany and Russia, was beyond description. The total infrastructure of modern society was obliterated leaving the survivors to scrounge in the rubble for food. The whirlwind the Japanese reaped was equally severe. In one night our B-29s set fire to 267,000 buildings in Tokyo.[12] This was not a nuclear weapon attack. This was conventional bombs especially designed to set fire to the paper and wood construction of the Japanese homes.

While one can certainly debate the role of nuclear weapons in holding the peace, the observed fact is that in the 57 years since nuclear weapons were introduced to the world we have had no global-scale conflict. We have had bloody wars with great loss of life, property, and potential, but nothing on the scale of the Second World War. Did nuclear weapons pull us back from some brinks? Did we just get lucky? I think a serious student of post-war history will tell you "yes" to both questions. Let me offer two quotes on the subject of nuclear weapons and leave you to think about the difficult question of what role nuclear weapons have played relative to world peace.

> "Be careful above all things not to let go of the atomic weapon until you are sure, and more sure than sure, that other means of preserving the peace are in your hands." —WINSTON CHURCHILL

> "The creation of nuclear weapons may yet prove to be a blessing to mankind, rather than a curse. Their over-whelming destructive force has proved to be a sobering force that can compel all sides to 'come to their senses' before the world again experiences such losses as both world wars produced. I believe we all have an obligation to be certain this remains so." —AMBASSADOR G. PAUL ROBINSON

Now, returning to the connection, or lack thereof, between nuclear weapons and peaceful civilian nuclear power, let's first look at the nuclear weapons side.

---

[12]*Superfortress, The B-29 and American Air Power*, General Curtis E. LeMay and Bill Yenne. McGraw-Hill, 1988, p. 123.

If you are going to make nuclear weapons, you are going to need a supply of fissionable material. Although there are several isotopes that could be made into a bomb, for any practical consideration there are two that have been made to work well. One is uranium-235 and the other is plutonium-239. Back in the nuclear science and engineering chapters, you learned that uranium-235 is present in natural uranium that we can find in the ground, but only as a small fraction, only 0.7% in fact. If you tried to make a nuclear weapon out of the natural uranium, it wouldn't work. All that neutron absorbing uranium-238 gets in the way. If you are going to make a uranium bomb, you have to get pretty close to pure uranium-235. As we saw in the fuel cycle section, you can do that by the process of enrichment, of which there are several different processes available.

Now civilian nuclear power also needs some enrichment, except for the Canadian CANDU reactor design, which uses natural uranium but must have heavy water to work. Typically light water reactors run with fuel enriched to 3 to 5% in uranium-235. You can't make a bomb with this stuff. The physics just won't let you. Enriching from natural uranium at 0.7% uranium-235 to LWR grade fuel at 3 to 5% uranium-235 is difficult and expensive. Enriching up to the nearly 100% pure uranium-235 needed for nuclear weapons is extremely difficult and amazingly expensive. However, within a national nuclear weapons program, it has been done. Both the U.S. and the Soviet Union built truly massive facilities to do this, and the Soviets in particular made a lot of weapons-grade uranium. Happily a lot of that is currently being blended with normal uranium and sold as civilian reactor fuel supplies removing it from weapons usability and providing the Russian Federation with much needed income.

The point here is that while both civilian power reactor fuel cycles and uranium weapons use enrichment, the nature of the process and its operation is very different. In some ways you might compare it to the distinction between the weekend fisherman with his fly rod to an ocean-going commercial fishing trawler. Sure, both catch fish, but the weekend fisherman is only catching a few fish while the trawler is trying to get them all. The investment and processing are on different levels.

So, what is the plutonium route to weapons? Again, going back to Chapters 4 and 5, you will remember that the part of the uranium that is not uranium-235 is uranium-238. This isotope has the property that it likes to capture neutrons. When it does that, it is transformed into uranium-239, which quickly decays to neptunium-239 that in turn decays to plutonium-239. Plutonium-239 is both an excellent reactor fuel and an excellent material to make nuclear weapons. With a half-life of only 24,300 years, no plutonium made in ancient supernovas is still around. Therefore, we never find it in nature. However, with a half-life of 24,300 years, any we make is going to be around a lot longer than either you or I. The important thing about plutonium-239 is that it is a different chemical element than uranium.

When we had that valuable uranium-235 mixed in with the normal uranium-238, they were both the same chemical element. No form of chemistry is going to be able to separate one from the other. Therefore, we have to go to the very difficult mechanical means of enrichment using the slight difference in weight to essentially sift the uranium-235 out atom by atom.

Because plutonium and uranium are different chemical elements and will behave different chemically, when we make plutonium-239 within a bunch of uranium-238, we can use chemistry to get them apart. Now this is easier than enrichment, but it is far from easy. The difficulty comes in that the way we got the extra neutrons to be captured in the uranium-238 is we put the uranium into a nuclear reactor. Then, we used the self-sustaining process of nuclear fission to give us the zillions of free neutrons we needed. This then means that the plutonium-239 and base uranium-238 are also going to be mixed up with a lot of very radioactive fission products. That makes the chemistry really difficult and results in a chemistry plant with very thick concrete walls. It also means process equipment that, once used, you can never let humans near to fix with hands-on maintenance.

The way the weapons process works is you build a dedicated military production reactor. This is unlike a civilian power plant for several reasons. First, since you are not trying to produce power, you need not operate at high temperature. Second, the fuel cycle will be very different. When you are making plutonium-239 from uranium, you are also making plutonium-240 from plutonium-239. That is, plutonium-239 also likes to absorb neutrons. About three-fourths of the time this will lead to a fission event and keep the reactor operating. The other roughly fourth of the time the plutonium-239 will not fission but rather be transformed into plutonium-240. Now from the chemical point of view, plutonium-239 and plutonium-240 are essentially the same. However, in the nuclear world, they are very different, actually just about black-and-white different. Plutonium-239 likes to fission and produces a relatively large number of neutrons per fission, which is why it is a good reactor fuel and a good bomb material. Plutonium-240 doesn't like to fission and instead likes to suck up neutrons in non-productive captures. It also has the annoying property of causing the random generation of neutrons as part of its radioactive decay. To someone trying to make a bomb, this is very annoying. Recall that, to get a big boom, the nuclear weapons designer is going to try to get a lot of generations of fission out before the weapon vaporizes itself. If there is too much plutonium-240 in his mix, some of its extra neutrons will start the fission process going before the weapon is configured for optimal explosion. That is, instead of waiting for all the fissionable material to be compacted in the smallest space, the plutonium-240 neutrons will start the fission process before everything is in the right place. This means the vaporization will come earlier, and the bomb will be less powerful, potentially a lot less powerful.

Now remember that plutonium-239 and plutonium-240 are both the same chemical element, just like uranium-235 and uranium-238 are both elemental uranium. Therefore, you can't separate them by chemical means. However, the problem isn't as bad in plutonium as it is in uranium. Nature gives us uranium with 99.3% uranium-238 and 0.7% uranium-235. There is no choice, that is the way it comes. However, we make the plutonium ourselves, so we have some say about how much is what. The rates of uranium-238 conversion to plutonium-239 and that of the conversion of plutonium-239 to plutonium-240 are such that initially most of what you get is plutonium-239. However, the longer you keep running the reactor, the more plutonium-239 you make and with more of it around the more plutonium-240 you make. Again, based on the relative production rates, it turns out the longer you run the reactor, the greater fraction of your plutonium is going to be plutonium-240.

How much plutonium-240 is too much and what do you have to do to not get more than that? The exact details of nuclear weapons designs are classified. No surprise there. However, the generally accepted unclassified number is that for use in nuclear weapons the U.S. requires plutonium at or less than 6% plutonium-240. How do you make that? Well, you run your reactor a very short time, just a few weeks, before you shut it down and remove the fuel. The fuel then is sent to a chemical processing plant where the uranium and plutonium are separated. The plutonium goes off to be made into weapons and the uranium recycled to be used as more fuel. The latter is important because the fuel has seen only a few weeks in the reactor and very little of its potential value has been used. The chemical processing will also strip out the radioactive fission products, creating a pretty nasty waste stream in the process.

Now a civilian power plant can't operate economically if it has to shut down every few weeks to pull out fuel. Neither can it stand the cost of new fuel many times a year even if you recycle the uranium. In fact, commercial nuclear power has been working on new fuel designs and operational systems to extend the time between refuelings. Because a refueling can take many days, the owners of the plants are very much inspired to reduce both the number and length of refueling outages. The plant makes neither kilowatt-hours nor money while it is not operating. When LWRs first came on the scene, most runs between refueling were around one year. Currently these have been extended to 18 months, and some plans are being made to go to 24 months.

When you leave the fuel in a reactor even 12 months, the plutonium-240 fraction is getting up into the 20%-plus range. Does that make it impossible to use a material in nuclear weapons? Actually the answer is no, not impossible, just much harder.

What bothers a lot of people is that in the operation of a civilian nuclear power plant, you make plutonium. The plutonium you will make will be way out of the required specifications for nuclear weapon-grade material, but as we in fairness must admit, it could be used in a nuclear weapon, at least maybe if you had a set of very clever weapons designers. When fuel reaches its end of life, it is removed from the reactor and stored on site. At that point the plutonium will represent about 1% of the mass of the fuel material. Most of the fuel is still uranium. About 3% of the mass will be fission products, rendering the fuel very radioactive.

If at this stage you wanted to get the plutonium out to make a nuclear weapon, you would have to:

- Get access, which is not easy with the plant guard force on duty
- Take the stolen fuel to some secret location, and it is pretty hard to hide something that radioactive
- Go through the chemical separation steps to get the plutonium out, which is not easy especially with the intense radioactivity of the fission products
- Purify the plutonium and convert it to a metal form ready for the bomb guys.

Of course, you would have to do all this while not killing yourself or letting yourself be caught by the entire world who would be looking for you. Then all you would have to do is figure out all the really hard stuff about making the bomb work, which is made much more difficult by the high plutonium-240 fraction in the LWR spent fuel.

In the U.S., we only take the commercial reactor fuel cycle as far as spent fuel. That is, after we remove the fuel from the reactor, we store it at the reactor for a few years until the initial short-lived fission products decay away and then . . .

Well, the idea was . . . and then the U.S. government would take the spent fuel and put it in a repository. In fact, they promised the reactor owners they would do this by 1998. The government has been charging the owners a special tax of 0.1 ¢ per kilowatt-hour to pay for the repository. Now if a tenth of a cent seems like a pretty small deal, for a 1100 MWe plant operating with a 90% capacity factor, that works out to $8,672,400 per year. The government has yet to take any of the spent fuel it promised it would accept, but this is a subject for the next section. The point here is that the U.S. policy is to throw the spent fuel away. This is called a once-through or throwaway fuel cycle.

Other nations, predominately Russia, Japan, and France, have instead chosen a recycling fuel cycle. In this system spent fuel is sent to a reprocessing plant where the plutonium and uranium are separated from the fission products. The plutonium and uranium are combined as a mixed oxide, or MOx, fuel and used to run the reactors again. Again, this troubles some people. First, this looks a lot like the military fuel cycle. Second, the plutonium at some stage is separated completely and even when combined within the MOx fuel could be separated by only chemical means. Again, this is pretty crummy weapons material, but it is still plutonium.

What we are left with is the unhappy fact that in either a throwaway or recycle fuel cycle, if someone tries hard enough and has enough skills, they can get plutonium out of a civilian reactor fuel system. For what it is worth, the Japanese like to say that they think the idea of recycling the uranium and plutonium not only makes economic sense but also is a good way to prevent misuse of the materials. After all, where is a safer place to put plutonium than back in the reactor? It is really hard to steal it from there.

Stepping back a bit, let's look at the big picture. When you get worried about nuclear weapons, you get worried about three kinds of threats:

- Existing nuclear weapons states
- Emerging nuclear weapons states
- Sub-national threats.

For existing nuclear weapons states, the potential role of civilian nuclear power is simply not a player. If any of those nations have combined or found synergies between civilian and military programs, well there it is. It is done and penalizing future civilian nuclear power will have no impact on that history. Dealing with this threat is an assignment for diplomacy.

For emerging national threats, we certainly need to be looking for cases where they seek to use synergies in civilian programs to support military nuclear programs. In some cases, civilian programs may be a front for a military program. However, it is pretty easy to tell the difference once they get going down the military path.

It is important to stress that military and civilian programs need not be linked. Israel has no civilian nuclear program but are widely believed to have a military nuclear weapons capability. Canada has an indigenous civilian nuclear power program but no nuclear military program.

Perhaps the single strongest demonstration of the separation of military and civilian nuclear programs is the case of North Korea. In the early 1990s they had a strong nuclear program heading down the plutonium path. We and other nations saw this as a significant military program despite the North Korean claims of a peaceful civilian reactor effort. The deal that was worked out at that time was that we would supply them two modern pressurized water civilian reactors if they would stop the operation of their existing nuclear program. The point was that a civilian nuclear power plant, operating under international inspections from the United Nations International Atomic Energy Agency (IAEA), is not a credible threat to the spread of nuclear weapons. As I write this, the North Korean question is much murkier with revelation that they were pursuing a uranium path to weapons as well, and their expulsion of IAEA inspectors. However, the point that civilian power reactors operating under IAEA inspection are not a weapons proliferation threat is unchanged.

This is not to say we need not work very hard with the world community to control the spread of nuclear weapons. The point is that purely civilian nuclear power is not a path to nuclear weapons. For example, what we have learned after the 1991 Gulf War is that Iraq was pursing the uranium enrichment route which requires no reactors at all.[13]

O.K., now for the tough one. What about sub-national threats? We have all seen movies and read books where the villains are international terrorists armed with a nuclear weapon threatening the civilized world. And, to our shock and horror, we have seen all too recently that there are those who would not balk at inflicting huge civilian casualties. We have also seen these true-life villains use relatively sophisticated technology to carry out their attacks, although learning how to build a nuclear bomb is a lot harder than learning how to fly a plane.

Does that mean that we have to bury all nuclear technology lest it be used against us? I think not. I think not for three reasons.

- It won't work.
- Nuclear terrorism isn't the easier way.
- It is wrong to deny ourselves the benefits of nuclear energy.

It won't work because the genie is out of the bottle. Once it was shown you can fission uranium and that the energy released can be turned into a weapon, then it becomes impossible to uninvent that. Even if we tore down all the nuclear reactors and sent all the plutonium and enriched uranium into the sun, somewhere, someone would remember that we could make nuclear explosives and figure out how to do it. We can't get all the uranium off the planet, there is just too much, and it is too widely dispersed. The genie is out, and for better or worse, it ain't going back.

Now, I will not offer up a suite of alternative methods for creating mass civilian casualties, but you all have seen enough movies about other means that I hardly need to do that. I don't suggest that any of these means are easy, just that in any relative scale terrorism using nuclear weapons isn't the easiest way to do it. While the sub-national threats appear to have a lot of money and

---

[13]Hamza, Khidhir, *Saddam's Bombmarker*, New York: Scribner, 2000.

a lot of technical sophistication, both have their limits. Full-scale nuclear weapons capability takes a lot of money and a lot of skill. Nations can do that, but while we can't claim it is beyond their reach, it is going to be a stretch for sub-nationals.

Finally, I feel it would be a mistake to unduly penalize civilian nuclear energy because of the possible potential for malefactors to misuse its fruits. There may be little difference between the production of insecticide and the production of chemical weapons. There may be little difference between the creation of vaccines and the production of biological weapons. However, that most certainly does not mean we should let people go hungry while we let the bugs get the food, or sit and watch while millions die of diseases we could have prevented with vaccines.

In the same view I think it is vital that we let nuclear power do what it is capable of in providing environmentally responsible energy that can improve and sustain the quality of life across the planet. While it is hard to fathom the justifications within the mind of a terrorist, perhaps the most effective weapon we can have against them is to reduce poverty and improve the quality of life worldwide. One has to suspect that such poverty and its frustrated desperation can be breeding grounds for international terrorism. Nuclear energy can help mitigate these deplorable conditions. Certainly we must do it with care and provide every barrier to misuse we can effectively employ, but the greater good is achieved if we use nuclear energy to close the gap between the haves and have-nots.

Nuclear Waste

If we see the connection between nuclear power and nuclear weapons as unfortunate, then we have to call the controversy over nuclear waste the most tragic. The tragedy is that nuclear waste has been sold to the public as a massive pollution problem and a problem without a solution. Both claims are just not true.

I have been told many times that what I am about to attempt cannot be done. What I am going to try to do is convince you that nuclear waste is not a big problem and that there are sound solutions to the small problem that nuclear waste represents. Many of my friends within the nuclear industry tell me that this is a losing battle and that I will never convince anyone that nuclear waste isn't a big deal. They will agree with me that nuclear waste isn't a big deal if seen in context and also agree that, if there is a lacking of a solution, it is political not technical. However, they are convinced that all of you have been so totally brainwashed that you won't see the light no matter what I say.

I disagree. I think you are all smarter than that, but you just haven't been given the chance. Let's see.

Let me tell you what I am going to do to try to convince you. First, I will give you a quick overview on the types of nuclear waste coming from nuclear power plants. This will help you distinguish between the low-level wastes and the high-level wastes. Most of the excitement is about high-level waste and that is where we will spend the most time, but I want to give you the complete picture. After that brief introduction to the spectrum of waste forms, we will focus on high-level waste. To provide what I hope is a true perspective, I will try to answer three questions about civilian high-level waste.

- What is it?
- How much of it is there?
- What can we do with it?

So let's start with the different types of radioactive waste that come from a commercial nuclear power plant.

A nuclear power plant produces a variety of radioactive wastes. These are classified by the amount of radioactivity in them. Low-level wastes are contaminated rags, tools, clothes, and the like. This is the greater volume of radioactive waste, being about 100 cubic meters per year at an average reactor.[14] However, the level of contamination is small, and the hazard is less than a number of things that show up in city landfills. Low-level waste has only about 1% of the radioactivity of all the radioactive waste coming from a nuclear power plant.

Low-level waste is disposed of in special landfills, but in effect it is very much like normal garbage. For a little perspective, the U.S. Environmental Protection Agency lists the U.S. 1999 municipal waste, that is everyday garbage, at 230 million tons. Since paper and yard waste make up about half this volume, we might guess that on average municipal waste is about as dense as water; that is, it would almost, but not quite, float. So that gives us a density of 1 gram per cubic centimeter or 1.1 tons per cubic meter. That means that 230 million tons of garbage is something like 250 million cubic meters. Which means low-level nuclear waste has a volume of about 0.005% of municipal waste. Or, if you like fractions, it is 1/20,000 the volume. If a mental picture helps, this is something like the difference in terms of water displacement between a modern aircraft carrier and a moderate sized (perhaps 20 to 30 foot) pleasure craft.

---

[14]The U.S. Department of Energy's Office of Environmental Management lists the low-level radioactive waste received from utilities at the nation's commercial waste disposal sites as 10,768 cubic meters for the year 1995. Divided among 104 reactors that works out to an average of 103 cubic meters each. This amount varies by reactor and by year for each reactor. Also with waste disposal costs rising sharply concerted efforts are underway to reduce the volumes sent for disposal.

Intermediate-level waste is stronger stuff. The largest source of this will be spent resins from the water purification systems. Light water reactors—in particular—pump a lot of water through a lot of pipes. Even with corrosion resistant steels and very careful water chemistry control, there will be some corrosion. The grains of corrosion products, little pieces of rust, will end up going through the reactor. When they do they will see a lot of free neutrons. This gives these materials the opportunity to absorb some of those neutrons and get transmuted into radioactive isotopes. Examples might include iron-55, cobalt-60, nickel-59, nickel-63, and niobium-94. These guys are pretty darn radioactive. Because the more times the corrosion products circulate through the reactor the more of them will get transmuted to such radioactive isotopes, the owner of a reactor will want to get them out of his system. Leaving them in makes it bad for the plant staff doing maintenance and increases the potential problems if you leak any of the water.

So the reactors are designed with water purification systems. The heart of these are the ion exchange beds. These capture the corrosion products in their resin material. After a while the resin is spent and needs to be replaced. The spent resin also has a load of radioactive corrosion products. The spent resin beds have to be handled more carefully than the low-level waste, with some shielding required for handling, transportation, and eventual disposal. However, a few feet of dirt is all that is needed for the final resting spot. Intermediate waste will carry perhaps as much as 4 times the radioactivity as low-level waste but have a volume less than 10% of the low-level waste. Many listings combine the low- and intermediate-level wastes. The DOE-EM volume numbers cited previously for low-level waste include intermediate-level wastes.

Now the type of nuclear waste that folks like to make a big deal of is high-level waste. This comes in two types: military waste and civilian waste.

Military wastes come from the operation of the military production reactors used to make plutonium for the nuclear weapons programs, the operation of naval reactors, and the operation of the nuclear weapons manufacturing facilities. Since both the Soviet Union and we made a lot of plutonium and floated a lot of submarines, both the Russian Federation and we have a lot of military waste. This comes in generally three forms. One is the fission product waste from the production and naval reactors. Another is a limited amount of spent fuel from those reactors that has not been chemically processed. And the last group is equipment and facilities contaminated in both the weapons and naval reactor operations.

The last group is somewhat like the civilian low-level waste but has special problems in being contaminated with plutonium, which is more toxic than the average commercial contaminated waste. Some is also of higher levels of contamination than you would find in commercial facilities.

The spent fuel is just what was left from the military production reactors that didn't get processed before the major government facilities that used to do that got shut down. The vast majority of the spent fuel did get processed and that is where the fission product waste comes from. The production reactor waste, which is the larger volume, is stored at Hanford, Washington, and Savannah River, South Carolina.

Both Russia and the U.S. are working toward solutions to handle this waste, but the challenges are large. However, *and this is important to grasp*, this has nothing to do with civilian nuclear waste. The forms, volumes, nature, and responsibilities are completely different. So, if someone starts telling you about the huge volumes of military high-level waste as part of the "problem" with civilian nuclear power, you can tell them they are off-base. The "problem" of military high-level waste[15] will exist and be dealt with independent of what happens in the civilian nuclear power world.

O.K., so let's focus on civilian high-level waste. What is it? In a throwaway fuel cycle like we have in the U.S., it is the spent nuclear fuel. Remember that, in a large modern nuclear power plant, we have about 100 metric tons of nuclear fuel. We run the reactor for about 18 months before we refuel. At that time we replace about a third of the fuel with new fuel. That means the new fuel coming in will see three cycles or up to 4.5 years before it comes out.

Now just what is the spent nuclear fuel? Recall that when we loaded the fuel it was uranium dioxide pellets stacked inside zircaloy alloy tubes. The uranium was 3 to 5% uranium-235 and 95 to 97% uranium-238. While the fuel was in the reactor for 4-plus years, we fissioned over half the uranium-235. We also converted a little bit of the uranium-238 to plutonium-239. We used some of the plutonium-239 as fuel and fissioned that portion along with the uranium-235. At the end we have a composition that is about 95% uranium-238, 1% uranium-235, 1% plutonium, and 3% fission products. We also have absorbed some neutrons in the heavy elements, which did not lead to a fission event. This means we get some new heavy elements, such as americium and curium, in addition to several plutonium isotopes. Collectively these are called actinides. There are not a lot of them, but they do represent a special problem for high-level waste. You will note a lot of "abouts" in that description. These are rough numbers, and the specific composition will depend on the exact nature of the fuel loaded and the specific operational history of any given reactor. But for typical large light water reactors, this is pretty close.

If you look at that composition, what should strike you is that it is very much like the stuff we started with over 4 years back. I mean, most of it was uranium-238 before and that is what most of it is after we are done. Only 5%,

[15]You will sometimes see military waste referred to as defense wastes. The terms are synonymous.

one-twentieth, or so of the fuel undergoes any change at all. However, what has changed is that we have consumed half or more of the good stuff, the uranium-235. Making some plutonium-239 along the way has been a help, but when the amount of good stuff gets too low, the fuel is "spent" and we can't make the reactor sustain fission any longer. In fact, if you tried to load a reactor with only the fuel that has already seen two cycles, it would not go. You need the one cycle and new fuel to "pep up" the older fuel and make the whole reactor run.

The other change we have made, of course, is we have made fission products equal to about 3% the original fuel mass. If we were using a recycling fuel cycle, like France, for example, we would take our spent fuel to a chemical reprocessing plant to separate the unused uranium, useful plutonium, and waste fission products. In that cycle this 3% of the original mass would be the high-level waste stream, the rest being used to make new fuel. In the U.S. throwaway fuel cycle, since we don't separate the fission products from the spent nuclear fuel, we have to treat the entire volume of spent fuel as high-level waste. To be fair, if we had separated the fission products, we would have dispersed them in some media such as glass, so the final volume would not necessarily be that much different than the spent fuel still in the form of the original assemblies.

In either separated high-level waste or spent nuclear fuel, the radioactive part is the fission products, right? Well, almost. Remember we also get some actinides. We will get back to fission products in just a bit, but let's look at the actinides first.

Along with fissioning uranium, we have also captured neutrons in uranium to make heavier isotopes. Remember, that is how we got the plutonium. Through neutron captures in higher isotopes, decay of those isotopes, and captures in the resulting isotopes, we get quite a mix of heavier isotopes. The actinide group has isotopes that range from some with half-lives that are quite short, and therefore very radioactive, to those with very long half-lives, which then are only slightly radioactive. Table 6.3 gives you a sample of some of these.

TABLE 6.3 Sample of actinides.

Isotope	Half-Life
Uranium-236	23 million years
Plutonium-239	24,000 years
Plutonium-240	6540 years
Plutonium-241	14.7 years
Plutonium-242	376,000 years
Americium-240	51 hours
Americium-241	432 years
Americium-242	152 years
Americium-243	7380 years
Curium-241	33 days
Curium-242	163 days
Curium-243	28.5 years
Curium-244	18.1 years
Curium-245	8500 years

Now back to fission products, which is the major radioactive part of high-level commercial nuclear waste. Recall that these are the isotopes that have resulted from the fission of the uranium and plutonium, which was the source of all the energy we got out of the fuel. Fissions don't always result in a uranium or plutonium atom breaking into exactly the same two parts. Uranium-235 has more than 40 possible pathways leading to over 80 primary fission product pairs. Back in Chapter 4 we learned that as we move toward heavier elements, the ratio of neutrons to protons increases. If we break a heavy element in to two parts, it is clear the result will have more neutrons than the "new" lighter elements need. The fission product isotopes are neutron rich relative to the normal stable forms of the elements. Nature solves this through radioactive decay converting neutrons into protons giving off an electron in the form of a beta particle. Excess energy is drained off through the release of gamma rays as well.

So the primary fission products are decaying and forming new isotopes as they do that. This increases the types of isotopes in the fission product mix. In addition, because a lot of this happens inside the operating reactor, many of these fission product isotopes pick up a free flying neutron and get transmuted into a different isotope yet. The result is that fission products are a pretty complicated group of radioactive isotopes. Just like the actinides some are very radioactive with short half-lives and some are only slightly radioactive with very long half-lives.

This leads us to the public relations problem of high-level radioactive waste. Folks like to tell you that radioactive waste stays radioactive for tens of thousands, even hundreds of thousands of years. How can we possibly seek to contain this awful hazard for such enormous periods of time?

If I have conveyed anything to you, I hope you have captured the idea that if the half-life is long the radioactivity is small. So if it lasts a long time, it is because it isn't very radioactive. It is a complex mix; however, the single most important point is that the radioactive hazard decreases continuously. In just 40 years it is about 1/1000 of what it started at.

Let's just look at a few of the components of high-level waste to get a feel for this. Remember that there are a lot of isotopes and this is just a small sample, but it should give you a feel for the general types of isotopes that make up the mix.

Let's first look at some short-lived isotopes. Two that draw attention are iodine-131 and krypton-85. These have half-lives of 8 days and 4.5 hours, respectively. In the Chornobyl accident in which the nuclear fuel was destroyed and the containment was breached, these guys were a problem. Their short half-lives make them very radioactive. Iodine is bad because it is mobile in the atmosphere and will move through the food chain and collect in the thyroid gland of humans. Krypton is a noble gas, which means it will not chemically combine with anything and, therefore, flows free in the air once released. The good thing about both of these is that they go away pretty fast. Iodine-131 is down to 1% of the original level in 54 days after the reactor shuts down. For krypton-85, you get to a 1% level in less than 30 hours. When people are talking about highly radioactive fission products, these two guys qualify. However, if you are talking about long-term storage of radioactive waste, these isotopes, and other similar short-lived isotopes, aren't in the picture. They have decayed away.

So what else do we have? Let's look at a couple of intermediate-lived isotopes. Cesium-137 and strontium-90 are good examples. Cesium-137, which has a half-life of 30.2 years, is a reactive chemical and can be mobile if you let it get loose. It also decays with a reasonably strong gamma ray, so it takes some shielding to contain that radiation. Strontium-90 has a half-life of 29 years. It decays with a beta particle, so you can shield its radiation with just thin materials. However, if humans ingest strontium, it seeks out bones and replaces calcium, somewhat like fluorine does in teeth. What would happen then is the strontium would be inside a person's bones and the beta particles would be doing damage to the bone marrow. With half-lives in this range neither of these are as radioactive as the short-lived isotopes we just talked about. In fact, if you had the same number of iodine-131 atoms and cesium-137 atoms, the cesium-137 would be decaying, giving off radiation, at a rate almost 1400 times less than that of the iodine-131. The same comparison between krypton-85 and strontium-90 shows the latter's rate over 57,000 times less.

But, with those longer half-lives, it does mean both of these guys are going to be around for a while. To get down to a 1% of what you started with will take about 200 years for cesium-137 and 193 years for strontium-90. That is quite awhile. However, humans have built structures that have lasted for thousands of years. Building something to contain these isotopes for a few hundred years ought to be a snap.

Well, what about the things that last a long time? What about those terrible things in the waste that are radioactive for tens of thousands of years? Two that trouble people are iodine-129 and technetium-99. These bother people because they have half-lives long enough that you can bet any structure we build to hold them won't last as long as they do. Iodine-129 has a half-life of 15.9 million years, and technetium-99 has a half-life of 213,000 years. Those are seriously long times. But again, if the half-lives are this long, the radioactivity is small. Using the same example of having the same number of atoms of two isotopes, iodine-129 is 526,000 times less radioactive than cesium-137 and is 722 million times less radioactive than iodine-131. That last number is essentially one billionth. Technetium-99 is only 7000 times less radioactive than cesium-137 and 9.7 million times less radioactive than iodine-131.

The other thing about these very long-lived isotopes is there isn't very much of them to start with. We will get into total waste volumes in a minute, but it is worthwhile here to capture some sense of just how small the long-lived fraction really is. While it will vary a bit with the particular reactor and how you run it, for each metric ton of spent fuel you will get about 890 grams of technetium-99 and 190 grams of iodine-129. As we will see in a bit, if all the reactors in the United States get their licenses extended and run a full 60 years, we will get about 105,000 metric tons of spent fuel. That means 20% of all the electric power for the entire United States for 60 years would give you 93 metric tons of technetium-99 and 20 metric tons of iodine-129. Isn't that a lot?

The density of technetium metal is 11.5 grams per cubic centimeter, and solid iodine has a density of 4.9 gram per cubic centimeter. That means the 93 tons of technetium would fit in a cube just under 2 meters, about 6.5 feet on a side. The iodine-129 would be a bit smaller at 1.6 meters or 5.2 feet on a side. That is it. I have seen walk-in closets that would be big enough for both. And remember, this is from all the waste from all the reactors in the United States assuming they all run for 60 full years.

But how radioactive would this pile of iodine-129 and technetium-99 be? The radiation would be so intense that the thermal heat coming off the combined accumulation would be about 800 watts. That is half a hair dyer. This is the "fission products radioactive for tens of thousands of years" problem, half a hair dryer of energy and a walk-in closet worth of space.

Let's just summarize a bit with Table 6.4. This recaptures the six example isotopes. Shown is their half-life and how long it would take to decay away to 1% of what you started with. Also shown is the relative radioactivity. This is

TABLE 6.4 Example fission product isotopes.

Group	Isotope	Half-Life	Time to Decay to 1% Original	Radioactivity Relative to Iodine-131
Short-Lived	Iodine-131	8 days	54 days	1
	Krypton-85	4.5 hours	30 hours	43 times more
Intermediate-Lived	Cesium-137	30.2 years	200 years	1400 times less
	Strontium-90	29 years	193 years	1300 times less
Long-Lived	Iodine-129	15.9 million years	106 million years	722 million times less
	Technetium-99	213,000 years	1.4 million years	9.7 million times less

the number of decays per second for a given number of atoms relative to the decays from the same number of iodine-131 atoms.

So, if the long half-lives bother you, look at the last column. Yes, a half-life of nearly 16 million years for iodine-129 is a long time, but a factor of nearly a billion less radioactive than iodine-131 means it is pretty dull stuff.

O.K., what about those terrible actinides, what about plutonium? Well, let's talk about plutonium a bit. You may have heard it said, plutonium is the most toxic substance on earth. That is simply not true. As we have seen demonstrated to our collective dismay, there are disease spores that in even very tiny amounts are very deadly. Plutonium is a very toxic material but it can't get close to the toxicity of the highly virulent disease spores.

But what about the statement that even one atom can cause cancer? Well, one spore could cause the disease but it is generally recognized that it takes a lot more than one. If a single plutonium atom could cause a cancer, we would all be dead. The atmospheric testing of nuclear weapons in the 1950s and 1960s released enough plutonium atoms into the atmosphere for there to be trillions and trillions of atoms for each of us. The concerns that made nations stop atmospheric testing had a lot more to do with iodine-131 than it did plutonium.

Plutonium is a toxic heavy metal as is uranium or lead. All of its isotopes are also radioactive. As we saw in Table 6.3 they range from a 14.7-year half-life for plutonium-241 to 376,000 years for plutonium-242. Because each plutonium isotope comes from the preceding one and non-fission nuclear capture isn't the only thing a plutonium isotope might do with a neutron, the higher the isotope the less of it there is going to be. You can think of it as the higher the isotope, the tougher it is to make. So there will be less of it, at least at the start. Clearly, after a fairly short time the plutonium-241 is going to be a lot less than the plutonium-242 because it is going away much faster.

Plutonium-241 is going to be like cesium-137 in that it is fairly radioactive, but in a hundred years, it is essentially gone. Plutonium-242 is like iodine-129 in that it is going to last for a very long time, but it is not very radioactive at all and there is very little of it. Plutonium-239 and -240 fall in-between. With half-lives of 24,000 and 6540 years, respectively, they are going to last quite awhile but are only weakly radioactive. Both decay with strong alpha particles and weak gamma rays. That means as long as you keep them outside of you they will be of little risk to you. However, if you ingest them, it would not be good. Although weakly radioactive, the alpha particles they do emit do a lot of damage inside of a person. And there is the fact that all plutonium regardless of the isotope is just plain chemically poisonous.

It is worthwhile just to mention that, if we in the United States were to use a recycling nuclear fuel cycle instead of the once-through throwaway cycle, most of the actinides would be recycled. Much of the recycled actinides would fission into isotopes with shorter half-lives and go away faster, reducing the long-term radioactivity of the resulting waste.

But let's back up just a step, and try to capture a little perspective. First of all, recognize that a lot of the environmental hazards that technology creates are actually the collection of **natural** hazards and concentrating them. All the arsenic on the earth has been here for 4 billion years. It is only a significant problem to humans if, in the operation of a mine and the processing of ores, we get a waste stream that is concentrated in arsenic or, of course, unless a nice old lady in lace decides it is just what we need in our tea. But how long does arsenic last? What is its half-life? FOREVER. It isn't radioactive, it doesn't decay, and the hazard remains for all time. That is a thing a lot of folks miss about radioactive waste. Eventually it goes away, and the most hazardous stuff goes away the fastest. Perhaps the time it takes for the slower decaying stuff to go away is longer than either you, I, or any of our great-grandchildren care about, but it does go away while other chemical poisons don't.

Now looking from the big picture view, what have we done in using nuclear fission to produce energy? If we left the uranium in the ground, it would decay over millions of years on its way through a long chain of intermediate radioactive materials to a stable isotope of lead. During all that time it and its daughter isotopes would be radioactive. In fact the radon from that chain is the largest single source of natural radiation we get.

What we have done in using nuclear energy is we have pulled a tiny fraction of that uranium out of the ground and converted it to fission products and actinides. These are all going to decay much faster than the original uranium. Therefore, they are going to be more radioactive than the original uranium. But remember, if it is more radioactive that means it is decaying faster and, therefore, going away faster. Figure 6.1 shows the relative toxicity of commercial high-level waste as a function of time as compared to the original uranium ore. It is given as both a total and what it would be if the actinides were removed.[16]

What this is saying is that in about 500 years, the waste is no more than 10 times as toxic as the original uranium ore we started with. If it doesn't look that way to you on Figure 6.1, you need to recognize that the time is plotted against a logarithmic scale (that is, powers of 10). Therefore, 500 is not halfway between 100 and 1000, rather it is a bit more than two-thirds the way toward 1000. At 2000 years, it is less than 4 times more toxic. That is, if we leave the actinides in. If we were to reprocess and reuse the fuel, thereby putting the actinides back in the reactor to get fissioned or transmuted to shorter-lived materials we could reduce the long-term hazard greatly. Without the

[16]*Transmutation of Nuclear Waste Using Thermal and Fast Neutron Energy Spectra*, Rodigues, Baxter, McEachern, & McGill; General Atomics, Los Alamos, New Mexico.

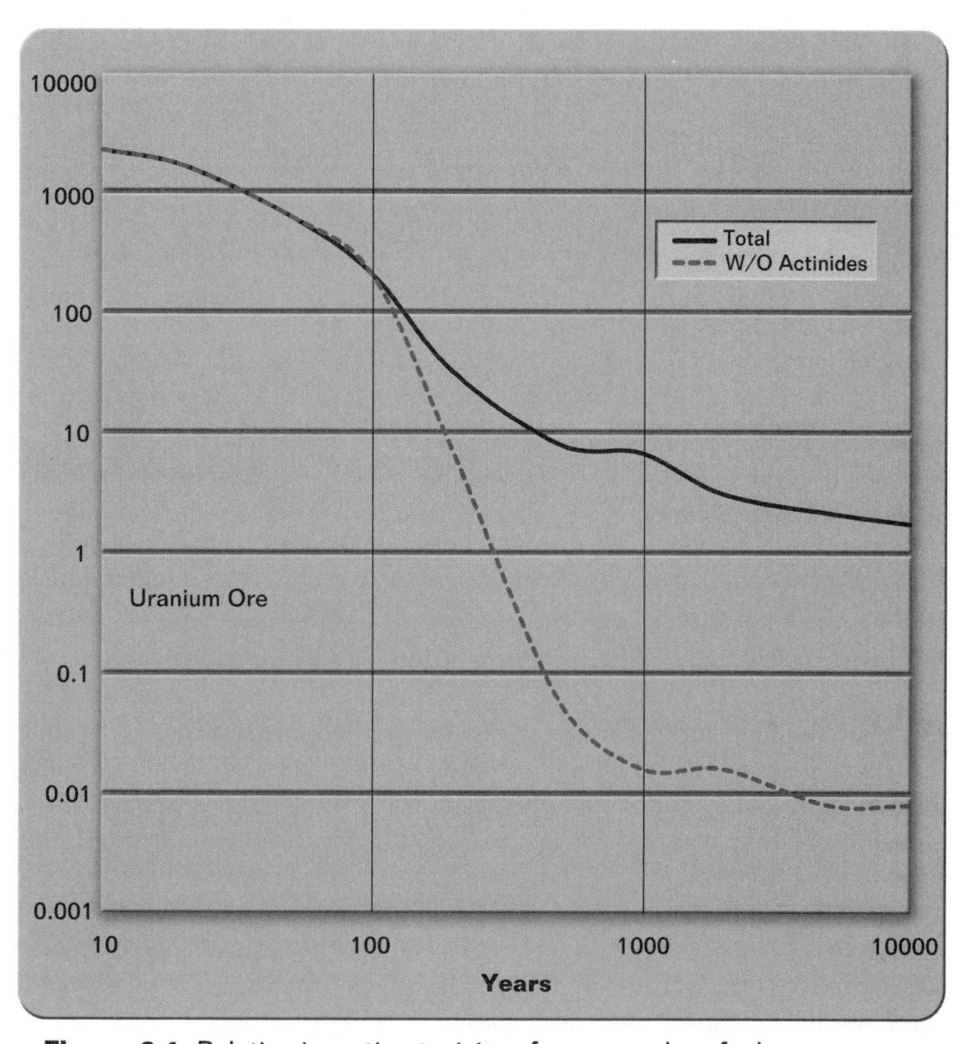

Figure 6.1. Relative ingestion toxicity of spent nuclear fuel.

actinides, the waste would fall below the toxicity of the original ore in about 300 years.

So that is what high-level waste is. Now, let's move on to the next question. How much spent nuclear fuel do we have?

The official number for the U.S. in 1998 was 38,500 metric tons. Since we add about 2300 metric tons a year, by the end of 2002, the total would be just under 48,000 metric tons. If all the currently operating U.S. power reactors ran to the end of their 40-year licenses; we would accumulate about 87,000 metric tons. Many plants are seeking 20-year extensions to their operating licenses. If all plants got such extensions, we would get up to 105,000 metric tons. Is that a lot? Well, that depends on how you look at it and what you compare it to.

To gain a sense of scale for what it means to collect all the waste for a nation into one pile, consider the following. Let's just assume that each of us uses one Kleenex® tissue a day. Now, if you add that up for 281 million Americans, you get 137,730 metric tons of spent Kleenex® tissues per year.[17] That is not 2300 added tons per year, but 137,730 tons *each* year. In 60 years it would be over 8 million tons, assuming the population held constant. This is not to compare the hazard of spent Kleenex® tissue with spent nuclear fuel, but rather just to make the point that 100,000 metric tons coming from 281 million people over a 60-year span is not very much.

Well, but how much is it? Let's look at it another way. An individual PWR fuel assembly is about 9.5 inches on a side and weighs about 500 kilograms. If we treat all the U.S. spent fuel as if it was PWR fuel of this type,[18] we would see we have the equivalent of 210,000 assemblies if all the reactors ran out to their 60-year life. Each assembly is 9.5 inches on a side or 90 square inches or about 0.63 square feet. The total area, if you put all the assemblies together, would be 132,000 square feet, or a square 363 feet on a side. To get a feel for that, this is about a football field plus both end zones long and the same wide. That is all the spent nuclear fuel if we run all the existing reactors for a full 60 years.

Now you really could not put them all that close together. What you might do is put them in dry spent fuel casks. I will explain more later, but these are big vessels designed to hold spent nuclear fuel. You can get about 32 assemblies in each one, and you put them on a concrete pad on about 10-foot center-to-center spacings. If you arrayed all 210,000 assemblies in that fashion, you would need 6563 casks, each needing 100 square feet for a total of 656,300 square feet. That is a square 810 feet on a side. That is two and a third football fields, plus end zones long and wide. That is it! All the space you would need for all the spent nuclear fuel if you ran all the U.S. reactors for a full 60 years each, producing enough electricity for about 50 million people during those 60 years. That space is for over twice as much spent nuclear fuel as we have now. Two and a third football fields long and wide. Surely there is that much space available someplace in all of America. If you build a whole bunch of new reactors and got twice as much spent fuel you would need a bit over 1.3 million square feet. That means you would have to add 340 feet to the length and width of your storage zone. The United States is something like 15,000,000 feet from Atlantic to Pacific coasts. Surely we can spare 1150 feet in there

[17] I worked this out by weighing a full and empty Kleenex® box (286 grams and 51 grams, respectively), taking the difference (235 grams), dividing by the number of sheets (175), then multiplying this per sheet weight (1.343 grams) by 365 days per year and 281 million people in the country.

[18] This is not a bad assumption because while BWR bundles are smaller in dimension, they weigh less so the math would work out very close to the same, even if we got the PWR to BWR fractions exactly right.

someplace. I wish I could give you a U.S. map and show the size of the space needed for this storage in perspective. However, even if printing technology could make the dot that small, you couldn't see it.

Again we need some comparisons. Recall we saw the EPA gives the mass of municipal waste as 230 million tons per year. We estimated that was about 250 million cubic meters. The space we just set aside for all the spent nuclear fuel coming out of all the nuclear power plants for all of their lifetimes was 656,300 square feet. The casks will be something like 18 feet tall so that is makes 11.8 million cubic feet or 335,000 cubic meters. So the space we need for all the spent nuclear fuel coming from 60 years is equal to 0.13% or 1/750 of the volume of the municipal garbage we generate every year.

Let's look at it yet another way. When you make electric power using coal, you get two forms of waste. One is the combustion gases, and the other is ash. A 1000 MW electric coal plant will make about 300,000 metric tons of ash a year.[19] We have the equivalent of over 200 coal plants of this size producing about half the electricity used in the U.S. Therefore, we are making something like 60,000,000 metric tons of ash each year. At an average density of 55 pounds per cubic foot that comes out to be 2,400,000,000 cubic feet or 70 million cubic meters. So the space we would need for all the spent nuclear fuel over the entire 60-year lifetime of the existing reactors, including the spaces between the casks, is 0.5%, 1/200, of the volume we need for find for the coal ash we make **every year**.

Like a few other comparisons? I read an interesting article in our local paper about a new high-tech dairy operation. When fully up to speed, this farm will have 50,000 cattle. The interesting numbers are that this stock will produce 152,000 gallons of milk a day and 675 tons of waste. My almanac tells me that there were 98 million cattle in the U.S. in the year 2000. If 50,000 cattle produce 675 tons of waste a day, then 98 million cattle produce 1.3 million tons a day, which is a bit over 480 million tons a year. Assuming a density equal to water, that is 15 billion cubic feet or 430 million cubic meters per year. One more interesting tidbit from the article was that 50,000 cows produce as much waste as 250,000 people. Follow the math on that and you get 277 million tons of human waste in the U.S. each year, which using our as-dense-as-water estimate works out to 8.6 billion cubic feet or 245 million cubic meters.

Table 6.5 attempts to capture these comparisons. The point is that compared to other forms of waste, which we as a nation of over 280 million people create, high-level nuclear waste is vanishingly small, factors of a thousand, tens of thousands, and even hundreds of thousands of times less.

---

[19]I have seen larger numbers cited, but to allow the kindest comparison to coal, we will use this lower value here.

TABLE 6.5 Selected U.S. annual waste generation comparisons.

Source	Annual Waste	Units	Tons/ Year	Ratio Waste/ Nuclear Waste
Nuclear Power—high-level waste	2300	Metric tons	2530	1
Kleenex®*	138,000	Metric tons	152,000	60
Coal Power Plants—ash*	60 million	Metric tons	66 million	26,000
Municipal Waste	230 million	Tons	230 million	91,000
Human Waste*	277 million	Tons	277 million	109,000
Cattle Waste*	480 million	Tons	480 million	190,000

*See text for bases of estimate.

Well, O.K., the radioactivity goes away and gets a lot less right off the start, and, in terms of volume as compared to other forms of wastes, there is not very much of it, but what are you going to do with it? Isn't even this small amount impossible to contain?

Actually, it is not hard to deal with at all. While there are a number of things one could do to securely isolate the hazards of high-level nuclear waste from people and the environment, there are three ways that will see practical use. These represent stages of storage ending with truly long-term disposal.

Remember that we are focusing on the once-through, throwaway cycle of a typical U.S. light water reactor. Let's look at how fuel becomes spent fuel and then follow it from there.

As we have seen a typical fuel assembly in a light water reactor will remain in the reactor for about 4.5 years. After that time the fissile material, the uranium-235 and plutonium-239, has gotten low enough that the fuel isn't providing the fission source needed to sustain the chain reaction. The fuel is "spent." So every 18 months or so the reactor shuts down for a refueling outage. The control rods are put in to stop the fission reaction, and the system is cooled down. Once it is cool the pressure in the primary coolant loop is reduced and the top or head of the primary vessel, the big steel tank holding the fuel, is taken off. This allows the fuel to be accessed. The "spent" assemblies are removed. This is about one-third of the total number of assemblies. New fuel assemblies are put in to replace the ones removed and the first and second cycle fuel assemblies, those in the reactor for 18 and 36 months, will be rearranged.

The spent fuel will be transferred to a spent fuel pool. This is essentially a big and very deep swimming pool. The pools are typically 40 feet deep, which is enough that you can lift one assembly over another and still have plenty of water to shield the workers from the radiation coming from the spent fuel.

When the fuel is first discharged, it is still producing quite a bit of heat but much less than when it was in the operating reactor. If you go back to Table 5.1, you see that one month after shutdown one-third of the core we removed from a reactor producing 1100 MW of electrical power was down to 1.9 MW of decay heat. Now, if we go out to the spent fuel pool where this fuel is stored, we will see we have it placed in racks inside this 40-foot-deep swimming pool. If we assume a little bit of space between the assemblies for the rack, you get something like 60 feet on a side to store that one-third core. Because the water over the fuel is about 26 feet (40–14) you have a 60 by 60 by 26 foot volume of water over that fuel, which at one month is producing 1.9 MW. So what? Well, engineers live for this kind of stuff,[20] but if you do the math, 1.9 MW is 6.5 million British thermal units per hour and 60 by 60 by 26 feet of water weighs about 5.8 million pounds. Because a British thermal unit is the amount of heat that it takes to heat up one pound of water 1°F, all this math tells you that the one-month old fuel is heating up the water in the pool around 1.1°F per hour. Now, we didn't take into account the water within the fuel or in the rack spaces or the water associated with older lower energy producing fuel. So the real number would be less that one degree per hour.

The point is that it is not difficult to remove that heat. Going back to Table 5.1, we see that at one year the amount of heat has dropped off another factor of 3. So while we need the water in the spent fuel pool for shielding and cooling, the cooling chore is pretty easy. In fact, safety studies have shown that if you lose all the water in a spent fuel pool and the fuel is at least 3 years out of the reactor, nothing happens. That is, the fuel doesn't fail. If you had fuel with less decay time and you lost the water in the pool, then it would be a good thing to get water back. However, you have some time to do that because the fuel will heat up slowly because the energy production is so low even in relatively new spent fuel.

You could leave spent fuel in spent fuel pools for quite a long time. The water chemistry is carefully controlled; hence, there is little corrosion of the fuel cladding. Removal of the heat, as we have seen, isn't particularly tough, so the spent fuel is safe, secure, and sound.

So why don't we just let it stay there? Well, the original idea was the we were going to reprocess spent fuel, that is, dissolve it in strong acid and

---

[20]One of the joys of engineering is to be able to apply a little bit of physics and math to understand what is going on. In this case we can take the energy generation rate and turn that into how quickly that much energy heats the water which in turn gives us a feel for how small the energy generation rate is in this configuration.

remove the fission products to be disposed of as waste and recycle the other 97% of the volume into new fuel. The reactors and their spent fuel pools were designed with this in mind. The purpose of the spent fuel pools was then to hold the fuel for a few years to wait for the heat and radioactivity to go down to simplify handling. Therefore, the spent fuel pools were designed to hold only a few years worth of spent fuel.

Now as we will discuss more in a bit, reprocessing didn't happen in the U.S. and the spent fuel started backing up in spent fuel pools. Fortunately the pools were over designed, and we were able to rearrange the storage racks and put a lot more fuel in them than was originally planned. However, even with that rearrangement, many pools have been filled up to the limit. But even if you stopped running the reactors and making more spent fuel, one has to ask if spent fuel pools are the best way to store spent fuel for a long time.

After the spent fuel is a few years old, you really don't need the water cooling and shielding of a spent fuel pool to safely store spent fuel. The expense of running and maintaining the pumps and water treatment systems is really not justified for fuel over a few years old. In addition, while the corrosion in the carefully controlled water is low, if you don't need water, why take any corrosion at all?

The answer is dry cask storage. This is the next thing you can do with spent fuel. We mentioned that previously when discussing the space you might need to store the nation's commercial nuclear waste. So what is dry cask storage?

The idea is that after a few years the heat and radiation of spent fuel is sufficiently low that you could wrap spent fuel in some steel and/or concrete, stand it up outside, and let the natural convection from the air cool it. That would mean no more pumps, water treatment, and the maintenance thereof.

There are a number of different dry casks in use. However, in general, the design concept is pretty similar. On the inside, you have a rack for holding the spent fuel assemblies. The capacity varies from 20 to 40 PWR assemblies depending on the manufacturer. The rack is contained in an airtight steel vessel or tank. When the fuel is loaded, the air in this tank will be pumped out and replaced with an inert gas like helium. This will essentially eliminate any further corrosion of the fuel.

Outside the steel containment vessel will come the radiation shielding. This could be reinforced concrete, ductile iron, steel, or a steel and lead combination. This shielding portion could be from 60 to 100 centimeters (24 to 40 inches) thick. Outside the shield will be an outer container probably of steel. The entire cask will be something like 250 centimeters (100 inches) in diameter and 5.5 meters (18 feet) in height.

The shielding will reduce the radiation to a very low level. A report prepared for Greenpeace International gives the dose at the surface of a dry cask as 200 to 800 micro-Grays/hr.[21] Because most of this dose will be gamma rays, we can convert this to rems and get 20 to 80 mrem/hr. That is right up against the cask. If you stand back just a bit, the dose falls off quickly. In studies done for Wisconsin Power, it was estimated that the dose rate at the public access point would be about 1600 times less than at 3 feet from the surface. The difference from a point on the surface of a dry cask and at the public access point would be greater yet. What that all means is that if you stood as close as possible to an array of dry spent fuel casks all year long, you could just about add as much dose as you were already getting from natural background radiation. But then again, you would be outdoors and not as exposed to radon gas trapped indoors, so perhaps it would balance out.

Normal airflow around the casks would be sufficient to cool the spent fuel. The dry cask surface would only be a few degrees above ambient temperature. As you might imagine from the description of the construction, these casks are very tough. Tornado-driven missiles (telephone poles and cars) can't hurt them and earthquakes don't bother them.

The design life of dry casks is on the order of 100 years. However, independent assessments have suggested that dry casks could safely hold spent nuclear fuel for 1000, perhaps as long as 5000 years. More on this later. The point is dry casks are very strong, very robust, shield the radiation very well, can handle the heat loads in a totally passive fashion (that is, you need not run any machinery to make them work), and can handle anything nature can dish out.

A number of power reactor sites have started to use dry cask storage. So far the storage has been located at the power reactor site. All you need is a good firm concrete pad. You fill the dry cask, seal it up, and park it out on the pad. There are proposals to build one or more national sites where spent fuel from many reactors could be stored more efficiently under a single control and management operation. None of these ideas have yet gotten to the construction stage.

Well, O.K., dry casks are pretty neat, but even 5000 years isn't forever. What about the really long term? There have been a number of long-term solutions suggested. Frankly some of them are pretty far out, literally. Those include shooting nuclear waste into the sun or burying it on the moon. However, what essentially every country has decided on is what is called geological disposal, which is science talk for burying it deep underground.

Folks have been thinking about geological disposal for at least 40 years. The basic idea is pretty simple. You wrap your high-level nuclear waste up in some suitable robust container[22] that protects the workers during handling, ship

---

[21]*Dry Storage of Spent Nuclear Fuel: The Safer Alternative to Reprocessing*, Dr. Ian Fairlie, May 2000.

[22]Some dry cask units are designed to be the container for the geologic disposal. These "multi-use" casks

these containers to a site where you have carved out tunnels far underground, load the containers into the tunnels, and fill up the hole. This gives you a robust package around the waste that itself will function as a barrier for hundreds to thousands of years and a final resting place that is far removed from human access and communication to the environment. If 40 inches of steel or concrete is sufficient shielding for spent fuel a few years old, then how good is hundreds or thousands of feet of dirt and rocks for spent fuel even older and less radioactive? Well, pretty obviously, it is a whole lot more than you need.

The current U.S. plan for geologic disposal would place our high-level nuclear waste in an underground vault at Yucca Mountain, Nevada. We will discuss that a bit more shortly; however, this would be a good place to give you a sense of just how safe the disposal at Yucca Mountain can be. Recognizing that you can get a range of potential estimates of radiation exposure to the public depending on the extremes of good or bad luck you want to assume, we can look for a best estimate value, that is to say, most likely to occur. *Discover* Magazine covered the topic in their September 2002 issue in an article entitled "Welcome to Yucca Mountain." They cited a discussion with Patrick Rowe, a Department of Energy engineer, as follows:

> "Rowe looked over toward Lathrop Wells. The Environmental Protection Agency's Reasonably Maximally Exposed Individuals were presumed to live there, now and for the next 10,000 years. 'Today eight people reside at the compliance point,' Rowe remarked. 'If everything goes well and we have the natural failure, the peak mean dose will be 0.00004 millirems per year.' That amount of radiation is less than negligible."

While simple in concept, the idea of geologic disposal does have a set of tasking details that have engaged many in years of study. The problem, of course, is since we are talking about tens to hundreds of thousands of years of "containment," just what does it take to "be sure?" A lot of effort has gone into choosing the right type of geological formation in which to build the repository. Granite, basalt, tuff, and salt formations all were studied and each has their special advantages. The truth of it is any would work fine. What you are really looking for is a place that is geologically stable, that is not subject to a lot of earthquakes or volcanism that might spit the waste back out of the ground, and a region that is fairly dry so that groundwater won't leach the waste out of the site to show up some miles away in drinking or irrigation water.

Because there are a lot of dry places on the planet and we can find geologic formations that have been stable for as long as a billion years, in actuality the location of a scientifically suitable spot is not super difficult.

The challenge for geologic disposal is an extreme example of the challenge for all of nuclear energy. That is, how good is good enough? Now I can look at corrosion rates in stainless steel, leaching rates for uranium dioxide, and geologic information regarding signs of seismic activity and feel pretty good about telling you what is going to happen to nuclear waste buried at a given site for the next 1000 or even 10,000 years. However, you might ask me if I chose another site, or spaced the waste canisters differently, or if I built the canister out of a different material, would that reduce the potential leakage from the site? The answer will always be, well I might be able to make it a little better if I were to . . . And, of course, as we go to the extreme times of the very long-lived waste isotopes, the uncertainties of the behavior of the repository increase, so the "what if" games get harder to do and the satisfaction of the "could you make it better" question becomes more difficult.

In real life we make decisions every day about how good is good enough. I am not just talking about how thoroughly you wash out your favorite coffee cup before you use it again. It is also things like how much life insurance do I need, do I need to have the brakes on the car checked before we go on vacation, or do we need to get that old oak tree thinned out before the next wind storm? Perhaps more truly frightening, if we were to really think about it, are decisions like can I pass that slow-moving truck around this curve hoping that there are no cars coming the other way, or can I take this call on my ringing cell phone and still get through the rush hour traffic without having an accident? Such decisions are potentially life threatening. However, we continuously make them, trading a few minutes of time behind the truck against the probability of injury or death.

I am not suggesting that the decisions on how good to make high-level waste disposal should be taken lightly or that we should be as caviler as some of our "on the highway" decisions often are. What I am suggesting is that, like those everyday decisions, we need a balance between perfect safety and the cost and physical impositions of providing that perfect safety. The safest car of all is the one that is always parked in your garage. Probably if you never take your car out of the garage and just go out and sit in it there, you are going to beat the odds of 1 in 7000 that you will die on the road next year. But you won't do that. For all the convenience and real requirements of having and using a car in our modern society, you are going to accept that risk, drive as smart as you can, and try not to be one of the 40,000 people who are going to die on American highways every year.

If you look at just about every idea ever offered for geologic disposal you would see that they put nuclear waste far from people and the environment and the odds of any of it coming back to hurt anyone in any significant way is really remote. We have spent a lot of money on the topic and advanced the art little in comparison to the money spent. It is just not all that hard. We really need to stop asking, could you make it a little better, and start asking, isn't that good enough? Can't we just use that and get on with it?

I also need to say a few words about transportation of high-level waste. This is a topic that some folks like to get very excited about. What one needs to understand is that, to provide the radiation shielding required, the casks used to transport nuclear waste have thick steel walls that are very, very strong. The Nuclear Regulatory Commission has a set of tests that all nuclear waste casks must pass that include puncture, fire, and water immersion. There have been a series of very dramatic tests in which trucks carrying casks were slammed into immovable concrete structures at freeway speeds and t-boned by railroad locomotives. The toughest SUV out there would have been creamed in any of these tests, but the nuclear waste cask survived without any leakage.

There have been a number of high-level waste shipments moving spent nuclear fuel between reactor sites and to shared spent fuel storage sites. The loss of life so far as been exactly zero.

But if we start shipping all that stored spent fuel to a geologic repository won't that be a lot more shipments? Well, sure, but let's look at that. As we saw there would be 105,000 metric tons of spent fuel if we ran all the current U.S. reactors for a full 60 years. At 500 kilograms per assembly that comes to 210,000 assemblies. If we only get 24 assemblies in a shipping cask, that would be 8750 casks. If we guess and say that each cask has to travel 2000 miles to get to the repository that is 17.5 million cask shipping miles. Isn't that a lot? Well, no, it isn't. In 1995 there were 6.9 million freight trucks traveling 200 billion miles. The entire truck transport required to move all the U.S. spent fuel to the repository would be less than 1/10,000 of the annual truck traffic. The hazard this minor addition to the overall truck traffic would be very small. The casks are very strong, the radioactive material inside would be very well protected, and the overall addition to transportation risk would be slight. Again, if the benefit is electric power for a fifth of the country for 60 years, isn't this slight increase in truck traffic an acceptable risk?

O.K., if we know how to store spent nuclear fuel, first in water-cooled pools, then dry casks, and eventually in geologic repositories, why is the problem still a problem? Why haven't we just quietly solved the problem? Well, therein lays a tale.

Way back in the good old days (1950s–1960s), when nuclear power was just getting started, the idea was that you would leave the spent fuel in the spent fuel pool for a few years, and then, when the short-lived radioactive isotopes were gone, you would ship it to a chemical reprocessing plant. That plant would separate the 3% of the spent fuel that is fission products and is truly waste from the other 97% which is uranium-238 (about 95%) and fissile uranium-235 and plutonium (together the remaining 2%). The waste would be put in a stable form and sent to a safe place for disposal, and the other 97% of the fuel would be recycled into new fuel. At this time enriching uranium was very, very expensive and reprocessing was not thought to be very hard.

We were routinely reprocessing spent fuel from our military production reactors and converting the technology to a civilian fuel cycle seemed pretty straightforward.

Several things happened along the way. One small commercial reprocessing plant was completed at West Valley, New York, and operated for a time. A large commercial reprocessing plant was being built at Barnwell, South Carolina, and almost finished while a second large plant was being built at Morris, Illinois. This was about 1974. Enter politics. The idea that, in a reprocessing fuel cycle, you would separate plutonium from the waste and put it into new fuel became increasingly troubling to a lot of people. The concern was that because plutonium was the primary ingredient in atomic weapons, it would not be a good idea to ever separate the plutonium in commercial waste from the highly radioactive fission products. As long as the plutonium was with the fission products, it was protected from any misuse by anyone less sophisticated than a national nuclear weapons program. That is, you could not get to it unless you had a reprocessing plant and the shielded facilities needed to handle very radioactive material.

This became a rallying point in national political debates. Presidential candidate Jimmy Carter was very strongly against a nuclear fuel cycle that recycled plutonium. In perhaps an attempt to steal some of that thunder, President Gerry Ford imposed a ban on commercial reprocessing shortly before he lost the election to Carter.

Now clearly the government had imposed a serious problem for the commercial nuclear power business who had a clear path to "solve" the issue of high-level nuclear waste. In closing the reprocessing route, the government came back with a series of studies on a "once-through" fuel cycle where, instead of securing the fission products in some very resilient medium like glass, they would dispose of the waste in the form of the entire spent fuel assemblies as they came out of the reactors. While having greater volume and mass, this had the charm of bypassing the whole reprocessing business, even if it did mean we didn't get to use the 97% of the mass that was still good.

At this point in the nation's history, inflation was high, and the cost of reprocessing had been increasing. The cost of enrichment, while high, was still subsidized by the huge enrichment capacity built by the government for the nuclear weapons program. With the promise that the government would make the once-through cycle work for them, the nuclear power business accepted their new fate and went along with the new government plan.

We do need to stress that this was a United States decision only. The rallying cry was that, if we were to forego plutonium recycling, the other nations with nuclear power systems would follow our lead and, therefore, the international flow of plutonium would never come to be. What happened was, in general, anyone with the ability to conduct reprocessing either did or is still trying to develop it. The problem wasn't that they weren't concerned about potential misuse of plutonium. It was that they believed they could adequately mitigate that threat, and they saw very important benefits to a plutonium recycle system. The major benefit being you get to use some of the value of the 97% of the fuel we throw away. Countries like France and Japan with limited internal energy resources were particularly focused on getting all the value out of the nuclear fuel they could. The U.S. is sitting on a 400-year supply of coal, so resource efficiency was not as big a factor to us. We also need to note that in 1974, greenhouse gases and global warming were not yet big issues, so all that coal looked pretty good.

France has led Europe in reprocessing and routinely uses recycled plutonium in its power reactors and provides reprocessing services for other countries. Japan has contracted for reprocessing in France and Great Britain. They are also attempting, with some difficulty, to develop their own capability in reprocessing and plutonium-based fuels. All this has been done without any loss of plutonium to malefactors. The Japanese have stated that in their view, a plutonium recycle system is the best way to protect plutonium because the very safest place for plutonium is back in the reactor. In fairness, we have to point out that, while that is exactly true, the time the plutonium spends between when it is separated from the fission products until it is put into new fuel and that new fuel gets in the reactor again does cause people some worry. A strong system of international safeguards administered by the International Atomic Energy Agency has so far provided an effective global policing of commercial plutonium production and use.

But back in the United States in the 1970s and 1980s, our government did what governments often do, which is to try very hard not to do anything wrong. The best way to do that, of course, is, while making a large show of effort, to actually do very little. This may be a harsh and unfair portrayal of the earnest efforts of a lot of well-meaning and talented people; however, the productive result of their efforts was very limited at best.

After years of lack of progress toward the development of a credible once-through system, that is, the development of acceptable long-term disposal of high-level nuclear waste in the form of spent fuel assemblies, a frustrated nuclear power industry increased its pressure on the government. The result was the Nuclear Waste Policy Act of 1982. In that act, the government promised to take from the nuclear power plant operators their spent fuel by December 1998. This required the reactor operators to find storage for their spent fuel on site up to that time but gave them the limit of storage required. It also gave the government 16 years to get its act together. It also gave the government a source of funding to do their work, charging the nuclear plants 0.1¢ per kilowatt-hour produced. By the end of 1999 that came up to $16 billion collected. That $16 billion has been added to the price of nuclear-generated electricity.

Well, here we are some years after 1998, and how much spent fuel has the government taken from the commercial nuclear plants? Except for tiny fractions for studies, essentially none. They aren't ready yet. The 16 years wasn't enough.

We should explain here that a number of ideas were looked at, such as disposing of nuclear waste on the polar ice caps or in the deep ocean, but as already noted the method generally accepted worldwide is geologic disposal, that is to mine a cavity deep (a thousand or more feet) in the earth, put the nuclear waste held in thick containers into the hole, and seal it up. The questions are what type of geological formation to use and how well will that work. Clearly you want a geological formation that is stable over a long period, has limited communication (such as groundwater flows) with the surface, and is in an area of low population to further reduce the potential for the waste ever getting back to many people.

During the years the government worked on this problem, they went through an exhaustive study program that compared a series of potential geological formations to store the spent fuel. While the advocates of each type could find problems with others, the fact is any would have probably worked. However, the problem was never a scientific one but rather a political one.

Basically, it comes down to who gets stuck with the waste. No elected official worth his salt wants to be the individual who brought nuclear waste to their town, county, or state. This is what is called the NIMBY (not in my back yard) problem. I want the power, or the freeway access, or the jobs from the industry, but don't put the power plant, freeway, factory, or nuclear waste site in my back yard. This then for the politician leads to a NOMW (not on my watch) problem. If something bad is going to happen, make sure it is not on my watch; don't let it be my fault.

After working on all these possible geological forms, the decision was made that Yucca Mountain in Nevada would be a good place. There are a lot of good reasons why that actually made sense: low rainfall, stable geology, and so on. However, the fact that Nevada doesn't have a lot of people, or voters, and that we have already placed massive radioactive contamination under the ground in Nevada by setting off a large number of nuclear weapons there was suspected to have a significant influence on that decision.

So for years the government has studied the Yucca Mountain site drilling huge tunnels at great cost. At the same time the Environmental Protection Agency and the Nuclear Regulatory Commission engaged in a protracted debate on the criteria for the safe storage of spent fuel. This, of course, made it challenging for those in the Department of Energy trying to design the repository. You might compare it to going to a General Motors design team, telling them you need a vehicle on the road in 3 years and then refusing to tell them right up to the last minute if it was to be a 2, 4, or 40 passenger vehicle.

The hang up between the EPA and the NRC is over the allowable public dose from the repository. The NRC thinks that 25 mrem from all sources is adequate. Note that this is about 8% of natural background radiation for the U.S. and 400 times less than the demonstrable level of radiation harm. The EPA wants a 15 mrem standard for all sources; however, they also want a groundwater source standard that is based on 4 mrem from that pathway. The NRC points out that, for Yucca Mountain, the groundwater would be the predominate pathway, so the EPA standard is actually more like 4 mrem than it is 15 mrem, and there is no need for a separate groundwater standard anyway. There is also a good deal of controversy over just how the EPA gets from the various limits on various isotopes in the groundwater to the 4 mrem value. The EPA says they are working on that but stand by their limits all the same.

While the controversy continues, the National Academy of Sciences has said that the EPA groundwater standard is "scientifically unsupported" and "adding little or no public health benefit." In fairness it must be reported that the NAS found no fault with the EPA's 15 mrem all source limit.

We need to remind ourselves of the Department of Energy's best estimate peak public radiation dose of 0.00004 millirem predicted from Yucca Mountain.

I counted the number of zeroes in that citation twice. That is 1/100,000 of EPA's most restrictive value, less than 7 millionths of natural background, and 250 millionths of the level of demonstrable harm. Could it just be we are working a little too hard to make that even better?

As previously pointed out, the frustrating thing here is that you can always make something a little safer. You can as long as you don't care how much it costs or how long it will take to build it. It is a constant problem with every aspect of nuclear energy. How safe is safe enough? For nuclear energy, the answer has been as safe as you can make it. For automobiles, it has been: well, we'll kill about 40,000 people a year, yeah, that sounds about right. You could make autos twice as safe by simply limiting the amount of gasoline produced to half as much and, therefore, cutting the number of miles we drive in half. This would seriously impact our way of life, but you could.

As another example, we kill up to 10,000 U.S. citizens a year with poisons. If we spent $16 billion (the amount nuclear power plant operators have given to the government so far to fix the high-level nuclear waste problem), what kind of an education, packaging, inspection program could we put together to bring that number of deaths down? We might also recognize that a lot of those 40,000 highway deaths were due to drunk drivers, and use the $16 billion to work a little harder on that topic.

So, the current best estimate is that the Yucca Mountain repository will accept spent fuel sometime in the next 10 years. In the meantime some of the reactors are running out of space in their spent fuel pools. What they have done is to turn to dry cask storage at the reactor sites.

When the Bush Administration took office, they called for a National Academy of Sciences study on the status of spent nuclear fuel. The questions were how much of a problem was this and do we need to do something about it soon? The answers of the study were oddly dissatisfying to both anti-nuclear and pro-nuclear people. The anti-nuclear people would have liked the answer to be that spent nuclear fuel represented a terrible hazard. The pro-nuclear people would have liked the study to say that it was time to act and finally get on with the whole thing.

What the study said was that basically you have two options. Store the waste as you are at the surface or go to geologic disposal. Of these two, only geologic disposal was a true long-term solution. They looked at other long-term disposal methods and concluded that essentially all users of nuclear energy worldwide were working toward geologic disposal and they agreed that this was the most credible way to go.[23] The question then isn't really do you store at the surface or deep underground, but rather the question is when do you move from the surface to deep underground. The answer was that, while some countries were running out of room in their current surface storage facilities, there really wasn't much hurry to go underground for the long term. The surface storage was adequately safe and would remain so for a long time. They referred to a DOE environmental impact statement that looked at the option of just abandoning the spent nuclear fuel dry cask storage at the reactor sites. The conclusion was that the casks would survive from 1100 to 5400 years fully containing the fuel for that period. Clearly that is an endorsement of the idea that we need not rush into a long-term solution.

While those commercial nuclear power plants that are running out of storage in existing facilities are in a jam, the solution of building dry cask storage is being used now in several countries to relieve this pressure.

The National Academy of Sciences report said that the major issues surrounding nuclear waste were societal not technical. Furthermore, since there was no need to rush to a long-term solution, they suggested that all the nuclear energy using nations ought to work together to address the societal and political issues, as well as the technical issues to get the long-term solution right.

Here are a few sound bites out of the report.

> "Today's growing inventory of HLW [high-level waste] requires attention by national decision makers. . . . Although quantities are minor compared with toxic wastes from other industrial activities, the

[23]Geological disposal is being planned for both once-through cycles that dispose of all of the spent fuel as waste and for systems that separate the fission products and recycle the other 97% of the spent fuel. In the latter case the fission products are mixed with a very robust material such as glass and placed in thick canisters before they are buried in the geological repository.

> inventories, particularly of spent fuel, are increasing in many countries beyond the capacity that can be stored in existing facilities."

> "Practically, geological disposal does not represent a major construction challenge. All of the techniques required to build a repository, encapsulate the waste in a series of containers and barriers, emplace the waste, and close the repository are established or could be developed from established practices, and the associated cost is entirely compatible with the economics of energy production. . . . Our present civilization designs, builds, and lives with technological facilities of much greater complexity and higher hazard potential."

> ". . . the committee found that moving from surface storage to geological disposition is a societal, rather than a technical, decision for each country and this decision need not be rushed."

> "SNF storage, primarily at reactors, has not been seriously questioned on safety grounds."

So where does that leave us on high-level nuclear waste?

It remains a small problem with feasible solutions but with a lot of highly emotional and political charged "stuff" wrapped all around it. If one could peel off all the wrappings, you would find a pretty small issue inside. The National Academy of Sciences is correct in their assessment that the societal issues have not been adequately worked. What I hope we can achieve is to treat the public with the proper respect for their native intelligence. We can only do this by taking the wrappings off this problem and letting them see the true nature of the problem and its true size in some relevant context. If we can do that, I think that you smart people out there would realize we are not put at much risk from nuclear waste and we need neither stop nuclear energy to protect ourselves from nuclear waste nor spend vast sums of money on it, money better spent on other concerns.

So, there you are. You now know what nuclear waste is. You have some feel for how much, or little, of it there is and will be. You know something about the radioactivity of this compared to the ore it came from. You also have been introduced to the unhappy recent history of the efforts to "solve" this problem. Hopefully, you also understand something about how robust the current surface storage is and how good the eventual geologic repository can be.

The question left is one only you can answer. Did I convince you? Do you believe nuclear waste is a huge problem without solution? Or do you think it is a relatively small problem in the larger scope of things and something quite solvable, if we would just stop fussing with it and do it?

7

ENERGY, THE BIG PICTURE AND OUR CHOICES

I need to tell you a few things to help you get the best out of this chapter. First, why include a chapter on the "big picture" of energy in a book advocating nuclear energy? Well, the overall purpose of this book is to give you perspective, to help you see nuclear energy in context. Therefore, it is very appropriate to give you a general overview of energy, so you can see how nuclear energy fits within that. It is also important to be sure we all understand, wide-eyed advocates and extreme detractors alike, that nuclear energy is not **THE ANSWER**. It is part of the answer, but in our technologically rich modern world, the energy demands are too great for any one source to provide solely the answer to all our energy needs.

The other thing I have to warn you about is that, within this chapter, I am going to say some bad things about some of the energy sources available to us. It is very important you recognize that this is not a "technology is bad" message. This is a "roses have thorns, and you have to handle them carefully" message. Air pollution from fossil energy use kills people, but without that fossil energy use millions, perhaps billions, of people would starve and surely billions would be inflicted with a crushing degradation in their quality of life. However, we can't see the true context of energy use unless we look at the warts on energy use as well as the blessings. After all, we spent a whole chapter looking at the "bad things" about nuclear energy. In looking at the big picture of energy use, we need to address some of the "bad things" about other energy sources if we are to have a balanced perspective. However, I do ask you in advance to remember that even with some bad effects, the net advantages of the energy that drives our technological civilization are significant. Without that energy, despite its warts, we would be a much smaller population with a much reduced quality of life. Back to nature may sound romantic, but there is little romantic about being hungry, cold, sick, and seeing more than half the children die by age 6.

In our attempt to provide some context for energy use, I want to take you through four topical areas. First, we will look at the really big picture, that is, the use of energy from an international perspective. It is particularly important to understand the differences between the developed industrialized nations and the developing nations. Next, we will focus on the United States and take a more detailed look at energy in this country. Following that, because fossil fuels are the predominate form of energy being used today and will be for the near future, it is important that we spend some time looking at how we use fossil fuels, what the reserves are like and where they are, and what are the costs of both using fossil fuels now and depending on them in the future. Finally, for context we will study renewable and alternative energy sources. While there are many desirable features of these energy sources, it is important to understand the limitations and even the warts that go with these. I believe those who claim alternative and renewable energy source are going to soon save the planet are misguided and unfortunately such unrealistic promises divert us from facing the real tough challenges before us.

The Global Energy Picture

Probably, the single most important thing for anyone in a modern developed industrialized country to understand about how energy is used globally is to learn that many people on the planet do not have the same access to energy as we do. Well, you knew that, but what I fear many of us fail to recognize is the extreme nature of the difference. Table 7.1 gives you some raw data to work with. Let's look at this together and try to get a sense of what these numbers mean in the lives of real people.

You may want to spend a little time just looking at these numbers, but I want to point out a few things in particular. Figure 7.1 shows the 1999 per capita use of energy by region.

Now probably you knew that someone in an industrialized country had access to more energy than someone in a developing country in Asia or Africa. But did you realize that the difference was 10 times more than the average in Asia and over 14 times as much as the average in Africa?

What does the difference in the availability of energy mean to people in these populations? In Chapter 2, we talked about the availability of energy

TABLE 7.1 Energy use by region.[1]

Region	Energy Use in 1999 (Quad)[2]	Energy Use in 2020 (Quad)	1999 Population (millions)	2020 Population (millions)	1999 Per Capita (Q/billion)	2020 Per Capita (Q/billion)
Industrialized	209.6	270.4	942	1030	223	263
Eastern Europe and Former Soviet Union	50.5	72.3	413	414	122	175
Developing (subtotal)	121.8	264.4	4628	6088	26	43
Asia	70.9	162.2	3212	4015	22	40
Middle East	19.3	37.2	239	350	81	106
Africa	11.8	20.8	767	1187	15	18
Central and South America	19.8	44.1	410	536	48	2
Total	**381.9**	**607.1**	**5983**	**7532**	**64**	**81**

[1]"International Energy Outlook—2001," March 2001, U.S. Department of Energy—Energy Information Administration, DOE/EIA-0484 (2001).

[2]In talking about really big amounts of energy, you will often see the use of this unit. A Quad is a quadrillion BTUs. A quadrillion is 10^{15}, or a million billion, seriously a big number. A BTU, as we mentioned earlier, is a British Thermal Unit or the amount of energy it takes to heat up one pound of water 1°F. One Quad is about 293 million megawatt hours, if that helps you at all. If that was all electrical energy and used at the rate we use electricity in the United States, one Quad would be about the electrical energy used by 24 million people in a year.

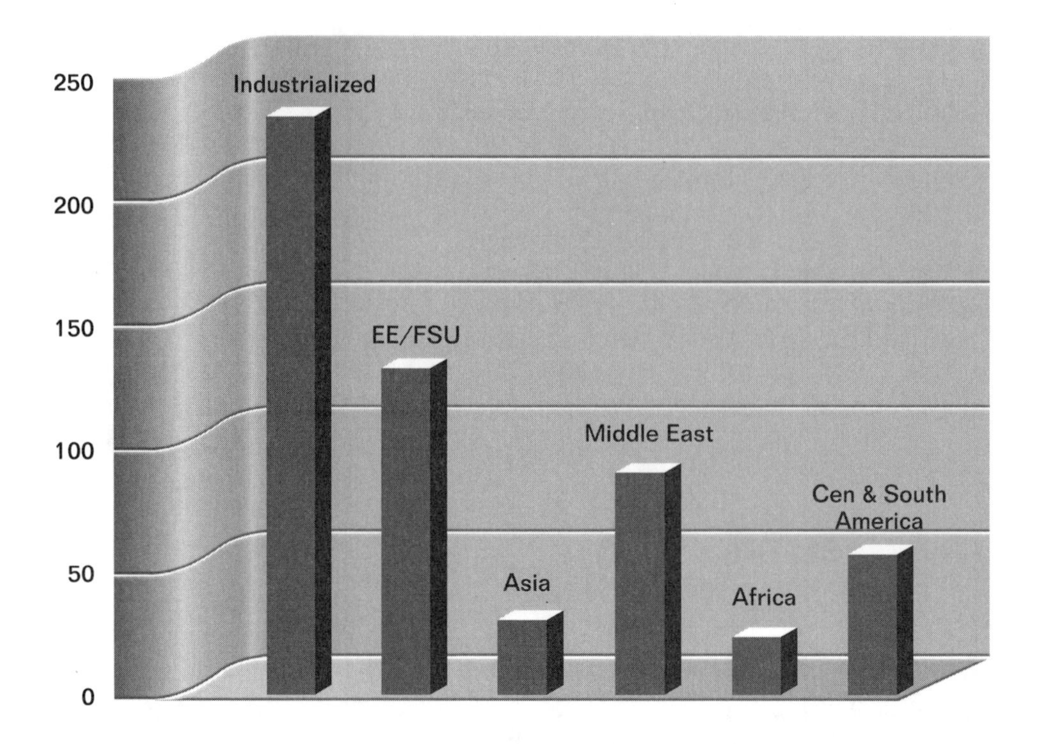

Figure 7.1. Per capita energy use by region (1999)—Quads per billion persons.

being the equivalent of servants to do work for us. This is a lot more that just a machine to wash the dishes for you or to avoid all that tedious toweling to dry your hair. This is running factories that produce goods at inexpensive prices that you can afford. It is powering a transportation system that moves your goods to distant markets and brings goods from afar to you. It is providing civic services like clean water, sanitation, and health care. It is refrigeration for food and aseptic preparation of meals. It means you have an economy that allows the majority of the population to do something other than manual agricultural labor.

There are 2 billion of the 6 billion people on the planet who do not have access to electrical power. There are places where people spend several hours a day searching for twigs and sticks to cook their food.

We use a lot of energy, but if we are to bring all of the earth's people up to the standards of an industrialized nation's quality of life, we will need a lot more energy.

That takes us to the other thing to point out in Table 7.1. That data includes the Energy Information Administration's predictions on energy use in 2020.

Figure 7.2 shows you the relative increases in per capita energy use that are predicted for each region. It would seem pretty heartening to note that the developing countries are projected to see much greater growth than the

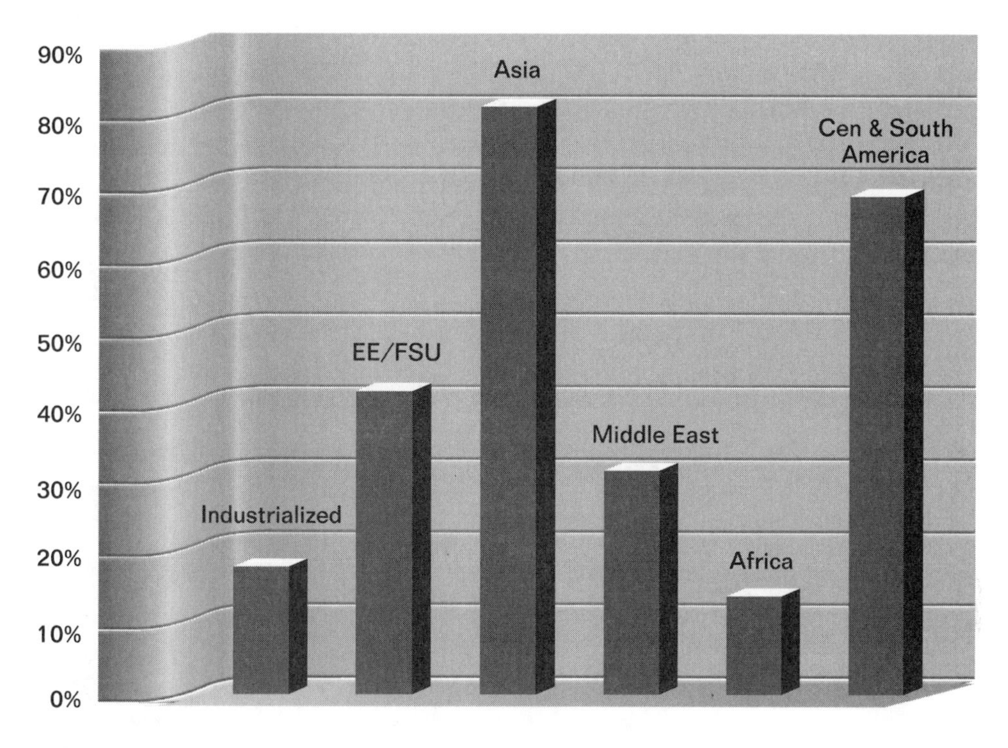

Figure 7.2. Increase in per capita energy use 1999 to 2020.

developed countries in per capita energy use. That is, the rich do get richer, but the poor move toward rich faster than the rich get richer. However, it isn't all good news. When you take into account that the industrialized countries had a large base to start with and, therefore, a small percent of that represents a larger fraction of the developing countries smaller base, it doesn't come out quite as pretty. Even with the significantly larger fractional growth in developing countries, the gap between haves and have-nots will still be there. The gap between industrialized countries and developing Asia is predicted to close from a factor of 10 down to 6.5, Central and South America closes from 4.6 times to 3.2. However, the gap between industrialized countries and Africa is projected to increase to a factor of 15 from the factor of 14 now.

While the per person energy numbers are important to understand the human impact of the disparity of energy use, we have to also look at the impact of population growth to get an appreciation for the total amount of increase in energy production required to meet just these modest projected enhancements to the 2020 quality of life in developing nations. This means we have to take into account the projected population changes in the regions as well as the energy use per person. Figure 7.3 shows the total projected increase in energy use by region.

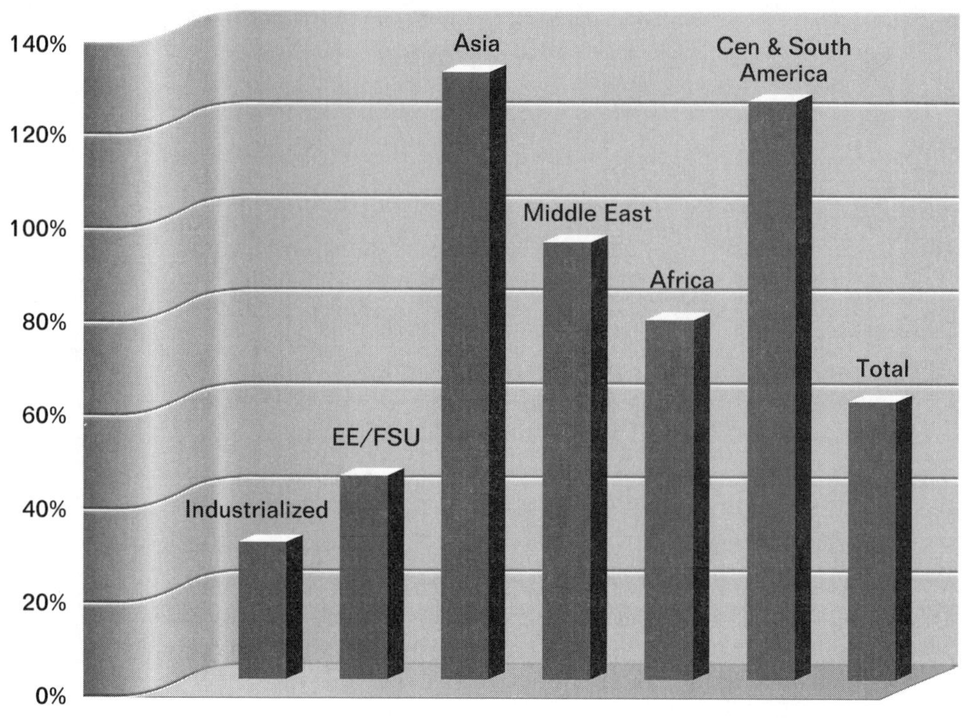

Figure 7.3. Projected total energy use increase by region 1999 to 2020.

What this shows is that we expect, in the next 20 years, very large increases in the amount of energy used in the developing world. This includes more than a doubling of energy used in Asia and Central and South America. Overall, this is saying humans are going to be using 60% more energy in 2020 than we did in 1999.

So where do we get that energy now?

When I started collecting data to answer that question, I thought it would be pretty easy to do. I soon learned it was not as easy as I had expected. Table 7.2 shows three different estimates of world energy use by fuel type. I have converted the original information all to Quads (Quadrillion British Thermal Units), so you can compare them. Please have a quick look, and then we can talk about the numbers a bit to help first understand the differences in the estimates, then what the numbers should be telling us.

In general, the three estimates pretty much agree with each other. The EIA estimate for oil seems a bit high compared to the other two, but other than that they are all in the same ballpark for oil, natural gas, coal, and nuclear energy. At that point, the estimates diverge. The EIA lumps everything else, that is, hydroelectric, renewables and whatever else is left, into "other." The International Energy Agency estimate breaks out hydroelectric as a separate

TABLE 7.2 2001 estimates of world energy use by fuel type—in Quads.[3]

Fuel Type	Energy Information Administration	International Energy Agency	British Petroleum
Oil	152.2	135.8	138.8
Natural Gas	86.9	80.3	82.6
Coal	84.8	91.2	86.4
Nuclear Reactors	25.3	26.4	26.1
Other*	32.7		
Hydroelectric Dams		8.9	9.1
Combustible Renewables & Waste		43.1	
Other**		1.9	
Total	381.9	387.7	342.9

*Hydro and all renewables
**Geothermal, solar, wind, heat, etc.

fuel type as does the British Petroleum estimate, and each provides some credibility in agreeing with each other. The IEA estimate lists a group of combustible renewables and waste as well as an "other" category, which is explicitly identified as geothermal, solar, wind, heat, and etc., while the BP estimate does not address any "other" group.

The question of combustible renewables is a tough one. This is basically burning non-fossil fuel combustibles. It is wood chips in forest industry, agricultural waste, wood in your fireplace, and animal dung in Afghanistan and Chad. That is a really hard category to get a good handle on. The EIA report takes a cut at the big pieces of its "other" category but states,

> "Although noncommercial fuels from plant and animal sources are an important source of energy, particularly in the developing world, comprehensive data on the use of noncommercial fuels are not available and, as a result, cannot be included in the projections."

We can get some sense that the amount of energy from the noncommercial fuels is somewhere in the difference between the 32.7 Quads of "other"

[3]EIA—"International Energy Outlook—2001," US DOE/EIA-0484(2001), March 2001. IEA—"Key World Energy Statistics from the IEA 2001 Edition," International Energy Agency. BP—"BP Statistical Review of World Energy June 2001," British Petroleum.

from the EIA estimate and the 53.9 Quads from the IEA sum of hydro, combustibles, and their "other."

This is also supported by the difference in the world total energy between the EIA, IEA, and BP estimates. The EIA and IEA are quite close, but the BP is significantly less. However, the BP leaves out all noncommercial fuels and "others" coming up almost exactly the 45 Quads the IEA estimated as combustibles and others.

The bottom line is that the IEA column seems like the most credible of the three.

So, now let's look at that a bit. The most important thing to see is that we are very much dependent on the fossil fuel trio of oil, natural gas, and coal. We will focus on the issues of fossil fuels more later, but recognize that 80% or so of our energy comes from fossil fuels. That means our technological society, which you will recall is needed to support any population greater than about one billion people, is very sensitive to the price and continuous supply of fossil fuels. To put it much more personally, if Safeway could not buy diesel fuel or had to pay 10 times as much for it, you would see food riots at your local grocery store.

The other thing I would like to direct your attention to is the combustible renewables. While we can agree with EIA that getting good worldwide information on this is very tough, we should recognize that it is a real component in world energy use. The significance is that much of this is non-sustainable and indicative of extreme poverty. There have been extensive portions of Africa where the search for wood and overgrazing by sheep, goats, and cattle have denuded the landscape. This is not something you can expand. The 2 billion people without electric power are not cooking over open fires because it is romantic. They do it because that is all they have. If we hope to provide these people with enhanced heat, shelter, refrigeration for food, and means to prepare meals that don't involve hours a day scrounging for twigs or dung, we are going to have to replace the fuel source with something else.

We will address non-combustible renewables and alternative energy later also, but just for now, note that our best guess from IEA cites these as 2.8% of the global energy, 0.49% if you take out hydro.

Let's now turn our attention to the United States. Here we will have better information and can be somewhat more definitive on where the energy comes from and how we use it.

Energy Use in the United States

The first thing to say about energy use in the United States is that with less than 5% of the world's population, we use a tad over 25% of the human-generated energy. That says a lot about the quality of life we have relative to the rest of the planet's population. It also says something about how dependent that quality of life is on energy.

Let's look at where the energy we use comes from. Figure 7.4[4] shows the sources of the energy we use.

As we saw for the total world energy use, the majority of our energy comes from the big three fossil energy sources: petroleum, natural gas, and coal. One of the more disturbing things about use of fossil fuels in the United States is that we are dependent on supplies outside our borders. Our single largest energy source is petroleum. Just over half the petroleum we use, about 56%, we import from other nations. About 16% of the natural gas we use is imported. The good news is that essentially all of that comes from our good neighbor Canada. We do not import coal and in fact actually export about 6% of what we dig up. We will examine fossil fuels more in the next section.

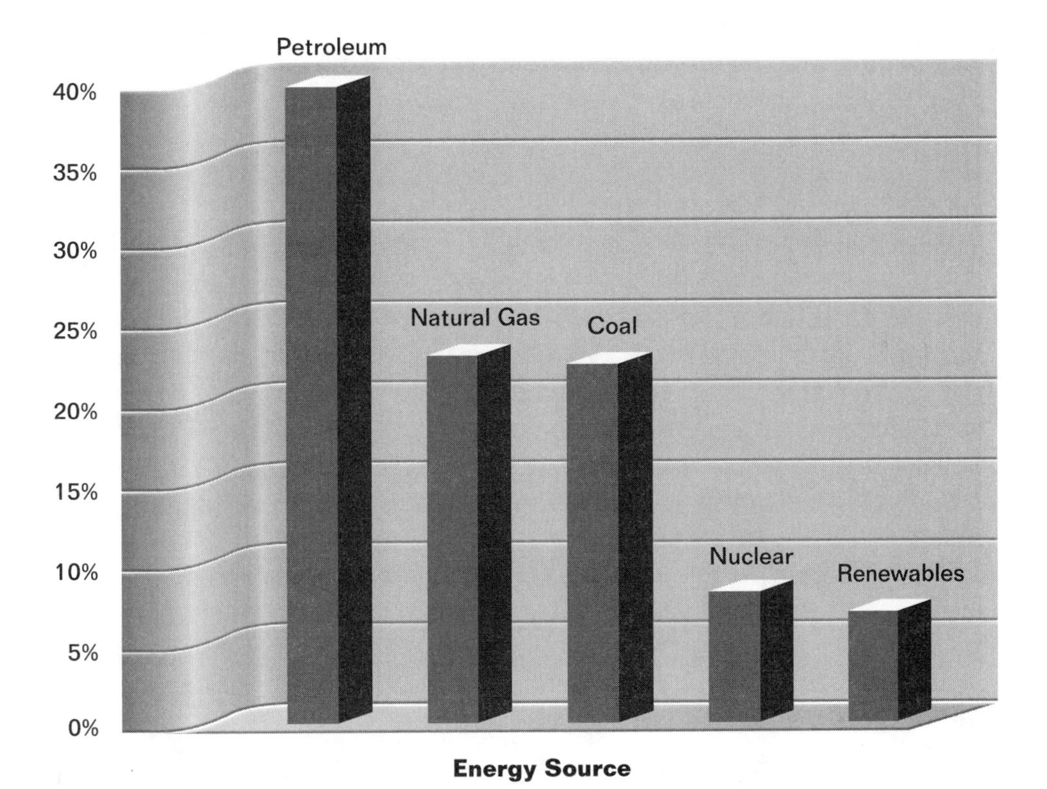

Figure 7.4. 1999 U.S. energy use by source.

[4]"Annual Energy Outlook 2001," DOE/EIA-0383(2001). Department of Energy—Energy Information Administration, December 2000.

Nuclear energy is the next largest source of energy used in the United States. This is followed by renewables. Much of that is hydroelectrical power and most of the rest is biomass, that is burning wood, waste, municipal solid waste, and landfill gas. We will look at renewables after we look at fossil fuels.

O.K., that is where we get energy from. What do we use it for? Figure 7.5[5] gives you that picture.

Here we see that industry is the largest user of energy. It is energy that drives the productivity of American industry. We have augmented the productive capability of people with machines and the energy that runs them. That means fewer people can make more stuff at lower costs, so more people can afford that stuff. Every time productivity is improved at a cost lower than the labor it replaces it makes "things" more affordable for consumers. That means you get more or better "things." Some would say that is excessive consumption. However, few I think would want to trade the comfort, ease, convenience, and pure enjoyment of a modern American home for that of 1950. I, for one, could not imagine trying to produce this book using a manual

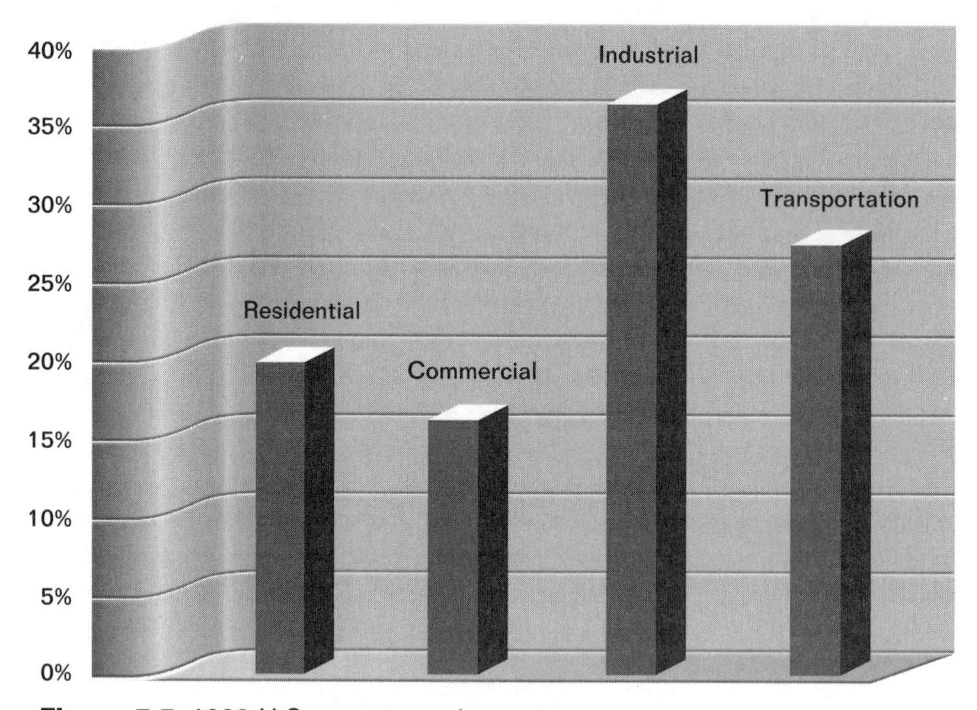

Figure 7.5. 1999 U.S. energy use by sector.

[5]"Annual Energy Outlook 2001," DOE/EIA-0383(2001). Department of Energy—Energy Information Administration, December 2000.

typewriter. The delete key is a super improvement over white-out. Do you remember whiteout?

It is instructive to see how much each of these sectors depends on electrical power to energize them. Table 7.3 gives you those percentages.

TABLE 7.3 Electrical percent of total energy by sector.

Sector	Electrical Percent of Total*
Residential	65%
Commercial	75%
Industrial	33%
Transportation	1%

*Includes electrical losses.

You need to appreciate that this is the total energy required to generate and transmit the electrical power to its end users, divided by the total energy consumed by that sector. If you recall, Mr. Carnot explained that in any process converting heat energy into mechanical energy, we could only get a fraction of the heat energy converted. There are significant losses. Only about a third of the heat energy ends up as electricity at the end use. However, we don't measure petroleum by the end energy in the motion of cars and trucks. We measure it by the energy content of the petroleum burned. If we measured it by the final energy of cars and trucks something like 80% would be "lost." So we are giving the total energy used in the production of electricity to keep our comparisons on an equivalent basis.

The message of Table 7.3 is that electrical energy is a big part of our energy use. It is very important in creating the comfort of our homes and drives a major fraction of our commercial enterprise. While electricity's role in industry is less, still at a third of the total, it is clearly a vital component in making American industry operate.

It is not surprising something like 97% of transportation energy comes from petroleum. That fact emphasizes the critical dependence we have on a continuing, stable, and affordable petroleum supply.

TABLE 7.4 1999 U.S. electrical energy by source.

Source	Percent of Total
Coal	51%
Petroleum	3%
Natural Gas	16%
Nuclear	20%
Renewables	10%

Let's look at electrical energy a bit closer. Where do we get electricity today? Table 7.4 shows you the numbers, and Figure 7.6 presents the same information graphically.

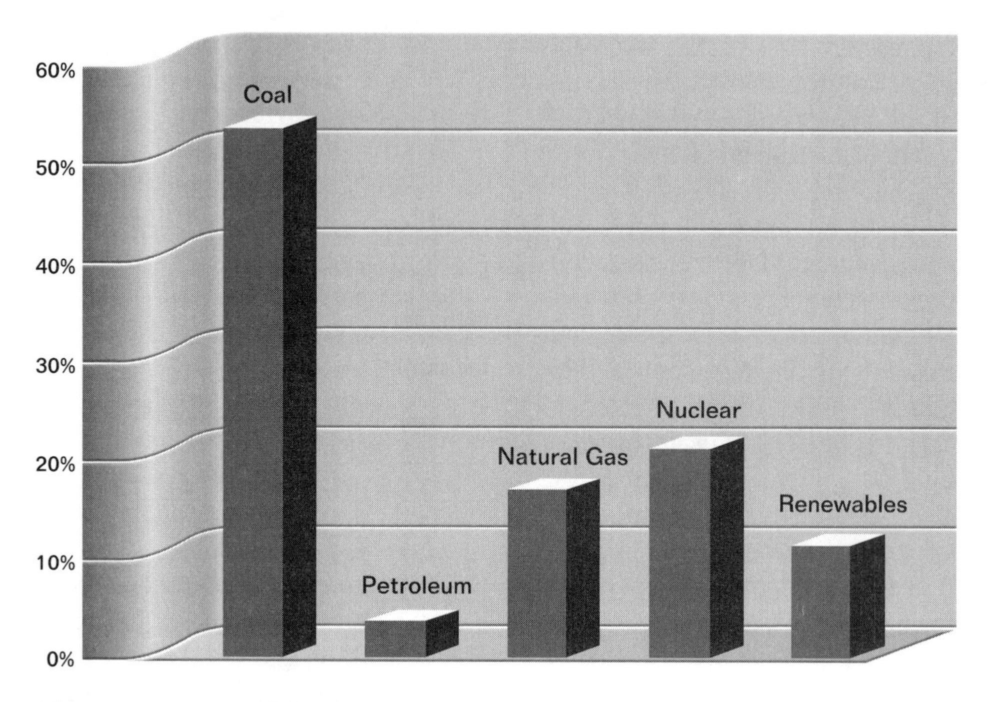

Figure 7.6. 1999 U.S. electrical energy by source.

From this it is pretty clear that coal dominates current electrical energy production in the United States. This is due to the fact that we have a lot of coal and it is pretty cheap. Not quite as "cheap as dirt," but not far from it. Petroleum used to be a much larger fraction but after oil rose from $2.50 per barrel to over $20 per barrel in the great energy crisis of 1973, oil-fired electrical generation went out of style quickly. There are a few special locations, like Hawaii, where oil-fired electrical energy makes sense. It was that same oil shock that set France on its campaign to convert its electrical generation to nuclear power. France has very little coal and now gets about 75% of their electricity from nuclear energy.

Natural gas is the rising star of electrical energy production. The price spikes of the summer of 2001 not withstanding, the price of natural gas has been reasonable and more importantly a gas-fired plant is the cheapest, easiest, and fastest to build. Modern gas-fired plants are essentially very large jet engines. You order a turbine, scrape off a flat spot, they bring you a turbine, you connect to a natural gas pipeline and an electrical line out, and you are in business. This is bit simplified, but the point is natural gas is significantly less hassle than a coal or nuclear plant to site and build.

A few quick words here on renewables. The renewable energy sources you are likely to hear the most about are wind and solar energy. Of the renewables contributing to electrical energy production, the vast majority is hydroelectric, a full 80%. Biomass and municipal solid waste make up just over 15%. That is a lot of wood chips from the lumber and paper industries and some innovative burning of garbage. Of the remaining 5%, about 3.5% is geothermal. Wind comes in with 1.2%, leaving about a quarter of 1% for solar.

I will let you think about the implications of that for a bit, and we will take it up again later.

Now that you have a feel for how we use energy from both a United States and global point of view and where we get energy, I want to look more closely at what it means to be so dependent on fossil fuels and also try to provide a frank and realistic appraisal of renewable sources of energy.

Fossil Fuels and the Problems Thereof

My friends have accused me of excessively dark humor in the following characterization, but I see only two futures for the human race if we continue to depend on fossil fuels. One future is that we run out of fossil fuels and civilization collapses with humanity reverting to medieval if not Stone Age conditions. The only other future is we do find enough fossil fuels and in burning all that we render the planet uninhabitable for ourselves and perhaps most other living things as well. That sounds pretty grim, huh? While it is very inviting to dismiss these as exaggerations, I think if you look at the logic of it, you will see that given the premise of continued dependence on fossil fuels, there really isn't an escape from one or the other of these futures.

Well, actually we will never see the "we run out of fossil fuels" future. However, that is not good news. The only reason we will never see that future is that, if we continue to depend on fossil fuels as those resources begin to get depleted in some regions but remain plentiful in others, civilized behavior is likely to go out the window. The resulting conflicts and wars will collapse civilization before we get to the "we run out of fossil fuels" state. That would hardly be an escape from the dark fate.

To try to appreciate this dark view of fossil fuel futures, let's look at each of these two potential paths. First, the "we run out" path.

Let's do a quick review of what fossil fuels are. You probably already know that fossil fuels are the remains of plants (and a few animals) laid down over millions of years. The significant start of the creation of fossil fuels began in the Carboniferous period some 345 million years ago. From the point of view of a 4.5-billion-year-old planet, this was pretty recently, but from the point of view of humans apparently ready to burn it all up in a few hundred years, it was a long time ago.

Once the planet got to the Carboniferous period, we had a lot of really lush plant life going on and in only short times of exception continued to have that lush growth. Where conditions were right, some of that plant life got converted to fossil fuels. For each type of fossil fuel, that is coal, natural gas, and petroleum, a different set of conditions was needed.

For coal what works is a nice large swamp with many generations of thick vegetation growing and successively dying. This leads to a bog with the current crop of plants growing on top of the rotting mass of the previous generations. After a while you get a pretty thick mass of rich organic material. If now you cover that up with an appropriate depth of overburden, you will compress and heat the organic material. With a little help from bacteria, this will break down the complex hydrocarbons in the organic material and give you something that looks more or less like pure carbon.

The various grades of coal are distinguished by how far down this path they get. Peat is still pretty much the plants you started with. Lignite is started on the path but still has a good deal of moisture and some organic materials. Bituminous coal is mostly carbon and fairly rock-like but still crumbly. Anthracite is hard black shiny rock, which is just about pure carbon. Sometimes you hear this called hard rock coal. My home town of Rock Springs', Wyoming, claim to fame in the first half of the twentieth century was that it was the only hard rock coal source in the West and the Union Pacific Railroad dug mines all under the town to fuel the coast-to-coast steam engines.

Petroleum is trickier. You start with the same general idea, that is decaying plant matter. However, to get petroleum, you need a more narrow set of conditions of heat and pressure. In the oil business, there is a thing called the "oil window." This is the temperature band in which petroleum can be formed from organic debris. Because there is a more or less uniform temperature increase with depth underground, the temperature window works out to a depth window of between about 7500 feet and 15,000 feet. Less than 7500 feet, the temperature is too low to provide the energy needed to convert organic materials to petroleum. Greater than 15,000 feet, the temperature is so high that you break all the carbon-to-carbon bonds leaving only methane, and as we will see in a bit that can be either good or bad.

Now creating petroleum is fine, but that doesn't mean we can get to it. A few other conditions are also required. First, the concentrations of the original organic material and the conditions for petroleum formation have to be right to allow sufficient creation of petroleum to allow collections. This is the value of the ore problem. It is hard to put a shovel in the ground and not get some iron oxide. However for economic recovery, you really want to find the places where the earth has already concentrated the ore so that you don't have to dig up many cubic yards to get enough iron oxide for one hubcap worth of steel. Oil works the same way. There may be trillions of drops of oil randomly scattered across the earth, but we really can't use them unless nature collects them for us.

The next thing you need then is an oil-bearing rock that is sufficiently porous to allow oil to flow. Even if you have a lot of little drops of oil in one place, it does no good unless you can connect those drops together to get a stream of oil out of the ground. The next thing you need is a capping geologic structure that is not porous to hold the petroleum in place until you are ready to come get it. If the critically important porous rock is not capped in some way, then the petroleum is going to get away from its underground home.

And that brings us back to the good and bad news about what happens if the organic debris get converted all the way to methane. If there is no capping structure, the methane comes to the surface and gets away. If it is capped, then we have a natural gas well. Often oil and natural gas are found together.

As an interesting little aside, let me mention that helium is a by-product of natural gas. Helium is pretty common in the universe but quite rare on earth because it is a very light gas and doesn't chemically combine with anything. Because helium is lighter than nitrogen, which makes up most of the atmosphere, helium floats in air. That is why you use it to fill party balloons and fill the Goodyear blimp. Any free helium on the surface floats to the top of the atmosphere and escapes to space. However, the earth is continuously making new helium. How does that happen? Helium comes from radioactive decay of uranium. That decay chain, as you will see if you look back to Table 4.1, has a lot of alpha decays. If you stick a couple of electrons on an alpha particle you get helium. Because there is a lot of uranium in the ground and it is widely dispersed, there is a lot of helium being formed all the time. Most is lost, but if you get the geologic conditions to trap natural gas, you trap helium as well. The interesting point of this is that the two protons and two neutrons in each atom of helium in your party balloon used to be part of a uranium atom and you only get to use it because uranium is naturally radioactive.[6]

Petroleum is a mix of organic chemicals, and natural gas is a range of organic chemicals in gas form. Petroleum can have a number of unfortunate impurities the most offensive of which is sulfur. Natural gas tends to be mostly methane, around 80%. While there are some exceptions, natural gas tends to be significantly less tainted with impurities like sulfur and, in any case, is much easier to clean up.

O.K., that is a bit of what fossil fuels are. Let's use that to look at the "we run out of fossil fuels" scenario, or that scenario modified to the "we get low on fossil fuels and beat each other up fighting over the last of it" version.

From the look at total energy use above, you saw that in the United States about 85% of the energy we use comes from fossil fuels. Globally the fraction

---

[6]And just to reassure you, the helium in the balloon is not radioactive. This is another example of the fact that once any given atom has undergone its radioactive decay that particular radioactive decay is over, and will not happen again.

might be a little less but still something near 80%. We also saw that the use of energy is not evenly distributed among the world's population and that the developed industrialized countries used a lot more. So here is the bad news. Because of the special conditions required for the formation of fossil fuels, the locations of large concentrations of those fuels are not uniform over the planet. And the really bad news is that the location of the greatest use of energy, hence demand for fossil fuels, and the location of the greatest concentrations of the resources of fossil fuels are not in the same place.

The following three figures show you the fraction of global use and the fraction of global resource in various regions. Figure 7.7 is for natural gas, Figure 7.8 shows you oil, and Figure 7.9 tells the story for coal.[7]

From these figures, you will see that the geographic location of oil and natural gas is very unevenly distributed. Realizing that you not only need the right conditions to form these fossil fuels but the right conditions to trap them, this should not be surprising. Coal is more evenly distributed. Again, this makes sense in that it is less demanding on special conditions to be formed

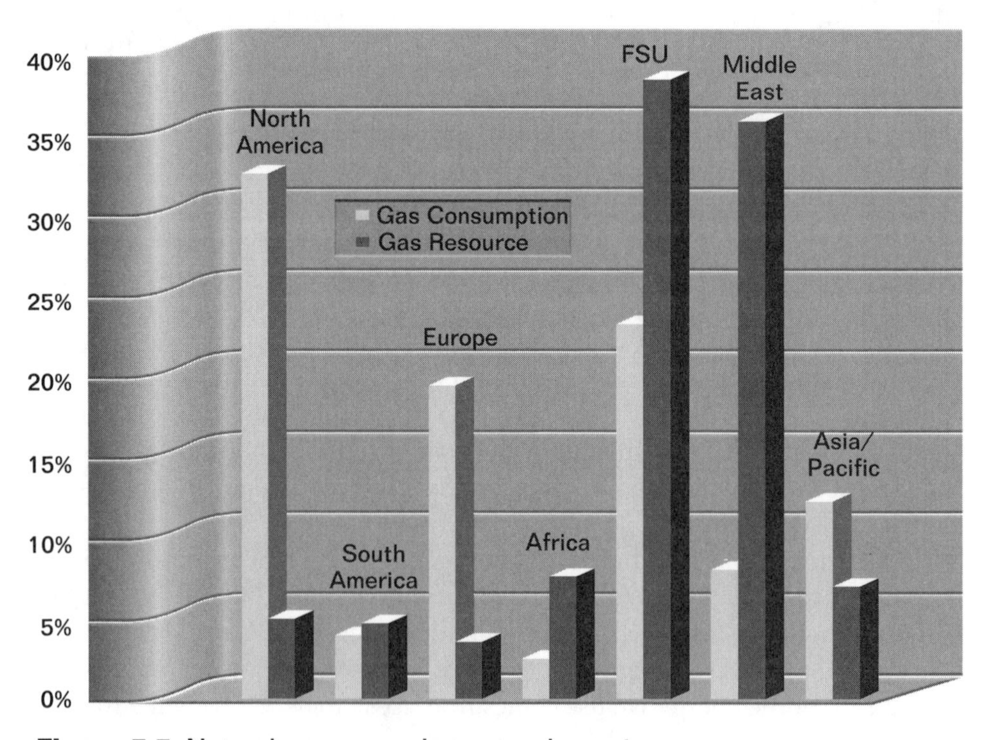

Figure 7.7. Natural gas use and resource by region.

[7]Data taken from "BP Statistical Review of World Energy June 2001," British Petroleum.

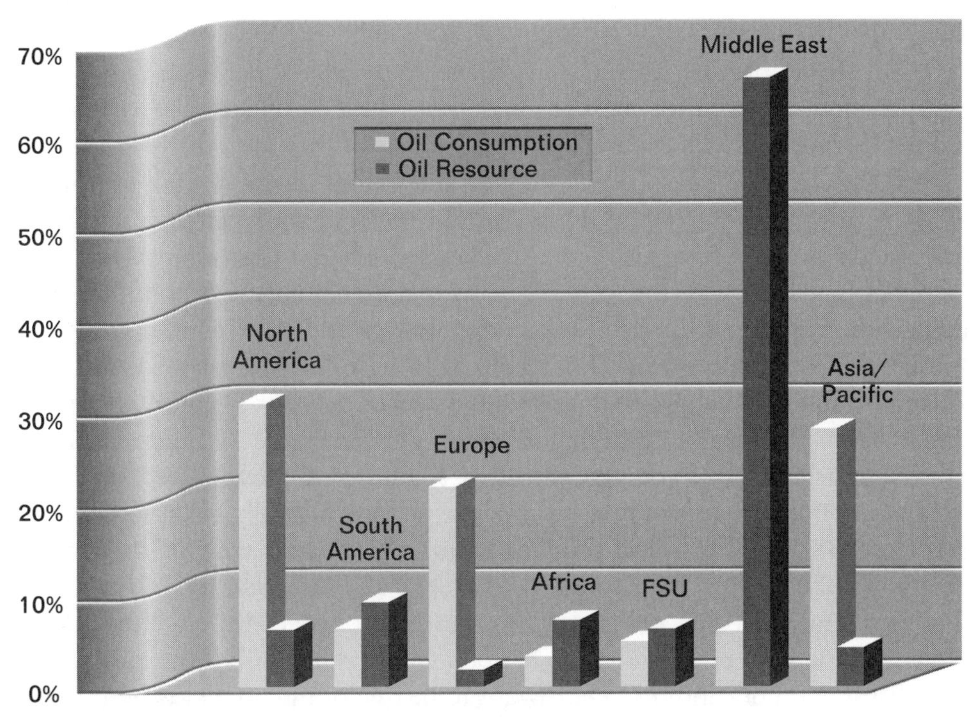

Figure 7.8. Oil use and resource by region.

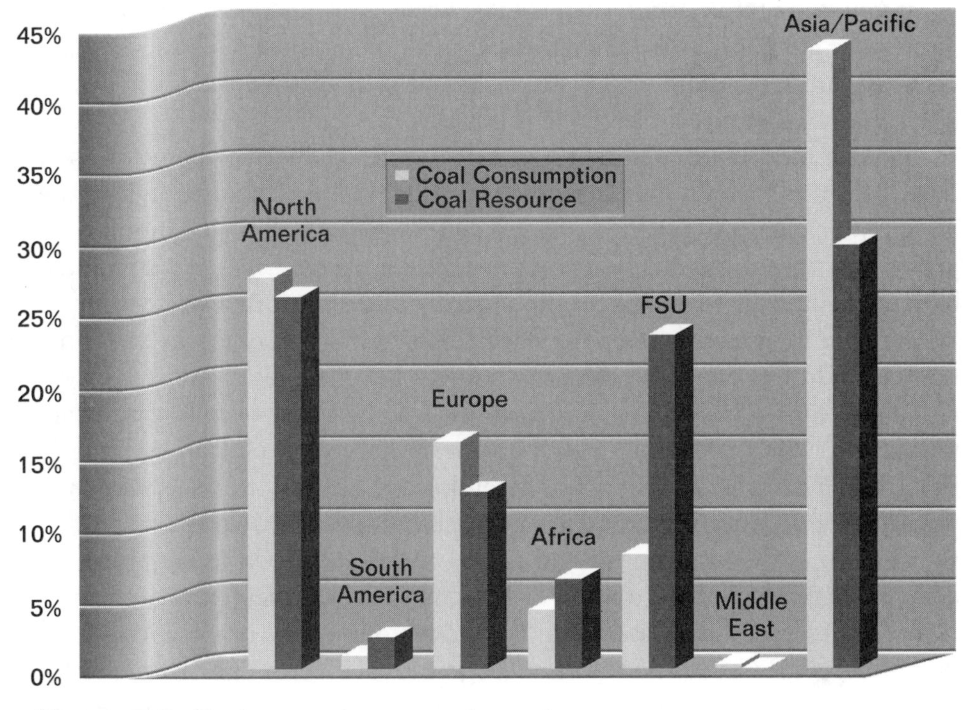

Figure 7.9. Coal use and resource by region.

and once formed stays put without special geologic features required to hold it in place.

The significance of the figures, however, is in the comparison between consumption and resource. Now it should not take a rocket scientist to figure out any region in which the fraction of world's consumption greatly exceeds the fraction of world's resource is a region that is going to be in a lot of trouble soon. If you live in North America or Europe and plan on using anything but coal for the long term, you should be getting a bad feeling about this.

Does that mean that the folks in the former Soviet Union nations and in the Middle East ought to be feeling on top of the world? Play a little game of thought experiment with me. Imagine you bring a half a dozen 6-year-old kids into a room. Now give them each a handful of really great candy. Happy times, right? O.K., do the experiment again. This time give 2 of the 6 kids 5 times as much candy as you give the other 4 kids. Happy times? Well, maybe for a while, but perhaps some tears are coming. Now to really test the experiment, we will carefully select the kids we use. The 2 kids we give the least candy to will be the biggest, toughest kids who also will have a really sweet tooth for candy. The kids we give the most candy to will be the smallest kids. Now what do you think is going to happen in that room with those kids and that distribution of candy? Do you want to be one of the little kids with the most candy? If I was a 14-year-old Saudi prince, I don't think I would have a really great feeling about my personal future.

If you don't think nations can behave like 6-year-old kids, I would like to direct your attention to 1991 when the United States and its allies kicked the stuffings out of one particularly stupid kid who threatened the candy jar by being mean to the nice kids who were sharing.

I will also direct your attention to the so-called California energy crisis of 2001. The crisis was not over the supply of energy; it was all about where the energy was relative to where it was needed. There is certainly enough blame to go around as to how that set of circumstances got established, but if you step back a bit you will see the crisis was caused because California had consumption greater than its resource. The Pacific Northwest was called in to supply what it could from its hydroelectric resources but that was a low water year and hydro capability was limited. However, Texas and other central states had plenty of gas-fired electrical energy and were happy to sell it to the energy hungry Californians. Well, perhaps gleeful is a better way to put it. Like a 6-year-old with a handful of candy facing another 6-year-old with a screaming sweet tooth, they saw an opportunity and took it. We need to be fair to the average Californian in the street. While we might ask if their excessive expectations on environmental issues and restraints on energy corporations were part of the problem, we need to also recognize that even after adjustments for the warmer climate, the Californians are among the most energy efficient in the land. They just found themselves unable to take care of their own and when they turned to their friends and neighbors they got help but at the price the suppliers demanded. And that price was high, cripplingly high.

That, scary as it is, should be the lesson of the California experience. If you find yourself in a big discrepancy between consumption and resource, it is going to cost you. You never saw the Californian National Guard getting ready to march across Arizona to liberate the natural gas wells in Texas, but if you had reached a point where California was really shutting down for lack of energy, what might one expect?

Now going back to Figures 7.7 to 7.9, we have to clarify a bit about the different ways to express resources. The data in those figures was based on "proven reserves." The British Petroleum folks define that as:

> "Generally taken to be those quantities that geological and engineering information indicates with reasonable certainty can be recovered in the future from known reservoirs under existing economic and operating conditions"

For the purpose of those figures, that is to give a feeling of who is using fossil fuels and who has fossil fuels, that is a reasonable definition of resource. But does that mean that is all there is? Well, no.

A game people often play is to get a past prediction of proven reserves, divide by the consumption rate per year to get a predicted date when that resource is gone. They will then point with some satisfaction that, using past predictions, the given resource should already be used up when in fact the current estimated proven reserve is greater than the old one. They then declaim that this proves all such estimates are worthless, but in actuality, it only proves they don't understand what "proven reserves" means. I presented the information in the form of fractions of the world's total to spare you from attempting to do that fallacious "how long will it last at this rate of use" calculation.

We have to look at the words in the British Petroleum definition closely. The words "from known reservoirs under existing economic and operating conditions" are particularly important. The United States' "proven reserves" took a nice spike up when we added 9 billion barrels of Alaskan oil to the books. As prices increase you can work harder to get more oil out of existing wells or mine more marginal coal deposits. Finding more and being able to get more if the price goes up are the major ways that proven reserves increase.

Another subtlety of "proven reserves" is that much of this data is not independently developed. That is, the owner, a company, or a country provides the data. It may be to that entity's advantage to overestimate the resource to attract investors or to underestimate it to sandbag the market at a later date.

So while "proven reserves" gives one some feeling for how much resource one might depend on, the data does have limitations. Worse yet, it doesn't answer the question we are most interested in which is "how much is there?" If we were to think about it a bit more, we would recognize that "how much is there" is not the real question anyway. Something like "how much that I can afford is there" might be a more useful question. Clearly one has to define how much you can afford before you can attempt to answer that.

One attempt to look at the issue of resource use is given in a very interesting book by Kenneth S. Deffeyes, *Hubbert's Peak, The Impending World Oil Shortage.* Professor Deffeyes provides a uniquely personal look into the business of oil production and the geo-science of finding oil. The key point in the book is the prediction by M. King Hubbert in 1956 that the United States oil production would peak between 1965 and 1972. There is a lot of statistical theory involved. Let me try to simplify it. He looked at the production rate of a finite resource. He noted that it is slow at the start but builds as uses for the resource are expanded and the technology for finding and extracting the resource are improved. As time goes on, the easy sites are used up and more difficult sites of the resource are tapped. As long as the technology for finding and extracting keeps ahead, the resource production rate continues to increase. At some point, all the easy and just-a-little-bit-hard sites are falling off in production and technology can't keep up. Production starts to fall. This increases the price, which allows you to work harder to get it, but eventually even that can't keep up and either the consumers simply can't afford it or they find some alternative that is less expensive. This logic said that the production of a finite resource would then follow a predictable shape over time. It starts small, increases up to a point, starts decreasing the rate of increase until it comes to a peak, and then starts to fall off. The curve looks like a speed bump. Hubbert's big leap was that he said that if you looked at the first part of the production curve, you could predict the rest of it, including when the peak would happen.

Using that premise Hubbert made two estimates. His pessimistic one predicted a peak in United States oil production in 1965 and a more generous one predicted a peak in 1972. So how did he do? United States oil production peaked in 1970.

The thing to appreciate about Hubbert's method is that it addresses the subtler question of "how much that I can afford is there" rather than the simpler but ultimately less helpful question of "how much is there."

Now the really bad news comes when Professor Deffeyes applies Hubbert's method to global oil production. Again, using a lot of math and statistics, the answer he comes up with is a peak oil production in 2005, no later than 2009. Well, that is not too bad is it? After all we have the whole downward side of the production curve to go. Reaching the peak is just when we have used up half of "what we can afford."

But remember, we are expecting a 60% increase in global energy use in just the next 20 years. The single most disturbing thing about the future of petroleum is that China is now a net importer of oil. As a billion-plus people fire their economy up and try to fuel it with Middle Eastern oil, things are going to get interesting. We are going down the backside of the curve a lot faster that we went up the front side.

The really, really bad news is about what we are going to have to pay for oil on that backside of the curve. During the 1930s, the price of oil in the United States was determined by the production rate in Oklahoma and Texas. Price gets high, turn up the pumping rate, price comes down. Price gets too low, turn down the pumping rate, price comes back up. That worked until the demand exceeded the capability of those oil fields. As soon as they could not keep up, they just ran flat out and someone else set the price. Saudi Arabia and their OPEC friends figured this out in 1973. Now they set the price.

Just to make the point on how much control the Saudis have, let's look at a little of the data Professor Deffeyes provides. The peak decade for oil discoveries in the United States was the 1930s, 60-plus years ago. How about this, the number of wells you have in production gives one a sense of where you are in your production curve. That is, if you are on the up side, you are still getting the easy sites and need few wells to pump a lot of oil. If you are on the backside and going down, you have a lot of smaller producing wells. The Saudis are pumping from 1560 wells; the United States is pumping from 563,160.

A little similar bad news on the natural gas front is that in 1971, United States well productivity peaked at 435,000 cubic feet per day. That is the average production per well for that year. By 1999 that rate had fallen to 157,000 cubic feet per day, just over a third of the 1971 rate.

In the winter of 2001-2002, OPEC called on non-OPEC states to join them in reducing production to help bring oil prices up. Several nations, particularly Russia, declined to join that effort. Saudis said, fine, you want production, watch this. They turned their valve and the oil flowed and the price dropped. The developed nations cheered, and Russia saw their oil revenues drop. The Russians got the point and agreed to restrict production along with the OPEC nations. The Saudis can do this up to the point when they can't produce more than people want to buy. As soon as demand exceeds their capability to just turn open the valve then the price is truly controlled by the market and can no longer be set by the supplier.

In Economics 101, we learned that price is the key to matching supply and demand. Given a demand greater than a supply, the price will increase giving the supplier the incentive and capability to increase supply to the point it matches demand. But what happens when the supply is finite? What happens when a doubling or tripling of price doesn't give the supplier enough money to increase the production to match demand? Well, it gets ugly. Simply economic theory says price increases without bound. However, we know that the social structure will fail somewhere along that path. Economics would also offer the opportunity to find alternatives that at some price can replace the unsupplyable and very expensive product. While there are some rays of hope there, the real question is the timing. Can those alternatives come into force in the very short time line Professor Deffeyes offers us?

It is a pretty scary question, and Professor Deffeyes doesn't offer us a lot of hope. Others offer more cheerful views.

In June 1999 *Discover* had an article entitled "Why We'll Never Run Out of Oil." In that article they start out reminding us that in 1973 we had been told the end was near. Oil prices of $100 per barrel were predicted by the year 2000. They pointed out that this had not happened and that, in fact, oil was selling for under $15 per barrel at the time of publication. So what happened?

In a simple answer, we figured out how to at least stem the rate of our increased use and we found more oil. The EIA data shows that world consumption of oil actually decreased by 7% from 1979 to 1983. Furthermore, we have significantly decreased how much oil we need to fuel a given economy. The barrels of oil per dollars of gross domestic product worldwide have decreased by 30% over the last 30 years. However, even with these efficiencies, the population growth and improving economies have caused us to resume our increases and now we use about 114% as much oil as we did in 1979. The predicted future growth rate is 2.3% per year, which is about 140% of the growth rate over the previous three decades.

The *Discover* article lists a number of new sources of oil we can expect from traditional oil wells and provides a number of sources that might give us oil from unconventional sources. To give a little context to the numbers, the British Petroleum values for proven reserves used in Figure 7.8 come up to just over one trillion barrels. The use rate in 1999 was about 25 billion barrels per year. The *Discover* article quoted the EIA prediction of a 2020 use rate of 36.6 billion barrel per year. The more recent 2001 EIA estimate of the 2020 use rate is 43.6 billion.[8]

Now to make us feel good, let's look at *Discover*'s list of "new" sources of oil:

- Caspian Sea—Up to 200 billion barrels (potentially as big as Saudi Arabia)
- New Gulf of Mexico wells—15 billion barrels

[8]An increase of 19% in the 2020 prediction in 2 years ought to give you a sense of how hard it is to make these predictions and also some feeling for how much confidence you might want to have in the predictions.

- Offshore coast of Brazil—30 billion barrels
- Offshore West Africa—75 billion barrels.

Sounds pretty good, huh? But it gets better. They remind us that, in the extreme, primary (that is normal pumping) recovery of oil leaves as much as 70% of the oil in a well. With improved secondary and ternary means, we can get much more oil out of existing fields. Deffeyes will remind us that the reason we don't do more of this is that hydro-fracture, steam injection, gas injection, and other methods are expensive. They are employed only if that is the cheaper way to get oil. The fact that, in general, folks are willing to poke multi-million dollar holes in the ground looking for a new well rather than extract the tough stuff out of existing wells speaks to his point rather well.

Discover isn't done with us yet, however. They offer that natural gas can be turned into liquid fuels. They quote a price of about $20 per barrel. Existing natural gas resources they tell us could provide 1.6 trillion barrels of such fuels. They do admit that there are competing uses for that natural gas, however.

Still, on the upside, natural gas isn't the only other place to go to get petroleum-like products out of nature. The Athabaska and Cold Lake tar sands in Alberta can be converted to useable oil. This is currently being done at a modest rate. If put into really big production, this resource is estimated to be 300 billion barrels, bigger than Saudi Arabia.

Venezuela has similar pre-petroleum goo in the Orinoco sludge deposits that could be processed into oil-like products. This is estimated to hold up to a trillion barrels.

To top it all off, *Discover* reminds us that in World War II the Germans used a coal-to-oil process to fuel their army and that coal makes up 90% of all the fossil fuels on the planet. The British Petroleum data for proven reserves are a little more conservative with oil and gas reserves both coming in at about 5500 Quads and coal with a bit less than 26,000 Quads, giving coal about two-thirds the total. However, any way you slice that, if you wanted to convert coal to oil, you could extend the resource by at least 4 times, probably even more.

The *Discover* article doesn't have much to say on the United States reserves of oil shale. This is a type of rock found in the Green River Basin of Wyoming, Colorado, and Utah. The rock is essentially pre-petroleum material, which holds the organic material but never got buried to the right depth to cook up oil. This is my old backyard. My dad told me stories of pitching rocks into the campfire and having them burn just like coal. The problem is you have to force the oil out of the rocks, which is neither easy nor cheap. Deffeyes relates that through his entire career in the oil business the price of extracting oil from oil shale has always been just a little above the current market price, no matter how high the price has gone. However, if you can use them, the oil shales might hold as much oil as all the other conventional oil sources, something like 2 trillion barrels.

Are you impressed? We aren't done yet. Let me tell you about methane hydrates. A methane hydrate is a type of water ice that traps methane gas inside. The neat thing about this is that it can exist at temperatures above the normal freezing point of water or can also be found if you provide an overpressure to keep it solid. Where might either of those conditions exist for a long time? One is in the arctic permafrost and the other is in the ocean at depths greater than 500 meters (1600 feet). The really, really neat thing about this is that there is a really, really huge amount of this stuff. We will get back to how huge in just a bit.

The methane ice can be found as ice crystal deposits in permafrost. You just scoop them up, heat them up, and you get methane, which you will recall is the principal part of natural gas. In the ocean it is a little trickier in that if you put a bucket on a rope, lower it to the bottom, scoop up some methane ice, then pull up the bucket, the ice will melt on the way up and the methane will bubble off. Several underwater mining methods are under development.

You can get a range of numbers as to how much of this there is. However, it is a lot. Even the more conservative estimates cite methane hydrates as the largest source of carbon in the planet's fossil fuels. Those looking at greenhouse gases wonder a lot about the history of methane hydrates and the planet's temperature history. For instance, in an ice age a lot of water turns to ice. Ice collecting on land locks up water and lowers sea level which, therefore, reduces the water pressure over some of the seabed methane hydrates. This causes methane to boil out of the ice and be released. Methane is 20 times more effective as a greenhouse gas than carbon dioxide, which means a big release of methane would tend to counter the ice age and reheat the planet. Methane hydrates also act as a cement to seabed sediments. If the methane hydrate boils away, the sediments become less stable and even slight slopes would tend toward landslides. Underwater surveys have found locations where this has appeared to happen, giving credibility to the idea that such mass releases have happened. Well, that is good, right? It gets cold, seawater freezes, sea level lowers, methane ice boils, atmosphere warms, all balanced back out. This is good, right? Well, maybe. It depends on the magnitude of the swings, how much you over control. Correcting a slight drift toward the centerline while you are driving with a quick full rotation of the steering wheel to the right will put you in the ditch if not roll your car. If you recall, one theory on the Permo-Triassic extinction 250 million years ago was a massive greenhouse gas release event.

As an energy resource, however, methane hydrates are the ultimate fossil fuel. The estimates range over a factor of 1000; however, if you pick one of the more generous values from the Department of Energy's Office of Fossil

Energy, the worldwide resource could be as high as 400 million trillion cubic feet. That is 75,000 times as much natural gas as in the British Petroleum's estimate of current proven reserves. As I said, this is a really, really huge amount. Think about it. Even if the very lowest estimates are right, it is still 75 times more natural gas than in existing proven reserves.

Let's do a little summary here. Let's look at fossil fuel use rates and the various reserve categories. Please recognize that the farther you go down this list the more "approximate" the numbers get.

1999 Fossil Energy Use = 314 Quads/year

2020 Fossil Energy Use = 500 Quads/year (assumes 2.3% annual growth)

Total Fossil Energy 1999 to 2100 = 76,000 Quads (assumes 2.3% annual growth)

1999 Proven Reserves

Oil = 5700 Quads

Natural Gas = 5400 Quads

Coal = 26,200 Quads

Reasonable Additional Reserves

Oil = 5700 Quads

Natural Gas = 5400 Quads

Coal = (I didn't find a value for this, current resources are so plentiful folks aren't looking for big new finds)

Unconventional Sources

Tar Sands = 1600 Quads

Orinoco Sludge = 5700 Quads

Oil Shale = 10,000 Quads

Methane Hydrates = 400 Million Quads

If you do a little math, you will see that, even without the methane hydrates, we ought to be able to get to 2100 with these resources given the 2.3% growth rate and given just a bit of luck finding more coal. So, *Discover* got the right title on their article, and Professor Deffeyes worries too much?

Well, look at the assumptions you have to make for that to work. You have to find a way to use the tar sand, Orinoco sludge, and oil shale, and you have to do that pretty darn quick and at a price we can all stand. In addition, you also have to get a new form of transportation fuel either from natural gas or coal because the oil out of the ground is not going to make it. Remember,

Deffeyes never said we were going to run out. What he said was that the production of oil would peak before 2009.

Deffeyes and *Discover* could both be right. Oil production peaks in say 2005, the price of oil starts to grow sharply and the incentives for unconventional fossil energy sources become sufficient to cause large-scale development. The real issue comes back to timing. Can we get to the new sources and get their cost down before our economies crater?

It seems to me that we are Little Nell tied to the railroad track without any stalwart hero in sight. The good news is only our legs are tied and our hands are free. The bad news is the train is coming fast and it is getting awful close. As fast as we can might not be fast enough to untie the ropes and leap off the track.

But if we can master methane hydrates, then oil or coal pale in comparison. Who cares if we run out of either of those? Methane is a superior fuel to any of those, at least if you can get it into an easily transportable form. So let's just bank on that.

There are a number of assumptions needed to make that work for us. Not the least of which is the development of extraction technologies that work at an acceptable cost. But if we can pull it off, it leads to my second scenario on potential futures with fossil energy. What happens if we find plenty? Let's look at that "we don't run out" future.

Well, if we don't run out of fossil fuels, we will burn fossil fuels. And we will be burning more fossil fuels in the future than we are now. We will burn more because there will be more of us and we will be improving the quality of life for the developing nations by improving their economies through energy-fed industrialization.

If we used that 2.3% growth rate all the way to 2100, we would see fossil energy consumption almost 4 times our current rate. There are a lot of things that might happen in the next 100 years. Some think the global population will peak in about 2050 and come back down to about our current 6 billion by 2100. Some think the population will continue to increase and we would see a world population of 18 billion by 2100. Most agree that we will continue to become more efficient in using energy relative to our economic activity improving by factors of 2 or even 3. There is also a range of possible degrees of improvement in the economies of developing nations. The result of this is that the carbon use might be about the same in 2100 as now or as much as 7 times more. We will look at this more closely at the end of the section, but the bottom line is if we don't run out of fossil fuels, in the future we will burn a lot of fossil fuels and probably at a rate greater than we are now.

So, if we burn a lot of fossil fuels and don't run out, what happens? Well, two things happen. One, we produce a lot of pollution, and two, we produce a lot of greenhouse gases. Each of these is sufficiently significant that we need to look at them in some depth.

First, let's address pollution. Ideally when you burn coal you are combining carbon and oxygen to get carbon dioxide. Ideally when you burn natural gas you are combining methane and oxygen to get carbon dioxide and water.

The problem is that it isn't an ideal world. Coal isn't just carbon, air isn't just oxygen, and natural gas isn't just methane.

Along with carbon, coal will contain sulfur, arsenic, mercury, cadmium, selenium, lead, boron, chromium, copper, fluorine, molybdenum, nickel, vanadium, zinc, and as we learned earlier even some uranium. When you burn coal some amount of all these less-than-friendly substances are going to be put into the air. Natural gas will also have some fraction of bad actors, but generally it is much cleaner than coal.

The problem with the air, in which you burn either coal or natural gas, is that 4/5 of it is nitrogen not oxygen. If you get nitrogen and oxygen hot enough, they react and form nitrous oxides, usually referred to as NOX.

For coal, and remember with the exception of the potential methane hydrates coal is by far our greatest fossil energy resource, the biggest problem is sulfur. The sulfur content of coal varies quite a bit. In the United States, western coal seems to have less sulfur than eastern coal but less is not none. The Russians are cursed with some really nasty coal, which has not helped their air quality.

When you burn coal you are going to make sulfur dioxide. When that sulfur dioxide gets into the air, it does a number of very unpleasant things. As emitted from coal-burning plants, sulfur dioxide is transported as very fine particles. These are too small to be impeded by the natural defenses of a human's respiratory system. If they are in the air and you are breathing, you are going to get some into your lungs. There, these particles will damage the air sacks within your lungs leading to a range of respiratory problems like bronchitis, asthma attacks, and other long-term respiratory trouble. With that strain on the respiratory system damage can also be done to the circulatory system leading to heart trouble.

Nitrous oxides have similar effects on human heath. In addition, nitrous oxides interact with sunlight to create ozone, which is poisonous to humans, and good old photochemical smog. Toxins like carbon monoxide and mercury worry a lot of folks as well.

Of course, humans are not the only ones to suffer. The problem of "acid rain" and its impact on vegetation and wildlife has been widely discussed. Acid rain is what happens when rain is formed in an atmosphere containing sulfur dioxide and nitrous oxides. The rain will contain sulfur and nitrogen in the form of acids, which are then deposited everywhere the tainted rain falls. Beyond the direct damage of the acid-laced rain to the plant life, the accumulation of the acids in the soils and groundwater changes the chemistry of the natural systems. Plants have a harder time getting nutrients, and lakes are becoming acidic. More that 50% of the canopy red spruce trees in the Adirondack and Green Mountains died in the 1970s and 1980s. Sugar maple has seen diebacks from 20% to 80% in the northeastern states, which is at least partly due to acidic deposition. Acid-sensitive fish species such as Atlantic salmon, tiger trout, bluegill, walleye, and minnows also took a hit with 24% of Adirondack lakes becoming acidic and fishless.[9]

Then there are man-made structures. In turns out acid rain is pretty tough on roads, bridges, and buildings as well. In Europe, structures that have stood for over a thousand years are literally dissolving in the rain.

But we know about this now and have got it all under control, right? After all we passed the Clean Air Act in 1970, and all is on the mend now, isn't it? Well, the best you might say is it isn't as bad as it might have been.

First, you have to understand that coal for electrical power production is not the only source of sulfur and NOX in the atmosphere. The other big actors are coal burned for industrial uses, natural gas for both electricity and industrial use, and motor vehicles. Various industrial processes and other non-industrial fuel uses make up the rest.

While we have burned more coal, natural gas, and gasoline over the last 30 years we have reduced sulfur emissions, down to about 60% of what they were in 1970. Nitrous oxides on the other hand have increased by about 20% in the same period, but recognize that the total fossil fuel use has increased quite a bit.

Table 7.5 gives you the Environmental Protection Agency numbers for sulfur dioxide and nitrous oxide emissions along with the EIA values on coal consumption. Most of the coal consumption was for electrical power production.

What should we learn from these figures? First, note that natural gas is a lot cleaner than coal. The industrial uses of natural gas were pretty big players in nitrous oxides, but that has improved a lot over the last 30 years. As for coal, well, considering the increase in coal use we have gotten nearly 3 times better on sulfur emissions per unit coal used and a little over 1.5 times better on nitrous oxides in the last 30 years. But in 1970, coal for electricity was 50% of our total sulfur dioxide emissions, and in 1999, it was 63%. Coal-fired electricity has continued to be about 19% of the nitrous oxide source. So, while better, one has to recognize that it is still a very large source of bad stuff.

So how bad is that amount of bad stuff? Well, there is a lot of debate on that. One source of analysis is the Harvard School of Public Health. They estimate that air pollution kills something like 60,000 Americans per year. In addressing sulfur dioxide alone, they estimate that 2400 people die for every

[9] "Acidic Deposition in the Northeastern Unites States," Charles Driscoll, et al. *BioScience*, March 2001.

TABLE 7.5 U.S. air pollution.

	1970	1980	1990	1999
Coal Production (million tons)	520	700	910	1060
Sulfur oxides (tons)				
Coal—electric	15,799,000	16,073,000	15,087,000	11,856,000
Coal—industrial	3,129,000	1,527,000	1,914,000	1,317,000
Gas—electric	1000	1000	1000	12,000
Gas—industrial	140,000	299,000	543,000	576,000
Motor Vehicles	411,000	521,000	560,000	363,000
Total	31,161,000	25,905,000	23,678,000	18,867,000
Nitrous oxides (tons)				
Coal—electric	3,888,000	6,123,000	5,642,000	4,935,000
Coal—industrial	771,000	444,000	585,000	542,000
Gas—electric	—	—	565	367
Gas—industrial	3,053,000	2,469,000	967,000	985,000
Motor Vehicles	7,390,000	8,621,000	7,210,000	8,590,000
Total	20,928,000	24,384,000	24,170,000	25,393,000

million tons of sulfur dioxide emitted.[10] That would mean we are "saving" nearly 30,000 people a year due to the reduction in sulfur dioxide emissions compared to 1970, but still killing over 45,000. Some people have criticized the Harvard work suggesting that air pollution only "kills" people who were sick and going to die soon anyway. Harvard responded to this in a March 2000 press release with data showing that just wasn't so but rather it was the air pollution that was making people sick in the first place. Furthermore, Harvard updated their study suggesting their earlier health impact numbers were too low and that it is 70,000 Americans dying per year.[11]

[10] It is important to appreciate how hard it is assign deaths to any given cause. There are a lot of ways to develop these estimates and let us not forget Dr. Cohen's First Law of Information, "97.6% of all statistics are made up." However, the Harvard School of Public Health appears to be fairly credible if not totally without controversy.

[11] If you think I am being tough on fossil fuels, let me direct you to *The Environmental Case for Nuclear Power: Economic, Medical, and Political Considerations* by Robert C. Morris. Dr. Morris is not a nuclear engineer or involved in any part of the nuclear industry, but rather a retired chemistry teacher. His private study into energy sources led him to the conclusion that fossil fuels were, well, bad and nuclear was a lot better. If you want to see the warts on fossil fuels, read Dr. Morris' book.

Let's do a little check here. I am telling you that, according to the Harvard health study, 1999 U.S. sulfur dioxide emissions are estimated to kill 45,000 people per year and, of those, we could tag coal-fired electricity production with about 28,000. But, at the start of the section, I told you that while saying bad things about fossil fuels, I was not trying to convince you that overall technology was bad. So just to re-establish balance, let's recognize what those nearly 12 million tons of sulfur dioxide coal-fired electrical plants brought us along with 28,000 deaths. That electricity was fully half the electrical energy we have. Without that, the U.S. economy would have collapsed. Remember it isn't just hair dryers. It is providing the electrical power that provides the energy enhanced industrial and commercial productivity that make modern life possible and to a large degree comfortable. Just try to imagine what would happen if you and everyone else could only get electricity 12 hours a day and someone else chose which 12. Twenty-eight thousand people each year is a heck of a price, but then again we trade 40,000 each year for the convenience of personal automobiles. The point is, it is a trade. My point is that it ought to be a conscious, thinking, reasoned trade, not an unconscious one.

Which brings me to a point that really, really bugs me about the comparisons of nuclear power to other forms of energy. This is a bit of a digression, but it is tied up in the idea that using fossil fuel energy has impacts people don't always include in their thinking about the choice of energy sources.

The thing that bugs me is that the single most important reason why nuclear is not used more is that it is too expensive. We will talk more about that in the next two chapters, but let me explain a bit here. Existing nuclear power plants can compete with other electrical energy sources, but this is due to the fact that the plants themselves are by now paid for. However, new nuclear plants, and therefore wider use of nuclear power, aren't being built yet because they "cost" too much.

See it is like this. The cost of power from a plant has three general components. These are the cost of building the plant, the cost of the fuel, and the cost of other operations and maintenance items such as staff salaries. With the incredible energy density of nuclear fuel, the cost of fuel for nuclear plants is much less than any fossil fuels, again, more on this later. However, nuclear plants are expensive to build. Worse yet, there is a good deal of uncertainty as to just how long it will take to build a new nuclear plant and how much it might cost in the end because of uncertainty in what the regulatory demands might be and how those might change during construction. It is hard to get people to invest in anything if you are unsure how much it might cost and, therefore, unsure if it will turn a profit. Operational costs are also high due to the requirements of nuclear operation.

Hang on, and we will get to the thing that bugs me. There are two kinds of costs of anything. One is the direct cost, that cost you paid to get the thing you buy. The other is the indirect cost that must be paid but doesn't come in the price. What does that mean? Think about it. You buy a car and pay a price. So you have transportation, right? Well, no, you don't. Not unless you also have highways, traffic laws, highway patrol forces, an insurance industry, and a medical capability to take care of some fraction of the carnage all our cars are going to produce. And you, my friend, and I get to pay for all those other indirect cost items. Your taxes go to the government provided services, and your insurance premiums go to support the insurance and medical establishments.

If you think about it, only part of the cost of that half-gallon of milk you bought last night was paid to the supermarket. Part of the cost came from your taxes for the highways that allowed the transport of the milk to your local store, part came from taxes to support dairy industrial subsidies, part came from taxes to support government inspections, and so on. These other indirect costs might be a bit hard to separate out, but they are very real.

So what is the point? What bugs me is that nuclear-generated electrical power's "price" carries essentially all its indirect costs, while fossil and alternative energy sources don't. When your local power provider buys a kilowatt-hour of electricity from a nuclear power plant, the cost includes the paying off of the construction of the plant, the cost of the fuel, and the cost of other operational items like salaries for the plant staff. That would also be true if the kilowatt came from a coal-fired plant. But the nuclear cost also carries the cost of regulation, decommissioning, and waste disposal, which to a large degree other sources are exempted from in their direct costs. Regulation by the Nuclear Regulatory Commission is on a 100% fee-based system. That is, the entire operating budget of the NRC is paid by fees it collects from the nuclear operators it regulates. The plant owner is required to maintain a decommissioning fund with money locked up to ensure sufficient funds, based on NRC analysis, to completely clean up the power plant site after it is finally shut down. The plant owner also has to pay the government 0.1¢ per kilowatt-hour to support a nuclear waste fund, which can be nearly $9 million per year for a single large plant and has cost the total nuclear power industry about $16 billion to date. All these things show up directly as the price, the direct cost of nuclear-generated electricity.

Some will tell you that nuclear energy has had tremendous indirect government subsidies in research, facilities, and insurance. It is true that the current light water reactor technology came from the U.S. naval reactor program and benefited greatly from that. It is also true that the enrichment technology and the current U.S. facilities for enrichment were developed for the weapons program. However, in each case, what has happened is that a commercial, peaceful, civilian application has been able to make constructive use out of a military investment that would have been made anyway. Other directed research in the commercial power area can be legitimately scored as an indirect cost, however, seeing 20% of our electrical power come from safe, efficient nuclear plants it seems one of our wiser federal government investments.

Some will get very excited about the Price-Anderson Act, which sets a cap on how much insurance a nuclear power plant operator needs to pay damages in the case of a major accident. The government then stands by to pick up a serious chunk of cost beyond that. For those who get excited about this, let me just point out that so far this has cost the government exactly nothing. No claims, none.

On the other hand, fossil energy sources do not charge the buyers of their power for the pollution impacts of their use. The costs of those 28,000 deaths, the costs of treating all the health impacts of respiratory and circulatory illnesses short of death (measured in the millions), the lost work time, the impacts on forests and wildlife, the acid rain enhanced degradation in roads, bridges, buildings, and structures, all are not counted in the cost of that power. The cost of all those things is paid, make no mistake about that. They are paid in the form of taxes, lost resources, and real human impacts, but those costs are not charged as part of the price of fossil-generated electrical power. If more of those costs were internalized, made part of the cost paid to buy a kilowatt-hour from a fossil-fired electrical generating plant, then fossil-fired power would be more expensive, and nuclear would be much more competitive. The frustrating thing to me is that people end up paying those indirect costs, paying more for fossil-generated power and avoiding the truly less expensive, if you look at all the true combined costs, nuclear power because the price is seemingly too high.

But wait a minute, all that is all about coal. The EIA forecasts that essentially all of the new fossil energy growth is predicted to be natural gas-fired plants. The numbers we just looked at show natural gas is much cleaner than coal, so beating up coal is unfair to the future fossil fuel use.

Well, there are several things to be said about that. First, remember that, unless we unlock the methane hydrates, in the long run, fossil fuel use is going to be coal simply because that is the larger resource. Then there is the fact that coal is just about as cheap as dirt. Speaking of price, I get very nervous about the price of natural gas. During the California craziness, the price of natural gas ran up to $10/million BTU. Everyone with a natural gas well in their back yard turned up the pumps, and everyone with an exploration drill rig started poking holes in the ground. Seven months later the price slipped to about $3/million BTU and has slipped more since. The pumpers and explorers are now backing off. There is a feedback going on. When supply is low, the price goes up; when the supply increases, the price goes down. The problem is the timing of the two is out of sync. Price can go up very fast; for example, it increased from $4.50 to $9.50 between November and December 2000. However, supply takes months or even years to respond. This allows a lot of craziness to happen, and if I am on the end of the chain paying for natural gas-generated electricity, a lot of that craziness is going to happen to me. This will occur not only in the price of my personal electricity but also in the cost of my groceries and, essentially, everything else.

The other thing is that while natural gas is a lot better than coal, in things like sulfur dioxide, it is only a little better in carbon dioxide, which is a greenhouse gas. And that takes us to the next topic, which is global warming.

Remember that if we had perfect coal, just pure carbon, and burned it in pure oxygen without any nitrogen, we would get carbon dioxide and heat. If we burned pure methane in pure oxygen, we would get carbon dioxide and water. Lacking pure coal and pure natural gas and having to burn either in air, which is four-fifths nitrogen, we get other pollutants. We did a lot of things to reduce those pollutants, and while a long way from perfect, we have reduced pollutants per unit fossil fuel burned. However, there is very little you can do about the carbon dioxide. You can make the total thermal efficiency of the process a little better and use more natural gas relative to coal and get the most electrical power out per unit fossil burned, but any time you burn a fossil fuel you get carbon dioxide.

So, why is that important? Well, remember when we were talking about the history of the planet all the way back in Chapter 2, we saw that carbon dioxide was very important in regulating the heat of the earth. Carbon dioxide and a number of other gases are called greenhouse gases. It works like this. The frequency of light that a hot body gives off is proportional to its temperature. The sun is very hot. It, therefore, gives off heat in a high frequency of light. The earth is warm, but nowhere as hot as the sun. It too gives off heat but with a much lower frequency of light. Carbon dioxide is transparent to light with the higher frequencies but absorbs the lower frequencies. So what happens is the high frequency solar heat bearing light gets past the carbon dioxide in the atmosphere and warms the earth. The earth tries to radiate some of that heat back out into space, but this is being done with much lower frequency light. The carbon dioxide in the atmosphere absorbs that light, which warms the air and hence the earth.

Glass in a greenhouse does exactly the same thing as does the glass in your car on a hot summer day. The energy from the sun gets in but can't all get out and, therefore, the greenhouse gets warm.

Now carbon dioxide isn't the only gas that does this. It is neither the most effective per unit greenhouse gas nor the greatest overall effector. Methane is something like 20 times more effective on a per unit basis, and nitrous oxides are over 300 times more effective. But you have to weigh the per unit effectiveness with how much of it there is. In that context, methane comes in about 10% the total effect of carbon dioxide and nitrous oxide at about 25%. The single largest actor is water vapor because there is much more of it than the other gases.

What makes carbon dioxide so important is that it is the largest component of the greenhouse gas mix that human behavior is influencing. Before we look at just how we are doing that, let's do a little more review of the history of earth, carbon dioxide, the impact of carbon dioxide, and life on earth.

If you recall the very early earth had an atmosphere that had a lot of carbon dioxide and no free oxygen. This was the way that it was for the first 2.7 billion years. Then about 1.8 billion years ago sufficient photosynthesis (green plants) was active to start releasing oxygen to the seas and atmosphere. Over time the fraction of carbon dioxide decreased significantly. Along the way there were at least two "snowball" events in which a super ice age almost froze the planet completely. Our buddy carbon dioxide came along in both cases and set up sufficient greenhouse warming to thaw the planet back out.

Along with these dramatic "saves" carbon dioxide has also helped balance the planet's temperature over the long haul. We like to think of our sun as a pretty stable provider. And to a large degree it is. As a moderate- to small-sized star, our sun should produce a roughly constant output for billions of years. But it is only roughly constant. The life of a star is a minute by minute balance of gravity pulling in and the energy of nuclear fusion pushing out. As long as these match, it just cooks along. But clearly over time things have to change as the nuclear fuel gets used up. The nice thing about a smaller star is that its fusion processes are relatively slow and this takes a long time. Bigger stars go faster; the biggest perhaps as much as a thousand times faster. However, slow as it is, in our star there have been and will continue to be some changes. Over the life of the earth, the sun has increased its energy output by about 30%. That should have fried us by now or should have frozen us back when it was 30% less. Why neither of those things happened is that our greenhouse gases, especially carbon dioxide, have offset the change. Back in the days of lower solar output, we had more carbon dioxide and hence a better greenhouse blanket to keep us warm. As the sun has heated up, natural processes in the rocks, seas, and living things have reduced the free carbon dioxide and shed some of the greenhouse blanket. So carbon dioxide has been very important over time to keep the earth's temperature in a relatively narrow range. It is worthwhile to mention that, in 95% of the earth's history it has been warmer than it is now. In the last 40 million years the planet has been going through a series of ice ages with some short spikes of warmer temperatures called inter-glacials interspersed. Over the last 400,000 years, the inter-glacials have been coming along every 100,000 years or so. We are in the middle of one right now. If you follow the graph of past experience, you might expect it to run back to the ice age at any minute.

However, the theory of greenhouse gases and their relation to global warming is not without controversy. There are some features of existing climate change that do not match the theory. While this causes some to question the idea that human-generated carbon dioxide is really influencing the climate, the majority of scientific opinion is that it is.

Here are a few numbers to give you a feel of what is going on. Over the last 400,000 years, the carbon dioxide concentration has been floating between about 180 parts per million to about 300 or 325 at the highest. At the beginning of the Industrial Revolution, say about 1776, it was just shy of 280 ppm. In the late 1950s, it was up to 315 ppm. That is high, but within the band of "natural" swings. But by 2000 it had reached 370 ppm, which is the highest value in the last 420,000 years. This data and noticing all the fossil fuels we have been burning leads one to believe human activity may be playing a role.

There is a technique in science called change analysis. If something changes and you want to know why, you ask what else has changed and try to postulate a connection. The rapid and seemingly "unnatural" increase in carbon dioxide seems to match the significant increase in the emission of carbon dioxide by the rapidly multiplying and industrializing human population. If you are tempted to ask why can't we just add up all the stuff we have burned and take some samples of the air and figure it out, the answer is that it is a bit more complicated than that. You also have to understand the natural sources of carbon dioxide—like ocean offgassing and volcanic sources—and the natural sinks of carbon dioxide—like growing plants, carbon dioxide absorption by sea water, and take up in minerals. The uncertainty in all our estimates of these things makes it tough to predict who is doing how much to whom.

Being that folks were getting excited and concerned about the whole topic of global warming and suggesting some pretty draconian measures, the Bush Administration commissioned the National Academy of Sciences' National Research Council to look into it and summarize the best scientific opinion.

We need to appreciate the importance of the question. If greenhouse gases are a big problem, then we need to make some massive changes to our entire economic system in order to "fix" it. There are two huge forces going on here. One, changes that would be required mean vast portions of our industrial infrastructure must be replaced at a truly staggering cost. It is a legitimate question to ask just how firm the science is before you embark on that path. On the other hand, if the concerns are legitimate, then you need to understand how big the problem is and that acting as fast as you can is probably going to be too slow to avoid some pretty nasty impacts.

The National Research Council's report was released June 6, 2001. It was pretty cautious and stressed that there remained a lot of uncertainty in both the scientific models of climate behavior and the natural variation out of which we are trying to separate the human-caused impacts. However, the press release giving the key summary reads in part:

> "We know that greenhouse gases are accumulating in Earth's atmosphere, causing surface temperature to rise. We don't know precisely how much of this rise to date is from human activities, but based on physical principles and highly sophisticated computer models, we expect the warming to continue because of greenhouse gas emissions.
>
> "Based on assumptions that emission of greenhouse gases will accelerate and conservative assumptions about how the climate will react to that, computer models suggest that average global surface temperatures will rise between 2.5 and 10.4 degrees Fahrenheit (1.4 and 5.8 degrees Celsius) by the end of this century."

Well, how bad is that? There is not a single simple answer to that question; however, while we might stop short of "the end of life as we know it," it would probably be pretty bad. One of the things you hear the most about is the melting of glaciers and the polar ice caps causing a rise in sea level. Because a lot of our major cities are also major ocean ports and hence very near current sea level, this would not be a good thing. Some estimates predict sea level rising as much as 3 feet by 2100. If that doesn't sound like a big deal, recognize that if the sea level rises only half that we still lose about 800 square miles of Florida to the ocean.

Melting ice caps could also have some large impacts on ocean currents and global weather patterns. One chilling scenario is that melting the Greenland ice cap would alter the Gulf Stream pointing it away from Europe. This would mean that Great Britain would get very cold even as the average temperature of the planet increases. If that seems odd to you, go look at your globe. You will see that London, in southern Great Britain is actually about 250 miles nearer the North Pole than the northern most tip of Maine. Without the warm Gulf Stream, you could see caribou grazing in Hyde Park.

Another unhappy theory is that global climate change is not so much about total average temperature increases but about changes in the extremes of weather. That is, the theory suggests that it would not be it just getting a little warmer everywhere, but rather the swings of cold, hot, dry, rainy, mild, stormy, and such would be wider. This would result in more damaging storms, more severe droughts, more floods, more damaging hail, and all the other bad stuff weather can do to us. A few years of no summer rain in the American Midwest would turn one of the planet's most productive breadbaskets into a dust bowl.

It does make some sense that if you mess up the current balance of the weather, the atmosphere and oceans are going to be doing some very different things in getting to the new balance. Remember the scenario where an ice age locks up ocean water in snow and lowers sea level, allowing methane ice to thaw and causing huge greenhouse gas warming? Well, not to worry, because now we are talking about heating things up right? Well, how about this: You can get the methane ice to melt two ways. One way is to reduce the overpressure of the depth of water, or another is to heat the water. Oops. Is it possible to get a runaway warming if we heat the oceans too much? Perhaps not, after all the rising sea level will help us. However, it is an example of how a little change might get the wonderful fortunate set of factors holding the planet's climate in a livable condition out of balance and make it unlivable.

A very real example is the magnifying effect of water vapor. It works like this, as we add carbon dioxide to the atmosphere, we heat the atmosphere up a bit. This warmer air evaporates a little more seawater. Water vapor is one of the most effective greenhouse gases so that heats the atmosphere up even more. The multiplying effect is estimated to be about 2.5. That is, the actual atmospheric temperature increase will be 2.5 times as much as the carbon dioxide alone would have provided due to the added water vapor that comes with it.

Even if the doomsday scenarios don't get into play, a temperature rise of even 2.5°F is going to do a lot of things to agriculture. Areas near the equator may go unusable, and areas nearer the pole might become useable. So, should Kansas wheat farmers buy land on Hudson's Bay? Perhaps, it depends on what happens to rainfall patterns.

While there is a good deal of uncertainty as to just what will happen, the consensus is that it is going to be bad, perhaps very bad.

So if we are pretty sure our additions to the growing carbon dioxide in the atmosphere are going to warm the planet and that would be tough on humans and most other living things, why don't we stop doing that?

We might answer that simply by referring back to the need for energy to support a level of technology that supports our current human population and noting that our current infrastructure is based on carbon dioxide emitting fossil fuels. However, we can look at the problem more clearly if we break it down.

There is an equation that shows the various components that make up our human-caused carbon dioxide releases. It goes like this:

$$\text{Carbon release} = \text{Population} \times \text{GDP/Population} \times \text{E/GDP} \times \text{C/E} - \text{S}$$

Where:

GDP = gross domestic product, a measure of economic activity

E = Energy use

E/GDP = the amount of energy we use to create economic activity

C/E = the amount of carbon released to the atmosphere per energy used

S = sequestration, which are methods to pull carbon dioxide out of the atmosphere

The beauty of this equation is it lets us look at how each of these factors are changing, so we can predict what will happen to the resulting carbon releases. It also suggests where we can and cannot have an impact if we want to change things.

Let's look at these one at a time. We have already talked a bit about population. If we look out 100 years, we will see there is a lot of uncertainty on human population with estimates ranging from coming back down to the current 6 billion to rising up to 3 times that to 18 billion. However, one thing is absolutely certain and that is, in the near term, the next 50 years or so, the population will increase. This is particularly true in developing nations where the average age is under 30, unlike the baby boomer dominated developed nations. This means there is a lot of child bearing locked up in these young people and that momentum will carry the world's human population upward. Even the more conservative predictions look to a 2050 population approaching 8 billion, while others suggest a population northward of 10 billion. The point is there are going to be more people, that factor in our equation is going up.

So how about GDP/population? Recognizing that fully a third of earth's people live without electrical power, that is to say, in conditions at best matching the early 1800s in America or Europe, we can hardly ask people to dial this back. Rather, for altruistic humanitarian as well as self-serving national security reasons, we should be looking for the global average GDP/population to increase, and hopefully increase a lot. This is not just making life for the average American a little more plush. This is giving the average Bangladesh citizen clean water, medical care, an education, and a job that doesn't entail hours stooped over in a rice paddy. This is closing the gap between the haves and have-nots.

While Dickens was warning the English about the abuses of its class system in the following from *A Christmas Carol*, the excerpt also speaks to us about the real need to address the social gap on a global scale.

> "From the foldings of its robe, it brought two children, wretched, abject, frightful, hideous, miserable. They knelt down at his feet, and clung upon the outside of its garment.
>
> "'O Man! Look here! Look, look down here!' exclaimed the Ghost.
>
> "They were a boy and girl. Yellow, meager, ragged, scowling, wolfish; but prostrate, too, in their humility. Where graceful youth should have filled their features out, and touched them with its freshest tints, a stale and shriveled hand, like that of age, had pinched and twisted them, and pulled them into shreds, Where angels might have sat enthroned, devils lurked, and glared out menacing. No change, no degradation, no perversion of humanity in any grade, thought all the

> mysteries of wonderful creation, has monsters half so horrible and dread.
>
> "Scrooge started back, appalled. Having them shown to him in this way, he tried to say they were fine children, but the words choked themselves, rather than be parties to a lie of such enormous magnitude.
>
> "'Spirit! Are they yours?' Scrooge could say no more.
>
> "'They are Man's' said the Spirit, looking down upon them. 'And they cling to me, appealing from their fathers. This boy is Ignorance. This girl is Want. Beware of them both, and all of their degree . . .'"

O.K., economic activity and quality of life aren't places to make savings, but how about energy use per economic activity? Finally, we get a little good news. From a peak in about 1970 to the current performance, we have gotten about 25% better at this. That is, for about 25% less energy, we produce the same amount of economic activity: steel produced, jobs created, salaries paid. A lot of this was forced by the oil shocks of the 1970s and early 1980s. Some of it is caused by a shift from industrial to a more commercial economy, especially in the United States, but that can only go so far. Even if we shift from making steel to selling insurance to each other, someone—the Koreans, the Japanese, the Russians, someone—still has to make the steel. However, globally we have gotten better in just about every way we use energy. Some suggest that we can get twice even perhaps 3 times better at this by 2100. But it is going to take time and investment. You could make a major improvement if you just replaced all the incandescent light bulbs with florescent. All you have to do is go down to the local supermarket and buy one or four billion florescent bulbs and start screwing them in. I could buy a refrigerator that would use half as much energy as the one I have. It would only cost about $1000 to get that $50 per year savings. Like most people I am probably going to wait until my highly dependable refrigerator breaks before I replace it.

So what else do we have? The next factor in the equation is carbon released per energy used. There is a little bit of good news here and some not so good news. First, the good news. The various forms of fossil fuels produce different amounts of carbon dioxide per unit energy. Coal is the highest, and natural gas, the lowest, at about one half the carbon release for the same energy production as coal. Petroleum is in-between. Some of this is basic chemistry; some is in the thermal efficiencies of the ways each fuel type is used. The good part of this is that for the short term at least we are shifting more toward natural gas and away from coal. That means we should be getting a little better in this factor.

The bad news has got three parts to it. One, unless we tap into the methane hydrates in the long run we are going to have to go back to coal just because there is more of it. Two, the potential improvements are limited by the fact that any fossil fuel is still going to create carbon dioxide and by the investment required to replace all the less efficient, from the standpoint of carbon release per unit energy, coal users with somewhat better but hardly perfect natural gas. Three, the real answer, which is using a fuel that doesn't release any carbon dioxide, is not getting the chance it needs. And, of course, what I am talking about here is nuclear energy.

While it may seem a small point in a bigger picture, that latter point is the whole purpose of this book. The single largest thing we can do to impact the carbon release equation is to use fuels that take the C/E term to zero. Clearly you can't replace all the fuel use overnight and make the entire equation go to zero. But every non-carbon emitting energy source you bring on reduces the average and helps bring the average C/E down. Replacing an old inefficient coal-powered plant with a modern natural gas-fired power plant could half the carbon released per energy output. But a nuclear plant replacement would take it to zero, nada, zip, none at all. You don't burn anything; no carbon release for the energy you get. You can't get any better than that.

A lot of folks will suggest that alternative and sustainable energy sources also would take the C/E term to zero. We will talk more about them in the next section, but for now let me just say as much as I would love to have such energy sources be successful, I think a realistic assessment would tell you these are not going to have a planet-saving impact.

There is one more term in the equation. That is sequestration. The idea here is that, if by using technology, we can take carbon out of the ground and put it in the air, can we not use technology and get it back out of the air? The simple answer is yes, we can. We can do a little chemistry and turn carbon dioxide exhaust back into some solid form. We can bubble that exhaust into the ocean which can absorb truly huge amounts of carbon dioxide, we can pump the carbon dioxide exhaust into underground vaults like old oil wells, or we can grow a lot more green plants and have them take up the carbon dioxide. The general problem with all of these, except the green plants, is that the problem is so large the solutions have to be equally large. While you might do something with coal power plant stack exhaust if that plant is next to an ocean or oil field, what do you do about the roughly 520 million metric tons per year of carbon coming out of the American transportation system? Green plants are great, but again, how much more can you add on top of existing green plants and how much effect can these have?

So what does all this mean? Population goes up, maybe a little, maybe a lot; GDP/population goes up, probably a lot; E/GDP goes down, maybe quite a bit; and carbon per energy goes down, probably not a lot; and finally sequestration might help, but perhaps not very much. What does that mean to the overall amount of carbon we might be releasing in the future?

As you might very well expect, there is a range of possibilities. The general consensus is, however, the "the things get worse" influences outweigh the "things get better" factors. It comes down to the fact that the increases in population and economic activity per person will outweigh the mitigating features. Clearly a lot of potential assumptions could be made in doing the assessment. An organization established by the United Nations called the Intergovernmental Panel on Climate Change (IPCC) has been looking at the possible outcomes since 1988. They have produced several potential scenarios and charted the greenhouse gas concentration and the resulting global climate impacts of each of those. In their "middle of the road" scenario, they predict roughly a doubling of the carbon dioxide concentration up to 700 ppm by 2100 with a resulting 5.4°F average global temperature increase. This assumes a doubling of the human population and a moderate rate of economic growth. But listen to the list of things required from human technology to hold the carbon dioxide to a mere doubling:

- 75% of electricity will be from non-fossil sources[12]
- Overall fossil's share of our total energy will be down to 57% from the current ~85%
- End use efficiency will improve 1% per year suggesting a halving of the E/GDP term by 2100, twice as much product for the same energy.

Those are a lot of changes. It would require a massive change to the way people make electric power and an even larger change to how we transport our goods and ourselves. It also requires a lot of technology innovation in improving energy efficiencies and a huge economic investment to deploy those enhanced technologies. It should be noted that as I write this on March 14, 2002, yesterday the U.S. Senate defeated a proposal to require an improvement in the fleet averaged gasoline mileage standards for U.S. automobiles. The current standards have been in effect since 1989. In those standards SUVs were classified as "light trucks" and subject to a less strict standard. Because SUVs have become very popular, the American averaged gas mileage has hit its lowest level in 20 years. It doesn't look like we are tracking toward the energy efficiency technology challenges very well at the moment.

The point is that it is going to take a lot of work to get us down to a rate of greenhouse gas generation that only results in a 5°F increase and there is little time to get on with it. One of the things we need to realize is that it takes about 120 years to get carbon dioxide back out of the atmosphere. That is, on the average it will take 120 years for a gram of carbon dioxide we put into the

---

[12]France currently does that with its nuclear power, but worldwide fossil is about 63% now, with nuclear at 17% and renewables, mainly hydroelectric at 20%.

air beyond the natural balance to be extracted by the natural processes. So if we were to stop all human-generated sources of carbon dioxide today, it would be more than a hundred years before the planet could balance itself back out.

So what are people doing about this? Well, you may have heard of the Kyoto Protocol. This was the result of an international meeting held under the auspices of the United Nations in Kyoto, Japan, in December 1997. At this meeting, 160 nations addressed the problem of global climate change and the attending nations provisionally accepted targets on how they would reduce carbon dioxide emissions. As in any international setting, the politics and interactions were complex. It was recognized that the developed nations were the chief current sources of carbon dioxide. The DOE-EIA values support this citing that the industrialized nations plus the former Soviet Union nations account for 65% of the 1999 carbon dioxide emissions. At the same time, it was recognized that developing countries need to energize their economies to improve the quality of life of their populations. The deal that was worked out was that 38 developed nations would accept specific individual targets for reduction as fractions of their 1990 emissions, averaging 5.2% reductions, and that developing nations would not be constrained. This was a landmark achievement representing united global cooperation toward a common goal.

The spirit of cooperation did not last long after the meeting. Critics, especially in developed countries, looked at the predictions in future growth and said the agreement was not only unfair but would not adequately address the global warming problem. Again, using the DOE-EIA values, we see that by 2020 they predict that developing countries will become an increasing source of carbon dioxide coming up to 47% of the total, nearly equal the developed countries. Therefore, it was argued, some constraint must be placed on them rather than ask the developed countries to bear the whole burden of reductions. The rice farmer in Bangladesh might be excused if he fails to have sympathy for the American riding in his 6000-pound SUV.

The Bush Administration has taken a lot of flak from the environmentalists over the U.S. withdrawal from the Kyoto Protocol. Surely that is burying your head in the sand to not see long-term impending disaster in order to pander to short-term industrial greed. This is technology, industry, and government all being "bad" exemplified at their worst. Well, let me make two points. First, under the Clinton Administration, the Senate passed an essentially unanimous resolution rejecting the U.S. commitment to the Kyoto Protocol citing unsupportable economic impact. Second, the impact would have been unsupportable.

What the Clinton team signed up for at Kyoto was a 7% reduction from the 1990 carbon dioxide emissions by 2010. Well, that doesn't sound so tough. Why can't we do that? Well, the 1990 U.S. carbon dioxide emissions were 1345 million metric tons of carbon. That makes the U.S. Kyoto target 1251 million tons. However, because of population increase and economic gains, the predicted 2010 emissions are 1809 million metric tons. That is nearly 45% more than the Kyoto target. If you fool with the numbers, you will see that, to make that target, all we have to do is shut down 90% of all the 2010 transportation in America. Don't like that? O.K., instead you could just shut down 75% of all the industrial capacity in the country provided you also reduced the residential heating to half the projected amount. Pretty ugly, huh?

This shows the real crunch on the "what if we don't run out of fossil fuels" scenario. Just cutting back to 7% less than 1990 seems too little to avoid serious global climate change impacts, but even doing that would call for unacceptable economic and resultant real human impacts.

This is compounded by the have and have-nots gap. The haves don't want to give up what they have, and the have-nots want to have. Both drive the carbon emission equation upward.

One more point and I will turn to the next subject. In the middle of the entire debate on the Kyoto Protocol, the concept called Clean Development Mechanisms (CDM) was introduced. These were things that would help reduce the carbon emissions. Sustainable energy sources like solar, wind, and biomass were offered up as examples. Also substituting a less carbon emitting source, such as a natural gas-fired electrical generating plant for a coal-fired plant would satisfy the concept. The idea was to recognize things that could be done to influence the carbon emission equation toward reductions. Furthermore, the beauty of this was that developed nations could also get credit if they invested in CDMs in developing countries. Everybody wins. Developed countries would score against their carbon reduction target and developing countries would get development. Some worked to expand this to sequestration ideas and wanted to include reforestation projects.

Clearly the big ticket was working on the C/E term; especially here, you could add energy sources where C/E was zero, energy without net carbon release. Cool, right? The nuclear advocates should be happy because if anything is a CDM, it is us. We provide big energy with no carbon, and that is the very definition and goal of CDMs, right?

Well, it was a political not a scientific process. Much of the effort to deal with global climate change is driven by "Green" parties. Many of these, especially European Greens, have a dyed-in-the-wool technology-is-bad aversion to nuclear power. All science and logic aside, they are convinced that substituting nuclear for fossil is a bad trade. They cite nuclear waste, proliferation, and potential accidents as reasons to declare nuclear power "unclean" and hence not a valid CDM. Hopefully in the previous chapter I have convinced you that those concerns are overstated. However, in the end the Greens prevailed and nuclear energy, the single clearest, most effective way we have to offset fossil energy use, was not allowed as a CDM.

Little Nell is tied to the tracks, and she sees a Swiss Army knife lying within reach. Does she pick it up and cut the ropes? The European Greens disdain the idea of sharp objects that require intelligent handling to be safely employed and sit there admiring the beautiful spring foliage as the train bears down on them. Many of them still expect the hero to dash up and pull them off the tracks at the last minute. Alternative energy sources are the hero they expect. Let's look at them next.

The "Renewables" and the "Alternative Energy Sources"

Renewables are any form of energy that doesn't rely on a finite resource, that is one that can be sustained pretty much indefinitely. Folks get just a little generous in the definition because in the extreme nothing is indefinitely sustainable. After all in a few billion years the sun will go out, but, hey, that is long enough for me.

Alternative energy sources are just means that are not in big use now. Between the terms renewable and alternative, they have come to mean environmentally friendly, lower technology, less big business dependent forms of energy, something nature-loving Green parties would support.

Let me say right up front that my maternal Scotch ancestry really resonates with using the "free" energy of the wind or the sun and the idea of self-reliance that such sources of energy offer. At the same time, my American pragmatic engineering sense tells me that these sources of energy have limited feasibility and relying on them to pull Little Nell off the tracks is inviting disaster and that those advocating that are something between naive and criminal. I am doing some damage to my no adjective promise, so let me try to show you how I came to that view.

First, what are our renewable-alternative energy sources? Well, there are a lot of potential ideas out there. However, only five seem to have real feasibility at making significant contributions to our energy use. These are hydroelectric, biomass, geothermal, wind, and solar. There are others such as tidal power and ocean temperature difference, but they appear to be pretty low on the practicality scale, so we will focus on the aforementioned five.

Let's first ground ourselves with a description of each of these then refresh our concept of their current contributions. I will follow that with a discussion of their charms and limitations and conclude with an appraisal of what we might expect from each of these in the future.

The most familiar to us of the renewables is hydroelectric power. This is the classical hydroelectric dam. Anyplace where you have reasonable rain or snowfall and some difference in the height of the terrain, you have a real opportunity to make use of hydropower. You just find a nice river with a steep canyon, build a big wall across the river, and wait for the water to back up in the canyon, and then divert part of the flow through a water turbine connected to an electrical generator.

While the original construction can be very expensive and is usually conducted as a government project, the operation is relatively simple and the electrical power is pretty inexpensive. The project often has other benefits in supporting flood control, irrigation, water-borne transportation, and recreation, all of which help justify the construction and offset the cost.

While perhaps less familiar, the other big actor in renewables is biomass. As we saw early in this chapter estimating the total global use of biomass as an energy source is very hard. The concept incorporates a wide range of activities, from sophisticated combine heat and power plants in industrial applications to a sub-Saharan shepherd burning goat dung, to a developmental methane-from-municipal-garbage facility, to you burning wood in your fireplace.

I was personally surprised to learn how large a fraction of our global energy use this was. Again, getting a good number is hard, but it could be as much as 11% of humanity's current total. However, as we move to more industrialized economies, the contribution declines. If we look solely at the United States, the total biomass contribution comes in at about 3% with industrial biomass making up about two-thirds of that.

Geothermal is getting heat from the earth—geo meaning earth, thermal meaning heat. The idea is pretty simple and the appeal is huge. The earth is a great big ball of rock and stuff in which is distributed a lot of slowly decaying radioactive material like uranium and thorium. This decay gives off heat. Because the earth is a very big thing, it is hard for the heat to get to the surface and escape into space, especially since there is a nice big friendly sun warming up the earth's surface. So what you get is the earth is warm inside. The further in you go the warmer it gets—about 14°F per thousand feet in the upper portions. Mr. Carnot, the clever French gentleman, told us that if we have a difference in temperature we could create a heat engine, pump the hot to the cold and extract some fraction of the energy. Well, the energy represented by the heat in the earth's crust is huge and the temperature difference is there so we got it made, right? Ah, no, I'm afraid not.

The problem is the distance between the hot and cold places. At 14 degrees per thousand feet, we need something like 2 miles to find temperatures hot enough to boil water. To find temperatures like you would get from the steam of a modern nuclear or fossil-fired electrical generating plant, you would need to go down 7 or more miles. The energy used in pumping water against the friction of 14 miles of pipe pretty much wipes out the energy you would get out of the system. However, nature does help us out.

Although the 14 degrees per thousand feet is the average, it is not the rule. There are places where the earth's crust is thin and the very hot molten mantle is not deeply buried. Such "hot spots" are found under volcanoes and hot springs. In a few places, nature is really good to us and provides its own sources of water down to the shallow hot spot and provides a path for steam to come to the surface. Neat, we get hot steam for free. The steam is free but getting electricity from that steam isn't free. The steam is at temperatures lower than you have in a modern power plant, so the thermal efficiency is lower. Also the steam is full of minerals that are pretty tough on the equipment, so maintenance is a problem.

Some have tried to beat the problems of naturally occurring steam fields by creating artificial ones by pumping water into shallow dry hot rock, heating it, pumping the hot water back to the surface, and then using a heat exchanger to boil clean water for a conventional turbine. Professor Deffeyes (remember him from the discussion on how long the oil supplies will last?) tells a story about visiting an experimental hot rock facility. The site used the latest in oil drilling technology to get down to a dry hot rock field in a major volcanic area. When Professor Deffeyes asked the tour guide if they got more energy out than they used in pumping the water around, the guide couldn't answer. After a few minutes doing back of the envelope calculations, Deffeyes concluded the answer was no. The energy extracted from the hot water was not enough to run the pumps.

Energy from the wind has a lot of appeal. Humans have been extracting energy from the wind for hundreds of years. Windmills for grinding grain were one of the first applications of non-muscle energy to human technology. Much of Holland's current cropland was taken back from the ocean using the famous Dutch dikes and windmills to pump the seawater out.

Energy from the wind is pretty simple. If you put a series of airfoils, that is, wing-like, blades on a shaft, and put those in a breeze the shaft will turn. The moving air creates a low pressure on the curved portion of the blade and the relatively high pressure on the flat side imparts a force on the blade. Because the blade is pitched relative to the shaft you get a torque, a twisting force, on the shaft and it turns. If we can make something turn, we can make electricity, which is a great way to transmit the wind's energy to distant places where we want to use it.

There has been a lot of work done on wind turbines in the last 30 years, and they have gotten a lot better. Likewise the electrical generators that are matched up to the wind turbines have also been improved. The difficulty of course is that the wind doesn't blow all the time. We will talk more about that in a bit.

Solar energy is the alternative energy the Greens love to love. What could be better than Mother Nature's own sweet warm sunlight to nurture us all?

All you have to do is turn your cheek to the summer sun to feel the energy that is there. Surely we can make use of that, can't we?

There are two general ways in which folks try to make constructive use of the sun's energy. These are solar photovoltaic and solar thermal.

Solar thermal is pretty straightforward in concept. You just let the sun heat things up for you. That works pretty good if what you want is water hot enough to swim in and you have a warm climate and plenty of sunshine. However, it doesn't work if you want to melt steel in Michigan in the wintertime.

There have been some attempts to use solar thermal to heat things hot enough to make electrical power. One used a field of mirrors to focus sunlight on a tank to heat a coolant to be used in a conventional boiler. One used a field of individual heaters in which a series of trough-like mirrors heated tubes running through the troughs. While the sunlight was "free," the cost of construction and maintenance made these impractical. An accident at one of these sites where the oil used as the heat transfer coolant caught fire didn't help the cause much either.

Solar photovoltaic is a neat idea in that there is no machinery involved. The semi-conductor materials in the "solar panels" make electricity directly from the sunlight. In simple terms, the materials in the solar panels allow electrons to be knocked free by the sunlight and collected as electrical current. While the panels are not eternal and do degrade with time, they do last quite a while and will produce electricity as long as the sun shines. Historically, the efficiency of photovoltaics has been low and the costs high reducing the applications to places where there is no other source of power, such as power to satellites in space. Efficiencies have improved over the years, and the cost of producing solar photovoltaic panels has been reduced.

O.K., that is short summary of what the major renewables are. Let's now look at their current contributions to our energy use.

Back in Table 7.2 we looked at global energy sources. At that time, we noted how hard it is to get a good handle on renewables, especially non-fossil combustibles. However, the International Energy Agency information took a shot at it. Let's repeat and do a little assessment of those numbers in Table 7.6.

If we look at just the United States, we can get firmer numbers. Using data from the Energy Information Administration,[13] we get the following for use of renewables across all energy uses (Table 7.7).

Finally, let's look at the contributions of renewables to U.S. electrical generation. Table 7.8 gives that data.

For now I just want you to hold on to those numbers for future reference. However, you ought to be getting the picture that when someone tells you

[13] Annual Energy Outlook 2001 DOE/EIA-0383(2001) Table A18.

TABLE 7.6 Global use of renewable energy.

Renewable Type	1999 Quads	Percent of Total Energy
Hydro	8.9	2.3%
Combustible Renewables and Waste	43.1	11%
Geothermal, Wind, and Solar	1.9	0.5%

TABLE 7.7 Renewables as energy sources in U.S. economy–1999.

Renewable Type	Quads	Percent of Total U.S. Energy
Hydro	3.35	3.5%
Biomass		3.0%
Industrial	1.97	
Residential Wood	0.41	
Municipal Solid Waste	0.25	
Ethanol	0.12	
Other	0.17	
Total Biomass	2.92	
Geothermal	0.40	0.4%
Wind	0.05	0.05%
Solar		0.04%
Thermal	0.04	
Photovoltaic	0*	
Total Solar	0.04	

*Apparently less than 0.01 Quad—shows as zero in reference.

TABLE 7.8 Renewables in 1999 electrical energy generation.

Renewable Type	Billion kW-hours	Percent of Total U.S. Electrical Generation
Hydro	312	8.4%
Biomass		1.6%
Municipal Waste	22.1	
Other	36.6	
Total Biomass	58.7	
Geothermal	13.1	0.35%
Wind	4.46	0.12%
Solar		0.03%
Thermal	0.89	
Photovoltaic	0.05	
Total Solar	0.94	

about a new renewable source with a 10 MW or even 100 MW capability, that it is a pretty small contribution to an electrical energy system that at 3693 billion kilowatt-hours annual generation has a capability in excess of 400,000 MW.

Let's look at some of the factors that limit renewable and alternative energy sources. Again, I am not trying to beat these guys up. I am trying to give you a balanced perspective that appreciates the rose's perfume but also recognizes the thorns.

Hydro has a lot of charm. After all once you get the dam built, all you have to do is sit back and collect the kilowatt-hours, besides you get a pretty lake to look at and play in. Well, it isn't all that perfect. It isn't free in that there is some maintenance. More importantly for a non-polluting energy generation system, hydroelectric facilities are not without some environmental concerns. The biggest thing you will hear about is the impact on natural fisheries.

Many fish species have a combined fresh and salt-water life cycle. The fish are hatched in fresh water streams and rivers, migrate to the ocean as young fish, then after years to reach maturity at sea they return to the rivers of their birth to spawn and lay their eggs. When we put a few million tons of concrete across one of these rivers, we screw up this natural cycle. Not only does the dam represent a barrier to the upstream migration, the presence of a big lake and the unnatural water falls of spillways changes the temperature and water chemistry of the river, all to the detriment of the fish. Fish runs have taken major hits. The degree of responsibility that should be assigned to the dams is subject to a great deal of debate. However, it is clear that the fish were doing a lot better before we started messing with their rivers.

Dams have two other unfortunate features. One, if you impound water you slow it down. Well, sure, so what? Water in a river is not just pure H_2O. It also carries a good deal of suspended silt, good old dirt. The amount the water can carry is proportional to its speed. If you slow the water down, then silt falls out. So over time what you get is the lake behind the dam is filling up with silt. How fast this happens depends on a lot of things, but in general it takes quite a while. The point is, however, we have to realize that once again our definition of "sustainable" has to be a little relaxed. Eventually our use of a hydroelectric site will fill in the lake with mud, and it will no longer be useable.

The other unfortunate feature of a hydroelectric dam is that a big lake represents a lot of stored energy. If that energy gets released in a short period of time, a lot of bad stuff happens. Now, big dams don't fail very often, and they are overdesigned to be robust against floods, earthquakes, and just about anything the designers could think of. However, dams have failed with horrible consequences. When folks play the "what if this happens" game with nuclear power, I wonder what they would do if they applied that paranoia to dams.

The biggest limitation to hydroelectric energy, however, is simply that we have used up most of the good sites. While there are a few big projects to be constructed throughout the world, most of the available sites in the United States and Europe have been developed. There is some opportunity in what is called low-head hydro, which are just small facilities with a relatively small fall of water. However, in the aggregate, there just aren't enough of these to have a major impact.

Biomass is pretty interesting. From the greenhouse gas point of view, it is a neat way to get away with setting fire to something without a problem from the resulting carbon dioxide release. The key is that, while you are burning some biological material, you are growing the same amount or more at the same time. Since the new material you are growing takes carbon dioxide out of the atmosphere, the net effect is zero release. That is, your new green plants take out all the carbon dioxide you put into the atmosphere burning the old plants. In some sense you actually get ahead because some of this represents biological materials that would have just decayed anyway releasing carbon dioxide without any energy benefit to humans. This has the appeal of doing better with what you have now. To the degree we can make better use of our waste materials, we get "free" fuel, get rid of wastes, and do so in a carbon dioxide neutral fashion. Now, of course, the fuel isn't free. Even if we are getting the fuel from municipal solid waste, garbage to you and me, the very fact we have to handle it, sort it, and get it into some type of burner means it costs us something.

There is also the problem that carbon dioxide is not the only gas given off, especially if you are burning garbage. One has to be careful about the sorting of the feed fuel, adjusting the nature of the burner, and applying appropriate pollution controls to mitigate this concern.

A lot of the energy in biomass comes out of forestry or agriculture by-products. These are certainly cleaner that burning garbage, but not totally free of non-carbon emissions concerns.

We also have to balance the production of new biomass fuels with the investments required to create them. It would be tempting to look at the huge acreage given to corn production and think of all the energy you could get burning all those dead corn stalks in the fall. However, plowing the decaying corn stalks back into the soil helps the farmer reduce his need for fertilizers. Well, we could always just add chemical fertilizers to make up right? Oh, right, those fertilizers are made with fossil fuels releasing carbon dioxide in their production.

There has been a lot of effort put into marketing ethanol made from corn as a replacement for gasoline. There are a lot of factors and debates regarding relative energy content and differences in the amounts and types of pollution between ethanol and gasoline. However, the real issue is whether you use more fossil fuel energy growing the corn, harvesting the corn, and turning the corn into ethanol than energy you get out of the ethanol. The detractors say you don't. They say if you tried to run the farms, fertilizer plants, and ethanol plants on the product ethanol, the whole thing would run out of energy quickly. I don't know. There is a wide range in the amount of energy it takes to farm. The Europeans use more energy than we do in the United States, but they get more crops per acre. Is there a balance out there in which ethanol is a net energy producer? Maybe, maybe not. But, if there is, it is going to be a lot like a charity organization in which the majority of the funds collected get spent soliciting for new donations. Some goes to the cause, but a lot gets spent feeding the organization itself.

We have already touched on some of the limitations of geothermal power. I sometimes wonder if those advocating geothermal as a pristine energy source have ever smelled a sulfur hot spring. If you are going to rely on natural steam or hot water, you are going to get a lot of dissolved minerals in the flow. In addition to sulfur, you might find such lovelies as arsenic, mercury, and antimony. This can be an environmental problem as well as just crudding up the piping with mineral salts. If you go to dry hot rock and keep your water separated from the minerals, you get the added challenge of getting your water down to the depth of the heat source and back up.

While we can look at the temperature difference between the surface and the deep earth and see tremendous energy potential, the true limitation is that feasible extraction is limited to few sites of exceptionally shallow hot spots. We have more spots to exploit than we have to date, but huge growth just isn't in the cards.

As I was beginning to write this section, the March wind was getting a good start, and we were having a beauty of a dust storm. Wind has a lot of appeal. Capturing energy out of the air just seems like a good idea. Wind power has been around for a long time, so one might think there is not much more technology to develop. However, recent technical advances have significantly improved the technical and economic feasibility of wind energy. Of the renewable concepts, wind is one of the more economically competitive. Having said that, the truth is that wind energy sits on the bubble of being competitive with conventional fossil fuel energy, or not competitive. It depends on a several things and that takes a little telling.

Obviously the greatest challenge with wind power is that the wind doesn't blow all the time or at every place. Therefore, the first thing you have to do is to find a good site. A good site is where you get a lot of strong but not too strong wind, with a lot meaning a lot of the time. To understand the significance of this, we have to introduce the concepts of capacity and generation.

Capacity is the maximum amount a generating source could produce if it were to run full out all the time. Generation is how much power that source actually did produce during any given period. Capacity factor is the generation divided by the capacity. So if I have a coal-fired fossil plant and it runs full out 345 days but shuts down for 20 days for maintenance each year, then its capacity factor is 345/365 = 95%. If I build a wind farm, I may have a capacity of 100 MW, but it is very unlikely it is going to run full out all year long. In fact, you probably shouldn't expect a wind power capacity factor over 30%.[14] That means a 100 MW wind farm probably isn't as valuable to me as a 100 MW coal plant.

It gets worse when you look at it from an electrical system operator's point of view. The system operator has to mix the output of a number of generating units to match the electrical demand of his operating area. The load he must meet varies during the day. There tend to be two peaks: one in the morning, one in the early afternoon. Clearly the time of year, which affects the time people switch on lights, or heat, or air conditioners, makes further differences. However, the difference between the peak and the average can be substantial. This means the system operator has to be able to reach a peak demand that is greater than his average load. He can't just look at the national average of about 12,000 kilowatt-hours per person a year, multiply that by the number of people and divide by the number of hours in a year. If he does that math, he gets the average load. He then has to first adjust that up or down for his part of the country, adjust for the time of year, and then multiply by the peaking factor for his region. He also needs to add a margin to ensure there are no short-term peaks that can destabilize his grid and bring the whole thing down if demand tries to draw more than there is.

Now if this system operator has the opportunity to add wind power to his system what does he do? O.K. the capacity is 100 MW, but what can he count on at any particular morning at 6 a.m.? Well, he really can't count on anything. Therefore, even if the wind power is relatively cheap, the system is still going to have to invest in coal-fired or natural gas-fired, or nuclear plant that they can be sure will be there when you plug in the toaster. What this means is the "cost" of a wind farm is effectively more than just the price you pay for it because you also have to build other capacity to ensure you have adequate generation when you need it. Sure, you save the cost of the coal or natural gas you would use during that time in which the wind farm is producing, but you don't save the cost of building that coal or natural gas power plant.

So how is wind power doing? As we observed, the financial success of wind power is just on the cusp. The truth of it is that, where and when wind power has been given government subsidies, it has done pretty well, and where and when those subsidies either didn't exist or were withdrawn, it hasn't done very well.

---

[14]The Census Bureau's *Statistical Abstracts of the United States: 2000*, Table 971, gives generation and capacity for non-utility wind power from 1990 to 1998. Doing the math gives capacity factors ranging from 18.0 to 23.7%.

We need to understand what these subsidies mean. To do that we have to step back a bit and look at the concept of human governments. The idea, at least since we have given up on the divine right of kings, is that collectively we will establish and empower a group to create and enforce a set of rules, policies, and programs for the benefit of all. We decide, for instance, that paved roads would be a good thing for commercial and personal transportation. We, therefore, empower state and federal highway commissions to build highways and allow them to levy taxes, which we pay to support that. The process of deciding just what are good things to put our collective support behind leads to the interesting human activity called politics and the wonderful saying that law and sausages are two things you really don't want to see being made.

So what happened is advocacy groups convinced a range of federal and state lawmakers that non-fossil energy sources were "good," and therefore, we should help support them. Now if we were strictly capitalists and interested solely in the least expensive, most efficient means of production, this would bother us. By subsidizing a non-competitive energy source, we are taking money out of more productive generation and increasing the cost of essentially everything. The socialist point of view counters that the "free market" does not adequately balance non-market concerns like air pollution, resource depletion, and global warming. Therefore, additional non-market corrections, like subsidies, need to be added to the market to get the more balanced, socially advantageous result.

We will return to this in the final chapter, but the point to be made here is that the political process is engaged all the time making decisions for you about things like this which a) will impact you in very real ways, b) never consider your interests if you just sit there and take it, but c) can be extremely responsive to your wishes if you do speak up.

So let's get back to wind power and subsidies. There have been several different state and federal programs designed to help renewable energy sources. The Public Utility Regulatory Policy Act (PURPA) of 1978 encouraged electric utility companies to buy power from smaller non-utility generations, some of which employ renewable sources. The nature of the law was that utilities were to pay these small generators at the highest rate of "avoided" costs. That means that, rather than letting the free market establish the price, one would look at what you think the highest price would be from the utilities "normal" sources and set that as the price the utility would pay the non-utility generator. The idea was to give the little guys a break and especially to help those out who had a "green" source of power but could not compete with the lowest cost providers. At that time we were in the middle of the oil shocks and folks were expecting $100/barrel oil to be appearing soon. Some of the PURPA contracts got written for prices reflecting this very high expectation of future costs. This means the average price paid to the PURPA sources was between 2 and 3 times the average price to the normal suppliers and often in excess of what the utilities actually sold the power to its customers.

We can see the direct effect of PURPA on wind power if we look at the growth in non-utility wind power generation in the 1990s. From 1990 to 1994, the generation grew by 55%, a whopping increase. However, about that time the 10-year contracts set up under PURPA started to run out. Generation took an 8.5% drop in the next year and by 1998 had fallen below the 1993 level. The 2001 data from the Energy Information Administration indicates that wind power is back on the growth track.

Helping to support that growth are a number of state programs encouraging renewables and in addition the federal production tax credit of 1.7¢ per kilowatt-hour was extended from its original termination in 1999 to 2001. It might be instructive to point out that in 1999, the U.S. nuclear power median generation cost, that is the cost of producing power once you have bought the plant, was 1.64¢ per kilowatt-hour. That means if you applied the 1.7¢ tax credit to nuclear power, then nuclear-generated electricity would have been free. Of course, you wouldn't do that because, while we could afford the $76 million subsidy for the 4.46 billion kilowatt-hours of wind power in 1999, the $12 billion it would take for a similar subsidy of the 730 billion kilowatt-hours of nuclear electricity would have been prohibitively expensive. The point is the reason you could subsidize wind power is because there wasn't very much of it.

However, it does appear that with modest continuing subsidies, wind power will be competitive and may become more so if fossil fuel prices increase. Wind power does have a few other problems we do need to acknowledge. The wind turbines have proven to be a bit hard on birds. Much of the 1990 development of wind energy has been in California. The environmentalists were a bit disconcerted to discover that between 1992 and 1998 the Altamont Pass wind project had racked up 1025 bird kills including 149 endangered golden eagles.[15] This has been mitigated with newer designs that rotate slower giving the birds a better chance to see and avoid the blades.

The other complaint against wind turbines is aesthetics. That is, they just don't look very nice. If you have ever driven east out of San Francisco, at one point you will be given a grand view of a row of majestic hills covered with hundreds of spinning windmills. The grace of the natural lines of the hills is not enhanced. As in all things aesthetic, it is in the eye of the beholder. Is it a coarse invasion of nature with man's contraptions, or an uplifting testament to man working with nature? Hey, I think snowy white steam clouds rising majestically from a nuclear plant's cooling tower on a cold winter's day are lovely.

---

[15]*Seattle Times*, March 25, 2001, "Silicon Valley turbines too perilous for condors."

Solar is a tough one to deal with. On one hand, you have to love free energy from the sun. On the other hand, the practicalities are just, well, not very practical.

You will hear that there is much more energy in the form of sunlight falling on the earth than humans could possibly use. And that is pretty much true. However, one does need to appreciate that you can't divide your energy needs by the solar energy per unit surface area to get the square feet of solar panels you need. Follow along on with me as I attempt to demonstrate.

We will start with what is called the solar constant. This is the amount of energy per unit time (that is, power) divided by the area of a sphere with a radius the average distance of the earth from the sun. That number is 2 calories per square centimeter per minute. Converting to more convenient numbers that is just less than 1.4 kilowatts per square meter. For any given spot on the earth's surface, you will have an average of 12 hours of sunlight a day, so that is 16.8 kilowatt-hours/day • meter^2. However, not all that sunlight gets to the earth's surface. Only about 40% actually gets down to where we live, so that is about 6.7 kilowatt-hours/day • meter^2.

Let's do a little check on that number. The National Renewable Energy Laboratory (NREL) operated by the U.S. Department of Energy has collected solar energy data for quite a while. One of their interesting products is a series of maps of solar energy for the U.S. presented in a variety of formats. If you look at the annual average taken over the period 1961 to 1990, you will see the nation generally gets between 4 and 5 kilowatt-hours/day • meter^2. That is the actual received solar energy as would be collected by a flat plate laid on the ground. This then adjusts for cloudy days. If we compare that to our 6.7 kilowatt-hours/day • meter^2, it implies it is sunny about two-thirds the time. That seems to make sense.

Now NREL also shows that if instead of a flat plate laid on the ground, we used a two axis sun tracking system that would keep our panel optimally pointed at the sun throughout the day and month to month, we can get more energy. The NREL data range from 8 to 10 kilowatt-hours/day • meter^2 for the desert southwest to 4 to 5 kilowatt-hours/day • meter^2 for the northwest coast. For the larger part of the nation, something between 6 and 7 kilowatt-hours/day • meter^2 seems about right.

So, let's continue the calculation using the 6.7 kilowatt-hours/day • meter^2 to get a feel for what solar can do for us. If we collected that all year, we would get 2450 kilowatt-hours/meter^2. The average use of electricity in the United States is about 12,000 kilowatt-hours per person, so that means each person needs about 5 square meters worth of sunlight to meet their electrical energy needs. But how many square meters of solar panels do you need to do that? Because solar panels do not convert 100% of the sunlight striking them into electricity, but rather a fraction of that expressed as an efficiency, we will need more panel area.

There are a lot of numbers on solar panel efficiencies. The Energy Information Administration suggests 14% for 2000–2004 technology increasing to 16% after 2005. I have seen claims beyond 14%, so let's guess at 15%. That means, to get the 12,000 kilowatt-hours, we will need 33 square meters of solar panels. But that is the annual average. We are going to need power in the "bad" solar energy collection months and really can't save up the amount above the average we get in the "good" solar collection months. If we look at the NREL monthly data, we see that really sunny places like Albuquerque have a minimum-to-average ratio of about 0.75. That is, the worst month gets about 75% as much as the annual average. Really cloudy places like Seattle have a minimum-to-average ratio of only 0.31. Just for round numbers let's use a worst month at 0.5 of the average. That means, if we are going to get our 1000 kilowatt-hours in that month, we will need 66 square meters of panels. If we add a 15% capacity-to-generation margin to handle at least some of the daily peaks and swings, that takes us to 75 square meters. That is about 810 square feet. That is not too bad. A lot more than the 5 square meters we started with but still not a really huge area. The first house I owned was slightly smaller. If you want enough electricity for a family of four, that takes you to 3240 square feet, which is a moderately large house. But then again, that really doesn't meet your electrical energy needs. You are going to need power at night and on those days when it is too cloudy to bring your panels even up to the worst month collection rate. That means, in addition to the investment to be made in the solar panels, you will also have to invest in a coal, natural gas, or nuclear plants to supply the energy you need when solar isn't available. You will save on the cost of fuel to those plants for the time solar is working, but you will have to build them all the same. That makes solar pretty expensive.

But just how expensive is solar if we ignore the little problem of buying all the backup capacity? After all the sun is free, so solar is cheap, right? Sorry, this is where solar really fails big time.

First, let's have a little lesson in how the cost of electrical power is figured out. If you own an electrical generating plant, you have three things that set the cost of your operation and, therefore, the price you have to charge to stay in business. These are the capital cost of the plant, what it cost you to build it in the first place; the operation and maintenance costs, things like salaries for your operating staff, spare parts, regulatory and administration costs; and finally the cost of your fuel. You might notice that only the cost of fuel and some of the maintenance varies with how much power you actually make. The capital and much of the operation cost are fixed whether you are running the plant or just standing by waiting on a cloudy day. Table 7.9 shows you some numbers from the Energy Information Administration on their

TABLE 7.9 Electrical generation costs, 2005 advanced facilities.

	Coal		Natural Gas	
Component	**¢/kW-hr**	**Percent**	**¢/kW-hr**	**Percent**
Capital	3.1	72%	1.2	29%
Operation and Maintenance	0.43	10%	0.2	5%
Fuel	0.78	18%	2.7	66%
Total	4.3		4.1	

estimates for these factors for a 2005 generation advanced coal and natural gas-fired plant, in 1999 dollars.

A quick look tells you why people want to build gas-fired plants. Their capital cost is low. You can build them relativity inexpensively and start making money without sinking tons and tons of money into them. A slightly longer look should also show you the scary thing about a big dependence on natural gas-fired electrical generating plants. Their big-ticket item is fuel, and as we have seen, the price of natural gas is something less than stable. If over the years you shy away from coal plants and replace the older ones with natural gas-fired facilities, when the price of natural gas doubles or triples as it did in the winter of 2000–2001, you won't be able to run anything but gas-fired plants because that is all you have and the price of your electricity is going to go way up. Doubling the cost of fuel for coal would not have anywhere near the same impact.

But let's get back to solar. The EIA information also lists what is called the overnight construction cost of various types of power plants. The term means the cost of labor and materials without any financing charges, as if you built it overnight and started making money with your investment the next day. The real cost of capital is going to be a bit trickier to figure because you have to work in the length of time for construction and what kind of a deal you can work out with investors. But overnight construction cost gives you a starting place comparison. The numbers EIA provides (in 1999 dollars) are as follows:

Natural gas = $450/kilowatt

Coal = $1100/kilowatt

Solar (2004 technology) = $7870/kilowatt

If we ratio that construction cost against the capital component given in Table 7.9 for natural gas, we get an equivalent capital cost for solar of 21¢/kilowatt-hour. If you use coal as your basis you would get 22¢/kilowatt-hour. That means that the cost of power from the solar system, even with free fuel and assuming zero operation and maintenance, of which there would be some, comes out about **5 times** as expensive as the fossil fuel options. That is it, that is the killer for solar.

But wait, isn't the cost of solar coming down? You bet. There is some really cool stuff going on. One of the most interesting things is the new thin film technology. Instead of the big thick fragile glass-like cells we are used to, this new technology is only a fraction of an inch thick. It uses a lot less material, and while the early versions were pretty poor in efficiency, the technology is coming along and seems to be matching the best of the older technology. And as you make more of these, the economies of scale will start to pay, and the cost will come down. The EIA predictions are that by 2015–2020, the efficiencies will be up to 20% and the costs down to $4158/kilowatt (in 1999 dollars). That is, between 11¢ and 12¢ per kilowatt-hour and only 2.7 times as much as the fossil options were in 2005, all calculated in 1999 dollars for consistency.

So while we can look for big improvements in solar technology and really hope to see a big breakthrough, the best look right now is it ain't going to happen. And remember, even if the cost of solar was the same as fossil, it wouldn't be. Did that sound confusing? Remember, even if the generating cost was the same, it would cost you more to have a solar power system, because you would have to build all that backup capability. If the generating cost was the same, it would be cheaper just to build the full fossil backup capability and pay for the fuel to run it and leave the solar in the laboratory.

O.K., that is a run down on the limitations of each of the major alternative energy systems. We have also touched a bit on the Energy Information Administration predictions. Let's look to see what they think will happen with renewables and alternative energy in the next 20 years.[16] To keep things simple, we will stick with just the United States. Table 7.10 is much like Table 7.7 showing renewables as part of the total energy picture but gives both the 1999 data and the 2020 prediction. Table 7.11 is similarly analogous to Table 7.8 focusing on electrical energy only.[17]

One thing that might surprise you in the table is that hydroelectric power actually is predicted to decrease. The rationale the EIA gives for this is that there is little potential for new additions of hydroelectric capability and existing facilities will be restricted due to environmental constraints, such as the added spill to help the young salmon get to the ocean.

---

[16]For full disclosure, I must say that I don't fully agree with all their predictions. I think their estimates for nuclear power are too pessimistic. The fact that their 2001 looking forward estimate for nuclear power is more optimistic than their 2000 estimate might support my view. However, I will admit that their nuclear estimate is a fair call as to what will happen if we stay our current course, which of course, I am trying to change.

[17]Both tables are data from *Annual Energy Outlook 2001*, Energy Information Administration DOE/EIA-0383(2001).

TABLE 7.10 U.S. renewable energy 1999–2020.

	1999		2020	
Source	Quads	Percent of Total	Quads	Percent of Total
Hydro	3.35	3.48%	3.24	2.55%
Biomass	2.92	3.04%	4.33	3.41%
Geothermal	0.4	0.42%	0.8	0.63%
Wind	0.05	0.05%	0.13	0.10%
Solar	0.04	0.04%	0.06	0.05%
Total Renewable	6.76	7.03%	8.56	6.74%
Total Energy	96.14	100%	127.04	100%

TABLE 7.11 U.S. renewable electrical energy 1999–2020.

	1999		2020	
Source	Billion KWhr	Percent of Total	Billion kWhr	Percent of Total
Hydro	312	8.44%	302	5.7%
Biomass	58.7	1.59%	103	1.94%
Geothermal	13.1	0.35%	25.8	0.49%
Wind	4.46	0.12%	13.1	0.25%
Solar	0.94	0.03%	3.48	0.07%
Total Renewable	389.2	10.53%	447.4	8.44%
Total Energy	3697	100%	5299	100%

O.K., what do the rest of the numbers mean? To show how it is important to be careful with your numbers, let me answer that question by telling you how some folks have used this information. I have seen in national publications words like: Renewable energy sources like clean wind and solar now make up more than 10% of our energy needs. Then a little bit later in the article, they will say something like: Solar is the fasting growing energy source expected to nearly quadruple by 2020.

Have they lied? Have they misled? Well, the answer to the first question might be ambiguous. The answer to the second question is an unqualified **yes.**

Let's look at what is implied in those words. First, they might have mixed total energy and electrical energy. They might like to quote the electrical energy numbers because they give the highest fraction for renewables (10.5%) versus using the total energy fraction (7%), and the highest growth rate, that being 370% over the 20-year span for solar.

Let's guess that you saw past the electrical versus total and looked at electrical only. If you assumed that because renewables **like** wind and solar were 10% and growing at 370% over 20 years, you would assume renewables were ready to save the day, at least in regard to electricity production. In fact, if you worked out the math, you would get an annual growth rate of 6.76% for renewables **like** wind and solar. If you looked at the total electrical growth predicted over the 20 years, you would see an annual growth rate of only 1.8%. Clearly then, renewables are gaining a big share. If you extrapolate at those growth rates, you would get something like 70% of all your electricity from renewables by 2040. Happy days. Forget about global warming. Lighten up on imported oil concerns. Why bother with a book advocating nuclear energy? Renewables are going to save the day. **NO, NO, NO!**

The misleading words are **like wind and solar.** Of the 10.5% of our current electrical power that are renewables all but the 0.5% is hydro and biomass. Are these **like** wind and solar? If by "like," you mean doesn't burn fossil fuels, then sure. If by "like," you mean wind and solar are economically competitive with existing fossil sources and, therefore, likely to contribute in a meaningful way as hydro and biomass will, then the answer is a great big no.

Table 7.12 does the extrapolation honestly keeping the individual terms clean without crossing the distinctions with the **like** terms.

TABLE 7.12 U.S. renewable electrical energy—extrapolated to 2040.

	1999	2020	2020/ 1999	Annual Rate	2040 Extrap	2040
Source	Percent of Total	Percent of Total	Percent	Percent	Billion kWhr	Percent of Total
Hydro	8.44%	5.70%	97%	-0.16%	292	3.85%
Biomass	1.59%	1.94%	175%	2.85%	181	2.38%
Geothermal	0.35%	0.49%	194%	3.45%	50.8	0.67%
Wind	0.12%	0.25%	294%	5.54%	38.5	0.51%
Solar	0.03%	0.07%	370%	6.76%	12.9	0.17%
Total Renewable	10.5%	8.44%	115%	0.70%	514	6.77%
Total Energy	100%	100%	143%	1.82%	7595	100%

The key thing to glean from all these numbers is that the larger contributors to the renewables, that is, hydro and biomass, are not expected to grow at the great rate of the highest growth rate renewable, that is, solar. In fact, hydro is predicted to decline. And while the growth rate for solar is indeed impressive, it is starting from a fraction so small that, even at the end of 40 years of this spectacular growth, it would still be only a fraction of 1%. Wind, while not growing quite as fast, does start with a larger base, but still, in this extrapolation, doesn't beat 1% of the electricity supplied in 2040. The result is the overall growth rate of renewables measured as total electrical energy supplied doesn't keep up with the predicted growth rate of electrical energy overall. The fraction of renewables actually goes down with time. Rather than reaching 70% by 2040, the renewable fraction falls from 10.5% in 1999 to 6.8% in 2040, using this extrapolation.

Do you have to follow that extrapolation? No, things could happen. New solar technology might blossom and drive costs way down. Natural gas prices might go through the roof and wind energy would get so competitive that we will put a windmill on every hill. It could happen. But if you are going to count on any of those things, I have a bridge in New York I would like to sell you.

Which brings me back to those who use the words paraphrased above to imply that renewables are just about to save the day. If they make these statements not understanding the implications and the erroneous picture given, then that is very poor journalism. If they understand the error and purposely mislead, then that is criminal. For if the general public is misled into believing that wind and solar are going to meet the country's energy needs, then they are not going to be ready to address the pollution, global warming, and national security challenges that the actual continued use of fossil fuels brings. And, of course, my point is such a misled public will not see any reason to support my alternative of nuclear energy.

Concluding Thoughts on Energy, the Big Picture

We have looked at how energy is used in the world and more specifically in the United States. We have looked at where we get energy and how much the use of energy is likely to increase in the near future. We have spent some time exploring fossil fuels and the challenges those energy sources present. We have looked at renewables and alternative energy sources and touched on the limitations associated with them. So what should we draw from all of that? What messages should we take away from it all? To me the key points are

- In the near future, global energy use is going to increase and increase by quite a bit. This is due to both the increasing human population and what we hope to be an improvement in the quality of life for the citizens of the developing nations. The latter is important to support both from purely altruistic, humanitarian goals of reducing suffering and the pragmatic self-serving national security goals of developed nations seeking to reduce destabilizing and conflict-generating forces in the world.

- Fossil fuels are our primary source of energy both in the United States and the world at large. There is only so much you can do to change that over the short term. We have had over a hundred years to develop the infrastructure that both provides and depends on the current fossil fuels. You can't convert all the cars to run on hydrogen in a year and provide the production and distribution systems to make that work overnight.
- Continued dependence on fossil fuels is a problem. There are several ways to screw up. We might run out or at least run low in regions with high demand. That will lead to economic disruptions and most likely to armed conflicts. Or we may not run low and instead trash the planet through pollution and global warming. Or most likely, we will do a little of both and still make life very tough for a lot of people.
- Renewable and alternative energy sources are very appealing, but in the cold hard light of day, they just are not likely to make a huge impact. The larger current contributors are hydroelectric power and biomass. Hydro is limited by the very few additional sites to be exploited worldwide and by increasing environmental restrictions on operation. Biomass can see some growth, but the resulting predicted increase will only slightly outpace the growth of energy use overall. Alternative energy sources such as wind and solar have significant problems in their low energy density and intermittent supply. With no way to save the energy from windy or sunny days to be used on calm or cloudy days, energy systems employing wind and solar need significant conventional backup capacity. Wind energy has proved commercially viable only when subsidized by governments but is close in cost to conventional energy sources. Solar is burdened with much higher costs, and widespread application requires massive and, at this point, unexpected improvements in efficiency and reductions in cost. The net effect is renewables' growth is not expected to keep pace with overall energy growth, and, therefore, the renewable fractional contribution is predicted to actually decrease.

So where does that leave us? As I see it, we are Little Nell, tied to the tracks. We can't see the train that is coming, but from the vibration in the rails and the sound of the whistle, we know it is coming and it is darn close. We don't know if it will be economic ruin, international conflicts, or environmental disasters, but the train of fossil fuel dependency is bearing down on us. Some see a rustle in the bushes and are sure that it is the hero of alternative energy sources getting ready to spring out and free us of our bonds. It sure looks to me that the rustle in the bushes is nothing but a rabbit hopping away from us—no help there. But remember, only our feet are tied. Our hands are free. We can act. And there is a Swiss army knife lying within reach. Don't touch that nasty old Swiss army knife; it might hurt you! After all, what about nuclear waste, proliferation, potential nuclear accidents? Well, be careful with the knife for sure, but hey, let's get sawing on those ropes, don't you think?

We do have to be careful here in this analogy. The Swiss army knife is not just nuclear energy. It is doing all you can with renewable energy sources, it is doing all you can to extend the supplies of fossil fuel sources, it is working the political consideration to reduce the potential for future resource constraint generated conflicts, it is making fossil fuel use as efficient and low polluting as we can. What I am all about is that nuclear energy can have a very significant part of that mitigation. The train is coming at us. As fast as we can cut using all the tools we can may not be fast enough. Leaving any tool out of our disaster mitigation program, especially a major one like nuclear energy, would be a very bad thing. We owe the rest of the world and our descendents a better world than one in which Little Nell is still on the tracks when the train arrives.

8

THE STATUS OF NUCLEAR POWER TODAY

So, how is nuclear power doing today? Actually, it is doing pretty well. To appreciate how remarkable that answer is you have to appreciate that just a few years ago the answer would have been much different and much darker. During those dark days, I was told more than once that nuclear power was just a dinosaur too dumb to know that it was dead. However, today nuclear power is more likely to be seen as an efficient, reliable, safe, and economically competitive source of a good deal of the world's electrical energy. Why wasn't that true a few years ago, and what happened to change things?

To answer that, we need to do a mini-history of nuclear power. What follows is really compressed and focuses solely on the high points to illustrate the changing fortunes of the industry, so please don't take this as a comprehensive study of the subject.

The first commercial nuclear plant at Shippingport, Pennsylvania, started up in December 1957. This was an outgrowth of the naval nuclear program. From that beginning a slow groundswell grew with a diversity of utilities and commercial firms signing on to the nuclear bandwagon. By the late 1960s and early 1970s, nuclear power was enjoying considerable popularity. By 1970, nuclear power plants were producing a hundred times more electricity than Shippingport did in 1958. It seemed a new reactor was announced every week. At one point, it was predicted that there would be 200 reactors in the United States by the year 2000. We ended up with 103.

So what happened? Oh, I know. It was the Three Mile Island accident in 1979, right? Well, clearly that wasn't a big help, but no, that wasn't it. That is a common misperception.

The trouble started years before the Three Mile Island accident. The trouble started in 1973 with the oil crisis. Well, gee, if fossil fuels were in trouble, wouldn't that help other forms of energy? The answer is yes, in some ways and some places, and no, in other ways and other places.

In 1972, oil was one of the principal fuels used to make electricity. By 2000, oil was only rarely used to generate electrical power and is now limited to places like Hawaii, where other fuels are impractical. As oil was replaced, some of the slack was indeed taken up by nuclear energy. Before 1973, France, in particular, was very dependent on petroleum for electrical energy and essentially all of that petroleum was imported. The French decided this wouldn't do and established a very focused nuclear power program. France now has the world's second largest nuclear power program with 59 power reactors, and they get over three-quarters of their electricity from nuclear energy.

However, besides creating a demand for alternative sources of fuel for electricity production, the oil crisis had other more impactful effects. These were a general depression of economic activity and creation of strong inflationary pressures.

In the United States, these negative impacts had the greater influence on the fortunes of nuclear power. First, the general economic recession reduced the demand for new electrical energy generation. A lot of the new planned power stations, including nuclear ones, were no longer needed. Nuclear power got a double hit because inflation pressures had a disproportionate effect on nuclear construction compared to other energy sources.

In Chapter 7, we looked a bit at the components of the price of electrical power and at the "overnight" cost of building power plants. We saw that for coal-fired plants, the capital cost, that is the construction cost, was a larger share of the cost of power as compared to a natural gas-fired plant. On the other hand, for natural gas-fired plants, the cost of power is more dependent on the cost of fuel. This difference in the components of cost is caused by the fact that coal-fired plants are more complex than natural gas-fired plants. Nuclear plants, with their redundant safety systems and massive containment buildings, are even more complex than coal-fired power plants and, therefore, their construction costs are even greater than coal plants. While the coal-fired "overnight" construction cost might be on the order of $1100/kilowatt, nuclear costs are on the order of $1700/kilowatt.[1] However, that is the "overnight" cost.

---

[1]There are a lot of possible numbers that could be cited. The various plants that have been completed range from under $400/kW to as much as $5000/kW. However, something above $1500/kW but under $2000/kW is representative of what most think it would cost to build a new nuclear plant equivalent to an existing modern PWR with U.S. safety standards.

You don't build anything overnight. It takes time. As anyone who has ever calculated just how much they really pay for their house after they pay all the finance cost knows, the longer you pay on borrowed money, the more you pay. Nuclear construction, being complex, takes longer than other power systems. Therefore, when inflation and interest rates took off after the oil shocks, nuclear construction became disproportionately expensive.

And it gets worst. Another hit about this time was our increasing environmental awareness. In 1969, we created the Environmental Policy Act, which basically said that any major federal government action must include a consideration of environmental and resource utilization issues. It didn't say you had to take the environmentally best option, only that you had to consider the issues. This led to the creation of the environmental impact statement, a study of the environmental and resource impacts of the pending federal government decision. One of the first really big applications was the Nuclear Regulatory Commission granting of operating licenses to utilities for nuclear power plants.

Now life was already tough for a utility trying to build a nuclear power plant. The NRC licensing process was a two-step process. First, after stating that you wanted to build a new plant, you requested a construction permit. To support that, you wrote a preliminary safety analysis report that explained what you wanted to build, why that would meet reams of NRC rules, and why that would work at the site you had picked. The NRC staff would review your submittal and ask you a ton of questions to which you would have to respond. Then you would have public hearings. At these hearings, anyone would be able to come in and have their say. Well, good, that is democracy at work, right? Well, yes, but it is also a great way for a few vocal people to cause a lot of grief, which they did.

Now, if you got past all of that and you built your plant, then you came back to the NRC with a request for an operating permit. That was supported by a final safety analysis report, a document often spanning 3 to 4 feet of bookshelf space. This was reviewed, questioned, and yet again offered to public hearings. At this point, you have poured a lot of concrete, welded a lot of steel, run a lot of wire, spent a lot of money, and paid tons of finance charges. However, folks could come in and claim there was a new safety concern, or that you didn't do this or that right, and just generally run you around and around. So with that and the dozens of other state and federal permits you needed, you got the added environmental impact avenues for "public" participation. Those opposed to nuclear power used these to work a legal morass that added more and more time to nuclear construction. Again, this was during the period of high inflation and high interest rates. The reviews and public interventions extended the construction times. All that time you are still paying finance charges on the money you borrowed to build the plant, but you can't turn the plant on yet and start making money to pay down your debt.

I know of one case where the NRC decided that one plant needed to strengthen the seismic restraints on piping. The plant was nearly finished, but since the NRC insisted, they went back and re-hung a lot of major piping. A little while later, the science changed a bit and it was recognized that what you really wanted is pipe that was flexible and would flex in an earthquake not super strong supports that would actually break the pipe. The plant then had to go back at great expense and return the pipe hangers to much like what they started with. While the regulators were certainly well meaning, such horror stories added significant, real costs. Furthermore, the time extensions they inflicted added to the financing costs and, therefore, total costs.

United States nuclear power plant manufacturers have built nuclear plants in Japan and other countries in 4 to 5 years. In the United States, the construction times were reaching 10 years and beyond. The situation became untenable, and many planned nuclear units were canceled. A number were abandoned during construction and never completed.

This sounds pretty depressing, huh? Well, it gets worse. The next big two challenges for U.S. nuclear power were deregulation and relicensing. Let's look at each of these.

We talked a bit about regulated utilities back in Chapter 3, and you are probably aware of the concept. Just to be on the same page, let's do a mini-review.

As the industrialization of the country progressed, it was realized that it was inefficient to leave certain functions totally to the competitive free market. Chief among these were utilities, like electrical power, natural gas distribution, or telephone service. While competition is normally a good thing and helps to limit costs and maximize service, it was clear that the cost to a utility would be impacted greatly by the cost of building the distribution system. If several competing firms all had to build essentially identical distribution systems, the cost to the consumer would be high. That is, if you had four sets of telephone poles, one each for companies A, B, C, and D running down every street, the cost had to be higher than if you only had one set of poles and wire. Well, how do you get to only one set of poles?

The government stepped in and said, to get the best cost for the consumer, we will have only one company distribute the utility service and we, the Government, will pick the company. Well, good, that reduces the cost of the distribution systems, but you lose the competitive pressure on cost and performance of the utility. They have a monopoly. As Lily Tomlin's telephone operator used to say, "We are the Phone Company, we don't care, we don't have to." But in actuality, the utilities did have to care because the continued operation of the monopoly was subject to the pleasure of the government. In general, utilities were regulated by state public utilities commissions (PUC). The various PUCs would grant companies a service region and set the rates they could charge. It was a constant struggle between the PUCs and the companies, with the companies trying to get the best return for their investors and the PUCs pressing for the lowest reasonable rates for the consumers.

Overall, it worked pretty well for a lot of years. Utilities were never a great place for an investor to make a lot of money because the PUCs would not allow that. However, it was a pretty stable place to make a modest rate of return, because it was not in the interest of the PUCs to bankrupt the utility companies.

Nothing, of course, ever stays the same, even if it is working pretty well. There arose an advocacy for a reassessment of regulation. Questions were asked. Do the big utilities have unfair control of their services? Are there ways to inject the cost and service improvements of a competitive system into regulated activities? As you well know, both the airline and telephone industries were subject to major changes in regulation and much more competition was introduced. While this was good for the consumer in many ways, it was very tough on a lot of companies and a lot of them could not adapt and died.

Folks looked at the electrical power industry and concluded that while the regulated monopolies for the distribution still made sense, it did not necessarily follow for the production of the electricity. If you recall, in the history of electrical power, we saw that as electrical power grew in the United States, we had large companies formed that both generated and distributed electricity. Big firms bought many small firms, and many big firms were bought by huge firms. This allowed them to build really huge power stations and massive transmission systems and do so economically. However, with the concept of deregulation, new generating firms were allowed into the market and the distributor was forced to route that power to the consumer, if the consumer wanted to buy the new guy's power. The idea being that this gave the market an incentive to find cheaper ways to produce electricity. Now there are a lot of details as to how the deregulation can actually be implemented. Some states have proceeded with deregulation of electrical power with reasonable success. Other states, notably California, messed it up pretty bad. At this point, while there is a general view that deregulation of electric utilities is a good thing and will benefit consumers, the California experience, which bankrupted companies, ran up consumers bills, and caused the state to temporarily run short of power, has caused many to go slowly and think carefully about further progress down that road.

Having said all that, a few years ago deregulation was the acknowledged future of all electrical power production. Many considered that the sure-fire death for nuclear power. Nuclear power was holding high costs because of the massive construction debt and the high operations cost from all the safety regulation. Before the Three Mile Island accident, an average 1000 MWe nuclear power station would have about 500 workers. By the mid-1990s, this had grown to nearly a thousand. Under the regulated market, a utility could come to the PUC, show them how much it cost to run their nuclear plant, ask for consumer rates a few percentage points more than that to pay their investors, and after a lot of shouting, usually get that rate. But under deregulation, that same nuclear plant would have to compete with Fred-the-New-Guy who whipped up a natural gas-fired plant in 18 months and could sell power for three-quarters that cost. Doom, doom, doom.

But it gets even worse! Nuclear power reactors in the United States are operated under an operating license granted by the Nuclear Regulatory Commission. That license has a set period, 40 years. The original idea was that, frankly, folks had no idea how long a nuclear plant might "last" or even just what "last" meant. Clearly, it had something to do with operating safely, since that is the bases of an operating license; that is, a demonstration that you can operate the plant in such a fashion to protect the public health and safety. They figured 40 years was certainly long enough for the original builder of the plant to recoup their investment but probably not long enough for anything particularly bad to happen in the way of aging of equipment and facilities. Especially if we hang a whole bunch of required scheduled inspections and direct staffed on-site overview by the Nuclear Regulatory Commission at every plant.

This worked out pretty well. We had a number of surprises along the way, things that we didn't know about until we operated a nuclear plant for some years. However, none of these led to any public harm, and we learned how to either avoid or correct the problems. As years have gone by, we have gotten better and better at running nuclear plants. We will look at some numbers in a bit, but the point is that the original idea that plants could "last" for at least 40 years seems to be working out to be right. Like a well-maintained used car, you replace worn parts, inspect for wear, do service on a regular schedule, and they run like a top.

Because folks were pretty sure plants could "last" for 40 years, they also expected that they could last for some span longer as well. The regulation was written to allow a request for license extension, that is an additional 20 years added to the operating permit. The idea was that the nuclear reactor operator would request such an extension from the NRC and provide sufficient information to show there was no reason to expect the plant could not operate for another 20 years. This would have something to do with the surveillance program—the operator had to keep on top of aging issues and how they maintained the facility. However, the challenge was no one had ever asked for an extension, and the NRC had never really said what it would take to convince them that everything would be O.K. In some licensing debates, you can get into some really ugly traps where you are being asked to prove the negative. That is, how do I convince you that nothing bad can *ever* happen? How do I *prove* that there are no polar bears hiding under the desks in the control room?

This gets really hard when trying to do something that has never been done before, such as operate a nuclear reactor for 60 years.

As an example, how do I convince you that a steel reactor pressure vessel can stand neutrons smacking into it for 60 years? Well, I can show you what has happened in the first 40 years and give you a lot of good scientific models of what is going on at the atomic level and how that affects the mechanical properties of steel. I can also take small samples of steel and put them in test reactors and run the neutrons way up until I get 60 years worth quickly. However, even with all that I haven't really reproduced exactly what my reactor pressure vessel will experience over 60 years. What if there is some other physical thing about the behavior of steel no one knows about?

In this case, it turns out we know a great deal about the behavior of steel under neutron bombardment and planned inspection programs will guard against surprises, so a convincing story could be made. But at the time plant operators were beginning to face the question of whether to ask for a license extension or shut the plants down, no one, including the NRC, really knew just how strange the relicensing justification debates were going to be.

With all this going against nuclear power, if a plant owner got just a little bit of extra bad news, it was likely to push them to shut down their plant even before the original 40-year period was up. Something like nine plants met this fate. A major repair facing them or a particularly tough local anti-nuclear advocacy group or often a combination just made it too hard.

With this grim picture, the Energy Information Administration predicted in March 2000[2] 13 additional early retirements in the United States and only 11 license renewals. By December 2000,[3] the EIA upgraded that gloomy prediction to only one additional early retirement and noted that 17 reactor operators had already stated their intentions to obtain license renewals. Since then, the common view is that most existing reactors will apply for license renewals.

So after all the bad news, what happened to turn the fortunes around?

Well, it was a number of things. Perhaps the biggest thing was just the passage of time. Time passing in which we learned how to run nuclear power plants better. Time in which those better running plants made a lot of electricity, which was sold for a lot of money, which paid off a lot of the debt of the original construction. With the specter of deregulation coming at them, nuclear power plant operators got very focused on cost of operations. They found ways to use their staff more efficiently, reducing numbers while improving operational reliability. Perhaps one of the more striking examples has been the improvement in capacity factors.

---

[2]*International Energy Outlook*—2000, DOE/EIA-0484(2000).

[3]*Annual Energy Outlook 2001*, DOE/EIA-0383(2001).

As you recall, the capacity factor is the amount of power a plant actually generates in a given time period divided by the amount of power it would have generated if operated at its full rated power throughout that time period. Because nuclear plants are expensive to build, but the fuel is relatively very cheap, ideally you would want to run it all the time you possibly could. However, there are three reasons why you can't do that. First, there is some maintenance and inspections you can't do when the plant is operating. There are areas you can't get to when the plant is hot and running. Second, you have to shut down from time to time to refuel the plant. Because you have to cool the plant down, unbolt the massive pressure vessel head, remove the spent fuel, add the new fuel, rearrange all the other fuel, and then bolt it all back together, this takes quite a while. And, lastly, things just sometimes break. So what you are left with is planned outages, for inspections, maintenance, and refuelings, and unplanned outages, for when something goes south on you.

Anytime you are in any outage, the plant is not making electricity and, therefore, not making money. Anytime I encounter a group of people just getting organized to undertake some task, I think about a reactor outage meeting. These are tremendously focused exercises. Everyone there has a job that everyone else is counting on them to do absolutely correctly and absolutely as fast as can be done consistent with absolute correctness. No one is daydreaming. No one is confused about the goals of the effort. Everyone is on the team and on focus. You can feel the very air vibrant with the energy of absolute dedication to the common purpose. It is a pretty incredible thing to experience.

Over time the nuclear power industry got better and better at making things run well and more efficiently. They figured out ways to do more inspections and maintenance while the reactor was running. They figured out how to make fuel last longer, so they didn't need to refuel as often. They figured out even better ways to manage refueling outages to reduce the time it takes to complete all the steps. They also did a better job in their inspections and maintenance, reducing the number of things that would break unexpectedly, driving the unplanned outage rate almost to zero. So how well did they do? Look at Figure 8.1.

Pretty impressive, huh? At the start of commercial nuclear power, the refueling schedule was once a year and refueling outages were pretty long. By the mid-1990s, a lot of plants had gone to 18 months between refueling and refueling outages were getting shorter. In 1990, the average refueling time in the U.S. was 100 days. By 1999, that time had been reduced to an average of 42 days. The best in the business are getting refuelings done in less than 20 days. Plants are now shifting to 2-year refueling cycles. All this means that the plants are spending more time running, and more time making money.

In 1990, there were 112 operating nuclear power reactors in the United States. Those reactors produced 577 billion kilowatt-hours of electricity. In 2000, the 103 remaining operating nuclear power reactors produced

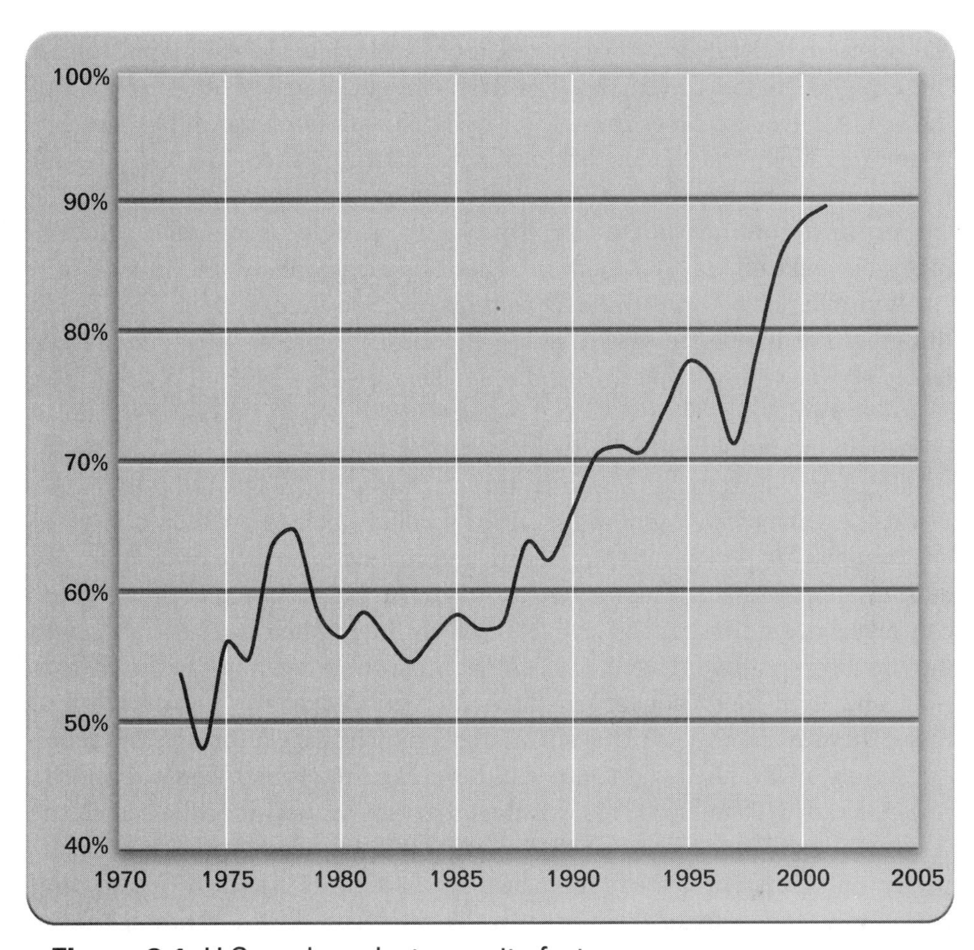

Figure 8.1. U.S. nuclear plant capacity factor.

754 billion kilowatt-hours, a bit over 30% more power with fewer power stations. How about this? In 2001, those 103 power stations pumped out 767 billion kilowatt-hours. It just gets better.

The financial markets have recognized this improved performance. With the original debt essentially paid off, the only cost to making power out of a nuclear plant is the fuel and operation and maintenance costs. Relative to fossil fuels, these are very competitive, that is if your plant is running. When nuclear had 50% nuclear capacity factors, the fossil guys looked like the better bet. At 90% capacity factors, a lot of nuclear operators are making a good deal of money.

Some years ago, the Vogel twin reactor station was coming in with the lowest operations and maintenance costs in the country. It was costing them about 2¢/kilowatt-hour to make electricity. They were selling that power for about 5¢/kilowatt-hour. And if you have got 15 billion kilowatt-hours a year to sell, that is serious money.

The electricity markets were helping out as well. In times of high demand, like peak winter heating or peak summer cooling seasons, electricity sold on the open market can get very, very expensive. It is a sellers' market under those high demand conditions. The first kilowatt-hour might be cheap, but the last one available goes for gold. It was calculated on one cold winter day in the late 1990s that if you had a nuclear plant, you could meet all your annual operating expenses if you sold your output for just 6 weeks at the highest prices on the Midwest spot market. You could have either taken the other 46 weeks off, or used them to generate pure profit. Sweet, huh?

A number of big firms decided that this meant nuclear power was a "good thing." What happened was a series of consolidations in which the larger, better run companies bought up the smaller, less efficiently run firms. The growing confidence of the financial markets in the value of nuclear power can be seen in the history of the industry consolidations.

Big firms were looking to buy the nuclear-generating assets of smaller companies. The idea was the bigger firms would be able to use their more flexible and more experienced engineering and management work force to more efficiently operate the stations they acquired. At the start, some of the smaller companies were more than willing to sell. To them, their nuclear experience had not been positive. They had been having trouble keeping their units running and with the vision of cut-throat deregulated competition coming at them, they saw no way to make any money on their nuclear units in the future.

In July 1998, a deal was set to sell the Three Mile Island Unit 1 reactor to AmerGen for $100 million. This was the other reactor at Three Mile Island, the one that was not damaged, but all the focus on its infamous sister made its life pretty unhappy. The $100 million deal may sound like a lot of money, but considering the value of the asset it was really, really cheap. The fuel in the reactor was worth about $70 million. It was like selling your used car for essentially the value of the gas in the tank.

A few other plants, notably the Pilgrim plant and the Clinton plant went for similar bargain prices. However, by 2000, things started to change. Folks begin to realize that these super cheap plants got turned into instant moneymakers. When the Indian Point Three and Fitzpatrick stations went for sale as a package in February 2000, they brought a combined price of $967 million. When the Milestone reactors were sold in August that year, the price was $1.3 billion, effectively $591/kilowatt. While less than the $1700/kilowatt or more it would take to buy a new power reactor, the fact that this was more than the going price for a new natural gas-fired station was a clear demonstration that the financial markets were assessing nuclear generation very positively. At the close of 2000, the sale of Nine Mile Point Unit 1 and part of the output of Unit 2 for $950 million brought the effective price up to $613/kilowatt.

Another very positive thing during this period was that, in March 2000, the NRC granted a license extension to Calvert Cliffs. The key to this was now relicensing had been done; not only done, but done on the pre-announced 22-month schedule. There were no weird detours, just a good solid review completed as scheduled. Existing reactors had a clear achievable path to a 50% extension in the time span they could be making money for an investor.

So where does that leave nuclear power in the United States today? We have seen the numbers on capacity factors and overall generation. The often quoted value is we get about a fifth of our electricity from nuclear power. While if you had to replace that tomorrow, it would be a very big deal, on a longer view, one-fifth may not be too impressive to you. However, I think you might be surprised at just how important nuclear electricity is in certain regions of the country. Table 8.1 shows you the number of operating reactors and the fraction of total electricity coming from nuclear power by state.

So what is the nuclear fraction in your state?

Were you surprised?

TABLE 8.1 U.S. 2000 nuclear generation by state.

State	Number of Reactors	Percent Total Electrical Generation From Nuclear
Vermont	1	67.3%
South Carolina	7	55.3%
New Hampshire	1	53.1%
Illinois	11	49.6%
New Jersey	4	49.3%
Connecticut	2	45.1%
Virginia	4	36.9%
Pennsylvania	9	36.0%
Arizona	3	34.2%
North Carolina	5	32.2%
Nebraska	2	29.8%
Mississippi	1	28.8%
Arkansas	2	27.7%
Maryland	2	27.7%
Tennessee	3	27.1%
Georgia	4	26.7%
Minnesota	3	26.3%
Alabama	4	25.3%
New York	6	23.1%
Kansas	1	20.4%
Wisconsin	3	19.6%
Michigan	4	18.2%
Louisiana	2	17.6%
California	4	17.0%
Florida	5	17.0%
Massachusetts	1	14.0%
Missouri	1	13.1%
Ohio	2	11.4%
Iowa	1	10.8%
Texas	4	10.0%
Washington	1	8.0%
Total	**103**	

Five states have essentially half or more of their electricity coming from nuclear power. Nine states are a third or more dependent on nuclear energy for their electricity, and 18 states are relying on nuclear for more than a quarter of their electrical power. How important is that? Well, if California had been in the 50% category during the 2000–2001 power crisis, perhaps they would have been much more cushioned from the run up in natural gas and out-of-state generated electricity prices.

That is a look at the status in the United States. How about the world in general? Let's look at Table 8.2, which gives international generation data to get a feel for the dependence of various nations on nuclear energy for electricity.

Any surprises for you there?

We knew that France was using nuclear power heavily. A few other countries might be a surprise to you. Lithuania is a special case because the former Soviet Union built two huge RBMK power reactors there, mainly to export power back into Russia proper. After the fall of the USSR, Lithuania was left with these two big rascals. With a lot of international help, they have them running pretty good and have done many safety upgrades. And so they have a lot of nuclear electricity for a small country. Armenia is another former Soviet state. It has two older generation Soviet-designed pressurized water reactors. Before the fall of the USSR there was a massive earthquake in Armenia and both reactors were damaged. They were shutdown safely but were deemed unsafe to

TABLE 8.2 Nuclear energy dependence by country 2000.[4]

Country	Percent Total Electrical Generation From Nuclear
France	76.4%
Lithuania	73.7%
Belgium	56.7%
Slovak Rep	53.4%
Ukraine	47.3%
Bulgaria	45.0%
Hungary	42.2%
Korea, Rep of	40.7%
Sweden	39.0%
Switzerland	38.2%
Slovenia	37.4%
Japan	33.8%
Armenia	33%
Finland	32.1%
Germany	30.6%
Spain	27.6%
Taiwan	23.6%
United Kingdom	21.9%
United States	19.8%
Czech Rep	18.5%
Russia	14.9%
Canada	11.8%
Romania	10.9%
Argentina	7.3%
South Africa	6.6%
Netherlands	4.0%
Mexico	3.9%
India	3.1%
Pakistan	1.6%
Brazil	1.4%
China	1.2%
Total	**16.0%**

[4]International Atomic Energy Agency Web site.

restart without repairs. With the withdrawal of Soviet support, Armenia was ill equipped to fix either plant. At the same time, they were facing a terrible energy shortage resulting in some awful conditions for their people. With international help, they got one of the units fixed, and it is now a true savior on cold winter nights.

What Table 8.2 doesn't show you is any sense of the size of the nuclear programs in the various countries. For instance, Armenia has a higher dependence on nuclear than does the United States. Does that mean Armenia is leading the nuclear power business and the United States is following? Ah, no. Table 8.3 gives you a view of the total size of the various nuclear programs giving the number of operating reactors, the total nuclear-generating capability, and the power generated in 2000.

TABLE 8.3 World nuclear power generation 2000.

Country	Number of Reactors	Capacity (MWe)	Generation (Billion kWh)
United States	103	97,411	753.9
France	59	63,073	395
Japan	53	43,491	304.9
Germany	19	21,122	159.6
Russia	29	19,843	119.6
Korea, Rep of	16	12,990	103.5
United Kingdom	35	12,968	78.3
Ukraine	13	11,207	72.4
Canada	14	9,998	68.7
Spain	9	7,512	59.3
Sweden	11	9,432	54.8
Belgium	7	5,712	45.4
Taiwan	6	4,884	37
Switzerland	5	3,192	23.5
Finland	4	2,656	21.1
Bulgaria	6	3,538	18.2
Slovak Rep	6	2,408	16.5
China	3	2,167	16
Hungary	4	1,775	14.7
India	14	2,503	14.2
Czech Rep	5	2,569	13.6
South Africa	2	1,800	13
Lithuania	2	2,370	8.4
Mexico	2	1,360	7.9
Argentina	2	935	5.7
Brazil	1	1,855	5.5
Romania	1	650	5
Slovenia	1	676	4.5
Netherlands	1	449	3.7
Armenia	1	376	1.8
Pakistan	2	425	1.1
Total	**437**	**351,327**	**2446.8**

The table is ordered by the amount of nuclear electricity generated in 2000. You will notice that in some cases countries with more capacity generated less electricity. This is due to a lower capacity factor for that nation. For nations with only a few reactors, one extra long outage at one reactor would be enough to reduce their rank in the generation ordered list. Likewise, you will see cases of countries with more reactors, such as India, producing less electricity than countries, for instance Taiwan, who have fewer reactors. While capacity factors influence that, the main thing going on there is simply that the Indian reactors are significantly smaller than those in Taiwan. You also see that effect in the numbers for the United Kingdom, where the individual advanced gas reactors and gas-cooled reactors are relatively small.

From the table we can see that the United States, despite coming nineteenth on the list ordered by fractional dependence, is by far the largest user of nuclear power. Likewise Japan is twelfth on the fractional use list but third in total nuclear electricity generated. This speaks to the size of the American and Japanese economies and how dependent those economies are on electrical energy.

We have learned a little about the status of nuclear energy in the United States. There is a growing acceptance of nuclear power in the financial markets, and operators of nuclear plants are doing pretty darn good. License extensions seem to be moving along well, and industry consolidation is putting plants under the control of well run, efficient operators.

As to the world picture, there are stories for every country in the table and each story is different. Some of the news is good, some is not. Let's hit some of the high points and start with the bad news.

The worst of the bad news is the influence of the European Green Parties. As we saw in the discussion of greenhouse gas emissions, the European Greens are dead set against nuclear power. They blocked nuclear as a Clean Development Mechanism in the Kyoto Protocol damaging the ability of nuclear to help mitigate the impact of global warming. The Chornobyl accident traumatized a lot of Europe as the radioactive cloud drifted to the north and west of Ukraine. Not surprising, Sweden and Germany became pockets of especially strong Green Party and anti-nuclear feeling. Both Sweden and Germany have passed laws to reduce and eventually phase out nuclear power. With 39% and 30.6% dependence on nuclear electricity in Sweden and Germany, respectively, this is not going to be easy. The Swedes have been working on it the longest. While they have shut down some of the older nuclear-generating capacity, the gains from conservation and "green" energy sources have not come up to expectations. There has been a lot of rumbling that they can't get to where the Greens want them to be from where they are. That debate goes on.

In Germany the story has had less time to unfold. Their existing plan stretches out over many years, but a similar debate on the impact to the economy and environment, since nuclear is being replaced with fossil sources, goes on.

Green Party assaults against nuclear power have taken place in other countries as well. The Swiss turned back a national referendum to close out nuclear power. Even in France, the poster child for nuclear electricity, the Greens are vocal, if not largely heard. Taiwan has seen an amazing political roller coaster ride tied to the issue of nuclear power. A few years ago a new party came to power whose platform included a stop to new nuclear construction. The economic and industrial leaders raised a huge cry against that. The new government undertook a "study" to examine the issue. Eventually they decided to stick with the campaign pledge and stopped construction. The legislature, still under control of the previous ruling party, toyed with a no-confidence vote to bring down the government. The national court entered the fray deciding that the executive branch could not legally stop the construction that had been authorized by the legislature. So for the moment, the construction is back on. Stay tuned.

In more neutral news, the Eastern European countries seem to be doing pretty well with nuclear. The legacy of the USSR was power plants that would not meet Western safety standards. With international help, especially from the U.S. Department of Energy, the equipment, components, operator training, and management practices at those stations have been significantly enhanced. Those countries are also making use of more and more Western technology.

Finland is a particularly interesting example of a blend of the Soviet-designed reactors and Western technology. The two Loviisa plants were based on the smaller version of the Soviet pressurized water reactors but incorporated a lot of systems from Western companies. They are sometimes jokingly referred to as "Eastinghouse" plants, a play on the Westinghouse company name and the terminology of referring to Russia as the East. The result was a darn good set of reactors that are very reliable and set operating record after record.

Russia itself still believes in nuclear power. They continue to have a large nuclear power program. The separation between their civilian and military programs has always been less distinct than ours in the United States. This meant that their overall program has been very big and, while much reduced from its heyday, there are still a lot of very talented folks working in the field. The Russian Federation continues to assert they will build new nuclear power stations. However, if their economy remains depressed, it isn't clear that they will need the extra power nor will they have the resources to finance new power stations. While Russia has a large resource of natural gas, it is difficult to transport natural gas across this very large country. Russia is also inspired to sell natural gas to Europe for hard currency rather than burn that gas themselves. This gives them a strong reason to employ nuclear-generated electricity rather than gas-fired generation. How this will play out it remains to be seen.

China is also an interesting case. Their goal is to import foreign reactor technology and use that to develop their own capabilities. They have done this with 1980s French technology and are looking to build state-of-the-art Canadian and Russian designs. American as well as French and other European vendors are looking for future sales. Some years ago China announced a massive nuclear expansion. That proved beyond their means, and they have since dropped that plan. Considering their rapid economic growth, resultant energy supply needs, and the vast negative global environmental impact if that energy growth was shouldered by abundant Chinese coal, we can all hope they return to a strong nuclear growth plan. However, like the United States, future growth in nuclear within China will depend on the economics relative to fossil sources.

Bright spots in nuclear energy worldwide continue to be in Asia, where nuclear power is continuing to grow in Japan and the Republic of Korea (South Korea). The Koreans have taken the same route as the Chinese intend. They imported American and Canadian technology to build the first set of their nuclear capacity. Building on that, they now have their own standard Korean design pressurized water reactor that they will use to expand their own capacity and also seek to export that to other countries. The International Atomic Energy Agency data showed they had four reactors under construction in 2000.

Japan is the third largest producer of nuclear electricity. They have a long-term view that includes reprocessing of spent nuclear fuel and recycling of the valuable parts of that spent fuel into new fuel. While not without its challenges, that program continues to progress. They had three more reactors under construction in 2000. They also have the world's most advanced nuclear reactors employing an advanced design from the American firm General Electric, which ironically has not yet been sold to an American power company.

Overall, the 2000 International Atomic Energy Agency data showed 31 reactors under construction throughout the world, with China, Korea, Japan, Ukraine, and Russia being the leaders.

Beyond those real examples of growth, there seems to also be a general revived awareness of what nuclear energy can do. I have seen many news articles with titles like "Nuclear Power, A Second Coming," or "Nuclear Power, A Renewal," or "Nuclear Power, Has Its Time Come?" This positive media attention is very welcome after so much negative press over the years.

There has also been a renewal of interest in nuclear power within government circles. Many have pointed out the favorable words given to nuclear power in the Bush Administration Energy Plan. However, the shifts started before that. Perhaps one watershed event was the Department of Energy's Nuclear Energy Research Initiative Workshop held on April 23-24, 1998, in Washington, D.C. This was a get-together of folks from the Department of Energy national laboratories, universities, nuclear industry, the Department of Energy itself, and congressional representatives. The goal was to figure out what might be done to return the United States to a position of leadership in the applications of nuclear energy and what constructive roles the government could play. I was there, and it was a very productive session. What it did was set the stage for a modest program of competitively chosen peer-reviewed projects to improve nuclear technologies. A lot of really fascinating things have been supported. Sadly the congressional support relative to the penned up creative energy is such that perhaps one in ten good ideas proposed is funded, but it is a start.

In 2001, the Department of Energy expanded this to include a small International Nuclear Energy Research Initiative in which American scientists and engineers would team up with their counterparts in other countries to explore technologies of common interest. France and South Korea led the list of these countries, and other cooperative arrangements are in the works.

The Department of Energy also has started a Generation IV reactor program. To explain, the first generation of power reactors were the first small plants built in the 1950s and 1960s. The current modern light water reactors, largely designed in the 1970s, are considered Generation II. During the late 1980s and into the 1990s, the Department of Energy had a cooperative program with three reactor vendors to design generic advanced light water reactors. The idea was to develop a small number of standardized designs, which could be pre-reviewed by the Nuclear Regulatory Commission. These advanced plants would take lessons learned from the operation of the Generation II plants, add improved technology such as improved electronics and computers in the control systems, and generally get a better, more reliable, safer, and more easily licensed reactor. It was a great idea. The designs were completed, the NRC did their reviews and approved the designs, and no one in the U.S. bought one. The Japanese bought the General Electric version and have built it. The Combustion Engineering design saw some application to the Korean reactors. But as far as the American market goes, DOE and the industry set the table, but no one came to dinner. Those advanced light water reactors are considered the Generation III plants.

Generation III plants are sometimes call evolutionary designs in that they are a modest modification to existing designs. The Generation IV plants are considered revolutionary in that they are expected to think outside the box and do things much differently and much better. We will take a peek at some of these ideas in the next chapter.

The Department of Energy's Generation IV program is hoping to focus the nation's creative efforts toward a revolutionary design that can leap over the Generation III concepts. The goals are to greatly reduce the cost of construction and operation while improving safety and proliferation resistance.

The Department has cooperating programs with other countries on the topic of Generation IV research. In addition, the International Atomic Energy Agency has its own advanced reactor efforts.

So, when we look at nuclear power today, we see that while there are some challenges in getting worldwide support, there is also a lot of good news. Power plant performance, both in terms of safe, reliable operation and economic performance, is going from very good to excellent. While there are no new plants being built in the United States, other countries are building new units, and worldwide concepts for new types of nuclear power reactors are being actively developed.

That takes us up to where we are today and sets the stage for tomorrow. Let's go look at that next.

9

THE FUTURE OF NUCLEAR POWER

In talking about the future of nuclear power, I want to approach the topic from two different angles. First, I want to look at the political and social possibilities. To do that, I will give you three different potential near-term futures of the human race. Now clearly, there are infinite shades of variation between just three potentials, but the idea is to examine the political and social context in which decisions about energy use will be made and how those decisions might impact our lives. Going through these will help prepare you for the final chapter in this book. Second, we will shift to a science and engineering mode to look at some of the very interesting new technologies that may play major roles in the future of nuclear energy, assuming one of the more happy political, social scenarios comes to pass.

Potential Futures—Political and Social Context for Nuclear Energy

I was once given the assignment to write a "vision" statement for the future of nuclear power. While this might seem a Dilbert-like exercise of debatable value, it turned out to be instructive. I was asked to consider three scenarios: 1) bad for nuclear, 2) neutral for nuclear, and 3) positive for nuclear. The instructive part was that after developing these scenarios, we then tried to come up with business plans to mitigate the impact to our firm if the bad scenario seemed to be playing out, how to take advantage if either the neutral or positive scenario appeared to be happening, and my favorite, try to come up with ways to make the positive outcome happen.

Let's start with a bad future for nuclear energy. In that potential future, the doomsayers of things nuclear will have won out. Either through apathy or lack of understanding the general public would have allowed the small vocal anti-nuclear community to work their will on the rest of us. The Greens will have shut down nuclear power in Sweden and Germany and gained ground in other European and Asian countries. China would have not found the will or the economic rationale to support nuclear developments, and both the French and Japanese programs would have stalled out and started a slide down as the older plants were retired. In the United States, no new reactors would be built, and as in all other nuclear countries, the fraction of energy gained from nuclear would continuously decrease as existing nuclear plants reach the end of their service lives.

Now some would speculate that this would be achieved through wonderful new methods of energy conservation and vast expansions of "alternative" and sustainable energy sources. As I have described in Chapter 7, I don't think either of those things is very likely. We will get better at the use of energy, and we will get greater use out of alternative and sustainable sources. However, the contributions will be small relative to the demands of a growing human population and hopefully improving human condition in developing countries.

Given a decision to step away from nuclear energy, then only fossil fuels would remain as the source of the energy needed to serve a 6- to 12-billion-person human population. That leads down one of the two dark paths I offered in the discussion of fossil fuels. Either we run short of fossil fuels and civilized behavior breaks down as we beat each other up over oil and natural gas, or we find enough oil, natural gas, and coal, and by using it render the planet uninhabitable. Now it doesn't have to go all the way to either of those end points before life gets ugly. Somewhere before civilization fails all together you *only* have major bloody regional wars. Somewhere before you get to a runaway greenhouse effect, you flood a major part of Bangladesh and set off killer winter storms in Europe and decade-long summer droughts in the American Midwest. Also, somewhere in-between the two paths, you set up a conflict between developing countries who are trying to establish an industrial economy that gives some portion of a healthy, hopeful life to their citizens and developed countries who can't tolerate the environmental consequences if all those billions of people in the developing countries start behaving as poorly as they themselves do.

The neutral-to-nuclear scenario is a bit more positive. In that scenario, the anti-nuclear forces have had some success, but only some. Europe remains divided as to its approach to nuclear energy. Sweden and Germany have retreated from a complete ban on nuclear but have limited future growth.

China has a growing nuclear program, but this hardly keeps pace with their overall economic growth, so much of China's energy dependence remains tied to coal. The French and Japanese programs remain robust. This includes a strong use of reprocessing and fuel recycling, holding out the promise for significant expansion of the energy content of uranium resources. The Korean program remains strong, and some use of nuclear energy is seen in developing countries. In the United States, most of those reactors operating in 2000 have been granted a 20-year license extension and some have even gained a second extension for an additional 20 years. Several new commercial reactors are built near the end of the first decade of the twenty-first century, and the government research and development programs are poised to give birth to revolutionary new reactor systems.

The world and especially the American public in this scenario remains a bit skeptical regarding things nuclear, but impressed with a continuing run of safe operation backed by excellent financial performance, they are willing to support nuclear development. Dissent to new nuclear construction is politely listened to, but the public at large has made its will known that such views are not mainstream.

Now as happy as that scenario might be, especially as compared to the first, it is still a pretty scary one in a longer term. This neutral scenario would result in nuclear power remaining as an option as an energy resource, but while a guarded reception is a whole lot better than a rejection, the guarded reception would probably limit nuclear's contribution to something close to the current fraction of total world energy production. Remember, we are looking at something like a 60% increase in worldwide energy in just the next 20 years. That means if nuclear energy expanded by 60% over its current level, say an increase from 437 power reactors to 695 worldwide, you would just maintain the fractional contribution. Considering that some existing reactors might be retired during those 20 years, you would need to build more than 258 new reactors.

And if you did build 258 new nuclear power plants, you would still see a 60% increase in the use of fossil fuels. The good thing is if you did build those 258-plus new power reactors, you would *only* need a 60% increase in fossil fuel supplies. If you built fewer than 258 new reactors, you would need even more than a 60% increase in fossil fuels. Yes, growth of alternative energy sources could take that down a bit, but again, it doesn't look to me to be more than a very small bit. The point is that in the neutral-to-nuclear scenario some enhanced use of nuclear energy would buy you some time relative to the twin evil outcomes of continued dependence on fossil fuels. To what end do we buy that time? Well, it might be enough to see me personally off the planet. Does it work for you? Does it work for your grandchildren? Probably not.

There is an old joke about a man brought before the king accused of treason. The king asks the man if there is any reason why he should not be put to death. The man answers, "If you spare me, I will teach your favorite stallion to sing, but I will need a year to do it." The king is impressed and grants him the year. Returned to the prison, the man is confronted by his fellow prisoners who ask him, "Are you crazy? How will you teach a horse to sing?" The man answers, "A year is a long time. In a year, the horse may die, the king may die, I may die, and you never know, the horse might learn to sing."

The neutral scenario gives us time to see if the horse can learn to sing. Not a likely outcome, but better than the quick trip to the block.

The positive nuclear scenario is, of course, my personal favorite. In this scenario even the most fanatic environmentalist has recognized that people are at least as important as snowy owls and snail darters, that energy is essential for the survival of humans, and that nuclear energy is a very environmentally friendly way to get energy, especially as compared to the truly feasible alternatives. The economic and political stars have been aligned such that nuclear energy is given a level playing field with other energy sources in which all the direct and indirect costs are fairly accounted. This has required an intra-governmental acknowledgment of the indirect costs of fossil fuels and the establishment of global programs such as Clean Development Mechanisms and/or carbon trading credits, which accept nuclear energy as a valid means to offset polluting and carbon-emitting fossil fuels and establishes the economic systems that incentivized clean nuclear use.

The result is a retreat from Green Party restrictions on nuclear energy worldwide. Sweden, Germany, and Taiwan return to a path of growth for nuclear energy in their economies. Japan, France, and Korea expand their efforts with all those nations increasing nuclear's contribution to their total energy use. China adopts nuclear as a key component of their energy-fed industrial growth. Power reactor systems are developed that have non-proliferation and safety features appropriate for developing nations. Those nations offset much of their otherwise fossil fuel-fed economic growth with nuclear energy. In the United States new commercial power reactors become an increasing component of our energy production. Most important, in the United States and across the world, nuclear energy applications expand in non-traditional economic sectors.

Currently the sole use of nuclear energy is in the production of electrical power that is used mainly in the residential, commercial, and to a lesser extent industrial sectors. Some small portion of electrical energy currently goes into electrified trains, but in general, the transportation sector is essentially totally fueled by petroleum. In my positive nuclear future, nuclear energy grows in all economic sectors.

Nuclear energy becomes a major source of electrical power, offsetting significant fractions of fossil fuels. This stabilized the cost of electrical power reducing social and economic instability worldwide. In addition, nuclear energy grows in the industrial sector with the development of dual-use power stations that supply process heat directly to industry while also producing electrical power for both industry and general use. Nuclear power is used in the transportation sector as a source of power for recharging battery-powered vehicles and a direct source for an expanded fleet of electrified trains. Nuclear energy is also used to extract hydrogen from water for use in industrial processes, such as production of ammonia-based fertilizers without the combustion of natural gas. Nuclear-generated hydrogen is also used in the production of transportation fuels. Such transportation fuels could include hydrogen for direct clean combustion, use in electricity-producing fuels cells, or the creation of hydrogen-rich liquid fuels such as methanol.

Another important feature of my dream scenario is that we have to develop a way to use the other 95% or so of the energy value of nuclear fuel that the current United States once-through fuel cycle throws away. What we are doing now is like running your car off the evaporating gas out of your gas tank, but throwing away 19 out of every 20 gallons you buy. Now reprocessing and recycling have some challenges, non-proliferation and economics being at the forefront. However, in the longer term, we have to face up to this, and work the challenges, or nuclear can't live up to its promise in displacing a meaningful fraction of fossil fuels.

Does this scenario mean the evil outcomes of fossil fuel dependence are banned from existence? Would we be home free, forever? No, of course not. While use of nuclear energy with recycling would give us a very large energy resource at the end of the day, nuclear energy is not a sustainable energy source. Once you use up all the uranium, it is gone. The key, however, is even if we shifted to this high-nuclear-use scenario we would get several hundreds or more years to teach the horse to sing (more on this in a bit). That gives us time to get the human population under control. It gives us time to find ways to mitigate fossil fuel pollution and figure out the global climate and mitigate our screwing that up. It gives us time to find truly effective and economical ways to use solar energy. It gives us time to continue the search for practical nuclear fusion, the power of the stars, which has fuel for many thousands of years. It might even give us enough time to get off the planet and find other resources in the solar system to support our human population.

The whole point is that we are in a race. On one hand, we are using up energy resources that we can't replace and doing so in a way that threatens our political and environmental equilibriums. And, on the other hand, we are headed toward a scientific future that holds the solutions to many of our current problems. The race is which will happen first. Will we find the solutions before we eat our seed corn and foul our nest? The race is on. My whole point is the degree to which we employ nuclear energy sets the length of the game and in this game an early finish is a guaranteed loss. The longer we can maintain livable social and ecological environments, the longer we have to come up with the technology that will allow the smart monkeys to beat evolution.

As I said at the start, there are an infinite number of shades of variation between my three scenarios. What it comes down to is that the fractional application of nuclear energy technology delays the forces of our destruction giving the potential forces of our salvation more time to get to the finish line first.

So how are we doing now, and how much delay is enough? I think that recently we have shifted from the negative scenario to something just slightly on the positive side of the neutral scenario. As to how much time we have, my fear is that all we can do might not be enough to win the contest. My sense is that we are late in the third quarter and down by two touchdowns. How hard do you play? I think the coach would suggest as hard as you can. He would tell you that if you pull together as a team, you can still pull this one out, but team, you got to dig hard and you can't wait any longer to get things going your way.

It is a lot more than a football game we are talking about here. We are talking about the future of the planet and all living things, human and otherwise, there-upon. Nuclear power cannot cure all the evils of the world nor solve all the challenges before us. However, if we use it enough, it might give us the time to do that before the forces of our destruction win the race. The clock is running, and we have no time outs left.

O.K., if we go for the positive nuclear future, what technologies do we have in the bag to pull this off? I'm glad you asked. Let's do a quick look at that.

The Technologies of a Nuclear Future[1]

Let's start off a review of future nuclear technology by looking at reactor systems. A good place to start is the Advanced Light Water Reactor (ALWR) program. We touched on that in the last chapter, but let's learn a little more about it here.

First of all, something like 80% of all the power reactors in the world today are light water reactors. The original concept grew out of the United States naval reactor program but has gained such worldwide favor because the idea

[1]Much of the information on future reactor system technologies comes from the excellent Web site information provided by the Australian Uranium Information Centre, www.uic.com.au; the International Atomic Energy Agency, www.iaea.or.at; the Nuclear Energy Institute, www.nei.org, and the Department of Nuclear Engineering at the University of California, Berkeley, www.nuc.berkeley.edu/.

works pretty darn well. Therefore, although there is much to be said about coming up with very different concepts, thinking out of the box as it were, there is also something to be said about taking a good idea and making it better.

In 1984, the Electrical Power Research Institute (a industry cooperative organization) and the Department of Energy teamed up with three major reactor vendors to develop a set of ALWRs. The idea was to take everything we learned about light water reactors from operating them over the previous 20-plus years and come up with a set of improved versions. However, it was to be only a limited set.

Although there were only a few models of current generation power reactors, the unfortunate truth was that, since the regulatory environment was changing rapidly and the vendors were trying to satisfy a lot of different utility customers, as these reactors were being built, each plant of that generation was essentially unique. This wasn't absolute, and you could find some twins and even triplets out there. However, there was also a lot of variation between plants even those of the same size produced by the same company. This meant the Nuclear Regulatory Commission needed to do plant-specific licensing reviews, and that we missed out on some opportunities to save money on standardized training, spare parts, and so on. The goal of the ALWR program was to develop just a few advanced designs and get these pre-approved by the NRC.

The program was a great success with limited impact. By 1997, the NRC had certified a design for a large advanced boiling water reactor, the ABWR, produced by General Electric, and a large advanced pressurized water reactor, the System 80+, produced by Combustion Engineering. A smaller but even more advanced pressurized water reactor design produced by Westinghouse, the AP600, got its certification in 1999.

These ALWR designs offered improvements in safety, easier maintenance, improved control systems, efficiencies in construction, and simplification of operating systems. Through a better understanding of the factors contributing to risk, they were able to make modest improvements that resulted in up to a factor of 100 decrease in the probability of a major accident. That is a big reduction on what is already a very small probability but still pretty impressive. By using standardized design, they felt they would be able to significantly reduce construction times. The French experience supports that idea. The first of their units took up to 7 years to build but by building essentially the same plant over and over, they reduced that to 5 years. In the smaller AP600, the goal is to reduce the work at the site by building a lot of the components in a factory, sort of like a manufactured home rather than a custom-built house. They think they can reduce the time to build their power plant to only 3 years.

The limited impact has come in the fact that when these designs were offered to the United States electrical production industry in 1997-9 with their pre-approved NRC certification, well, nothing happened. At least nothing happened in the United States. The Japanese liked the GE design and had been working with the ALWR program for some time. They bought two ABWRs. The Kashiwazaki Kariwa Units 6 and 7 came on line in 1996 and 1997, respectively. They cost about $2000/kilowatt to build and will produce power for about 7¢/kilowatt-hour. That would be a bit steep in a United States market. However, they expect that they can drive the construction cost down by 15% in future units. The other bit of good news is they built the first unit in 52 months, 4.3 years, beating their schedule by 10 weeks.

The Koreans had also invested in the ALWR project and used a good deal of the System 80+ technology in defining their Korean Next Generation Reactor. Two reactors are under construction at Shin-Kori and are expected to come in at $1400/kilowatt with later units coming down to $1200/kilowatt and construction times of only 4 years. This is getting into a range competitive with modern coal-fired plants in the United States.

The AP600 design was a greater departure from the Generation II reactor concepts. In addition to the types of improvements described previously for the other ALWR concepts, the AP600 employed a greater use of passive systems. In Generation II reactor designs, the continued safe cooling of the reactor depends on active systems. That is, pumps, valves, and control systems that have to do something, be active. And it takes some external power to do that. In Generation II reactors, we have designed in backup power systems and backup safety systems to reduce the probability that all the multiple systems stop working correctly at the same time. One of the goals of advanced designs is to move away from active systems to safety systems that don't require any system or component to be commanded to do anything or require external power to perform its safety function. The AP600 does that with its emergency cooling system. It places a large reservoir of water in the reactor containment, held above the reactor core. If for any reason the normal reactor cooling were to fail, the increasing pressure would cause this reservoir to flood the containment. The containment itself is designed to allow natural convention from the exterior air to remove the reactor's heat from the flooded containment. All that is required is that heated water expands, gravity remains pointed down, and hot air rises. Nature does all those things without human intervention. That is a passive safety system.

Unfortunately no one has bought an AP600 type reactor at this point. The designers are trying to make the concept more attractive by making it bigger. They are working on an AP1000 design. Using this larger size, they are shooting for a generating cost below 3.2¢/kilowatt-hour. If they can do that, it would be an instant moneymaker. The British are reported to be interested.

General Electric worked on a more enhanced boiling water reactor they called a Simplified Boiling Water Reactor. The SBWR took the passive concepts of the AP600 one step further. In addition to passive emergency cooling, they converted the normal cooling circuit to a passive system using only natural circulation. Again, relying only on nature retaining its laws about hot fluids being lighter than cool fluids and gravity remaining pointed downward, they could eliminate a lot of pumps and just make the whole thing much simpler. There is a cost to going to natural circulation. Natural circulation is not going to move the water as fast as you would with a pump; therefore, you can't remove as much heat, and, therefore, you can't get as much power out of the same space. The power density, power per unit volume, of the SBWR is about three-fourths that of the ABWR. That is still pretty darn impressive for a pumpless system. Although there remains interest in this system, GE withdrew its certification application to the NRC in the mid-1990s.

The Americans have not been the only folks working on advanced light water reactors. France and Germany teamed up to work on a 1500 MW electric pressurized water reactor they called the European Pressurized-water Reactor, the EPR. The idea was very much like the American ALWR program engaging the French Framatome and German Siemens reactor manufacturers to build on their existing sound designs and get the new design pre-approved by the French and German regulatory authorities. Siemens was also working on an advanced boiling water reactor, the SWR-1000, which would employ passive safety features.

One would assume the current Green Party restrictions on future nuclear power in Germany has put a damper on at least the German portion of these efforts. The French are pressing ahead and have plans to increase the EPR up to 1750 MW electric. They are predicting improvements in reliability and thermal efficiencies greater than have ever been achieved with light water reactors. They expect the EPRs to operate at a cost 10% less than their most modern current reactors.

We have mentioned the Japanese purchase of the General Electric ABWRs. They are also working on a next generation pressurized water reactor. This is a large 1500 MW electric design, which combines active and passive cooling systems. This design is expected to become the standard for the next generation of Japanese PWRs. It is being designed as a joint project between Mitsubishi and Westinghouse, a partnership that goes back to the start of nuclear power in Japan. Mitsubishi is also participating in Westinghouse's AP1000 efforts.

Russia also continues to have an active design effort in pressurized water reactors. Like other concepts, they build on their existing technology and add enhanced features including passive safety systems. They have proposed to build several of these advanced reactors, but financing remains challenging in

Russia. They have sold two advanced VVER-91 units to China. These will include Western control systems and represent the opportunity to demonstrate a world-class Russian technology.

The Canadians have also continued to look toward the future. Their next big advancement is to combine their traditional heavy water reactor design with a light water reactor. They will retain the heavy water moderator, but use light water for cooling. This maintains much of the excellent neutron economy from the CANDU reactors but uses much cheaper normal light water for the higher pressure cooling circuit. They also would reduce the size of the plant slightly, down to 650 MW electric. This allows them to build more of the plant in prefabricated modules. That and other simplifying features are expected to reduce the overall cost by 40%, bringing the operating cost into the range of 3¢/kilowatt-hour. If they can do that, it makes it a very competitive energy producer.

The light water coolant does mean they would no longer be able to run solely on natural uranium. However, moving up to a slight enrichment of 1.5% will allow them to keep the fuel in the reactor about 3 times longer. Not only would that reduce the refueling work, it would also reduce the spent fuel volume considerably. Also, they are looking at ways to reuse spent fuel that comes out of normal light water reactors. Such fuel has had too much of the fissile uranium-235 used up to continue to make a light water reactor go. However, in the heavy water moderated CANDU reactors, this fuel can still operate. It could turn out to be a pretty cool way to get more energy out of what would otherwise be waste.

One of the most interesting potential future nuclear technologies is gas-cooled reactors. As you recall, the British have been operating gas-cooled reactors for many years. However, the key to these newer gas-cooled reactors is the unique nature of their fuel. The basic unit of fuel is a kernel, approximately the size of a coarse sand grain. At the heart of that kernel is a tiny bit of uranium dioxide. That is coated with a layer of porous carbon that will absorb the mechanical deformations the fuel will see during reactor operation. That is, the deformations the fuel will see as it heats up, cools down, and as fission products are produced inside. That layer is in turn surrounded by layers of pyrolytic carbon and silicon carbide. These are very tough and are intended to form the primary containment of the fission products and other radioactive components of the fuel.

Now, the really neat thing about this fuel is that it can take a lot of heat. This means during normal operation, they can run at temperatures above that of light water reactors. Those, you recall, run at about 650°F (343°C). A high temperature gas-cooled reactor might run at 1740°F (950°C). Remember that the higher the outlet temperature, the higher thermal efficiency you can get. Instead of the 33% of normal light water reactors, the high temperature gas-cooled reactors should operate at up to 48% rivaling the best of the fossil-fired systems. Another neat thing about a higher operating temperature is that you can use the heat for direct industrial process heat. Creation of hydrogen or hydrogen-based fuels comes to mind as a possibility. But the really neat thing about the high temperature gas-cooled reactor fuels is that they can withstand even the very high temperatures you might see under accident conditions.

In a high temperature gas-cooled reactor, you remove the heat through the circulation of helium gas. If for any reason you lose the ability to circulate the helium, either you lose the fans or a cooling duct breaks, the fuel heats up. But is that bad? In a light water reactor, if you lose the ability to force coolant past the fuel bad things happen. Just ask the guys who spent years chipping the melted fuel out of the bottom of the Three Mile Island II reactor vessel. In a high temperature gas-cooled reactor, the fuel heats up, heats up a lot, and it just sits there glowing hot. It gets hot enough that the natural radiation and conduction of heat through the vessel and into the outdoors is sufficient to remove the heat. Once the fuel gets up to about 3000°F (1650°C), the driving force of that heat is enough to push the heat through to the outdoors and the fuel never gets any hotter, no matter what. That is without pumps, or coolant, or any human actions at all. And the fuel is stable up to something like 3600°F (2000°C). That means you can't fail this fuel by shutting off the cooling. All the heat gets out; all the radioactive materials stay in.

There have been several high temperature gas-cooled reactors built. The German 13 MW electric AVR operated for 21 years to great success. In the United States, a 40 MW plant at Peach Bottom worked well but the scale up to the 330 MW electric Fort St. Vrain was an unhappy experience. There was a long list of mechanical problems[2] that never allowed the plant to run well. None of those invalidated the high temperature gas-cooled concept, but it did show that basic mechanical engineering must be dealt with.

The two leading high temperature concepts today are the pebble bed modular reactor (PBMR) and the gas turbine-modular helium reactor (GT-MHR). The PBMR is an interesting example of international cooperation. Its primary sponsor is the South African utility Eskom. They have brought in European experience from Germany, France, and Great Britain. Until recently they had financial support from a United States utility, Excelon. However, Excelon withdrew stating that while they retained high confidence and interest in the idea, they had concluded that their core business was operating power stations not trying to design and construct new concepts.

---

[2]These included vibration of the fuel blocks in the helium gas flow and leaking water in the high temperature steam generators.

The magic of the PBMR is that they take the tiny kernels and glue them together in baseball-sized spheres. Then they put many thousands of these spheres along with a mix of carbon-only spheres into a steel vessel. The gaps between the spheres, the pebbles, become the gas flow paths. They then pump the helium gas through this bed of pebbles, and there you go. The carbon in the fuel kernels and carbon-only pebbles are the moderator, and the helium is the coolant. They will continuously sift the pebbles checking them for the amount of uranium-235 left. As pebbles reach the end of their useful life, they will be removed and new balls added. This means they keep the reactor operating with optimal fuel load essentially all the time and never have to shut down for refueling.

The other magic of the design is they would not boil any water to make steam. Instead, they will extract the energy from the hot helium directly in a gas turbine, much like a natural gas-fired fossil plant would do. It is the higher temperature of operation that allows them to do this. This greatly simplifies the design and avoids that nasty hot water whose leaking caused much of the grief at Fort St. Vrain.

The last bit of magic for the PBMR is that it is a modular reactor. What that means is the reactor itself is pretty small, perhaps 250 MW thermal energy producing about 120 MW of electricity. This small size means you can build most of it in a factory, where you can do it efficiently and cheaply. Then you put several individual reactors together, ten or more, with a common operations system to make up a power station. They think they can build such stations for $1000/kilowatt and generate power for 1.6¢/kilowatt-hour. If they can, that would be a construction cost on par with the best coal can do and the cheapest form of electrical power you can get.

The GT-MHR concept is the latest idea from General Atomics, the American firm that has been working high temperature gas-cooled reactors for many years. They built Fort St. Vrain and learned a lot there. Their design differs from the PBMR in that, instead of pebbles, they glue their kernels into pencil-like plugs, which they insert into blocks of carbon. They propose to refuel half the reactor each year and a half. Their reactor is a little bigger at 600 MW thermal and 285 MW electric but is still built in modules. It too uses a direct gas turbine concept for the conversion of heat to electrical power. They are working with Russian, French, and Japanese collaborators. The construction cost they are shooting for is $1100/kilowatt.

Other countries are also looking at gas-cooled reactors. Japan has a small high temperature gas-cooled test reactor that uses the carbon-block-type fuel, and China has a small test reactor that uses pebble-type fuel.

A lot of people think that high temperature gas-cooled reactors are the future for nuclear power. They may very well be right. The idea of failure-proof fuel and significantly cheaper construction and operating cost is very attractive. The key to gas-cooled reactors is getting all the technology demonstrated with all the little nagging mechanical problems worked out. While gas-cooled reactors have some advantages over water-cooled reactors, water-cooled reactors do have a much larger operating experience.

Beyond gas-cooled reactors lies liquid metal cooled fast reactor. We talked about these a bit back in Chapter 5. These liquid metal reactors (LMRs) have the advantage, like high temperature gas-cooled reactors, of operating at higher temperatures than current light water reactors. They can also be designed with a number of passive safety features such as passive shutdown and passive cooling. However, the really important thing about LMRs is they can operate with a fast neutron spectrum and, therefore, make use of the other 95+% of uranium that water- and carbon-moderated reactors can't. Now that is just a little unfair in that moderated reactors could make use of some of that, as we will see at the end of this chapter when we look at fuel cycles. However, LMRs are a lot better at it. Let's see why that is.

Back in our reactor physics lessons, we learned that by moderating neutrons, that is slowing them down, we got them to cause fissions in uranium-235 better. Because fissions are the way we get energy out of nuclear energy, that is a good thing. The bad thing is that uranium-235 makes up only 0.7% of natural uranium, the stuff you dig out of the ground. Most of that uranium is uranium-238. That type of uranium will fission but only if you smack it with a neutron that is going really, really fast. The spare neutrons that are released in a fission event are going fast enough at the start, but as soon as they start bouncing off atoms within the fuel and coolant, they will quickly fall under the speed needed to cause uranium-238 to fission. This is especially true if the coolant is normal water, which is very, very good at slowing neutrons down.

But recall, all is not lost in that we can get uranium-238 to absorb a neutron and convert to plutonium-239 which, like uranium-235, is a good fuel for slow neutrons. The bad thing is that in general you can't get ahead. That is, you use up more uranium-235 generating neutrons than you make of plutonium-239 out of neutrons absorbed in uranium-238. It may not seem that this would be fair. After all, on average you get 2.4 neutrons out of a uranium-235 fission. If I use one to cause the next uranium-235 fission and one to make a plutonium-239 out of a uranium-238, I still have some excess. Well, life isn't that easy. Not every capture of a neutron in uranium-235 causes a fission. Some just lead to uranium-236. Also some of the neutrons just get away from us and get out of the reactor. And, of course, the real problem is neutrons get absorbed by essentially everything else we build the reactor out of. Now we have tried pretty hard to pick materials that don't suck up a lot of neutrons, but you can only do so much. Normal light water, while it is great at slowing neutrons down, is also a pretty good way to suck up neutrons.

The result is that if you use normal water as a coolant and moderator in a reactor, you can't make more plutonium-239 than the uranium-235 you use up. You can reduce your loses, and current light water reactors do that using uranium-238 captures to build in some plutonium-239 to extend the useful life of their fuel. But you can't get ahead of the game.

However, if you give up on the idea of moderation and work only with fast neutrons, you can beat the game. First of all, fissions with higher energy neutrons lead to more neutrons released per fission, so you have a greater number of neutrons to work with. Second, you can get more fissions out of uranium-238 itself. Third, with higher energy neutrons, you are less likely to get neutrons absorbed in coolant or other reactor component materials. The result is that if you design your reactor right, you can make more plutonium-239 than uranium-235 you use up. This is called breeding, that is, making new fuel. Now, it isn't a free lunch; you aren't getting something for nothing. What you are doing is making plutonium-239 out of uranium-238. But the really cool thing about that is that you are getting to use the form of uranium that is most common. Again, at the end of the chapter, we will look at the significance of this, but for now I think you can see how important it could be to be working with all of the uranium you dig up, not just 0.7% of it.

The reason LMRs can keep all their neutrons in the fast range is they use liquid metals as a coolant. Remember the lighter the atom, the better it is as a moderator. That is why hydrogen is so very good as a moderator. It is the lightest atom there is. If we use a metal as a coolant, we will get a much heavier atom that will be a poor moderator, which in this case is just what we want. The metal should not capture neutrons, it should transfer heat well, it should melt at a low temperature, and it should be easy to pump. Now it is very hard to get anything that does all these things super well, but most folks have settled on sodium. It is very good in the heat transfer and pumping departments and ignores neutrons just about completely. It does have the unfortunate chemical property of exploding on contact with water, which makes operation of a power plant using a steam cycle, well, interesting.

The LMR promise of extending the fuel supply has brought this concept to the attention of many countries. LMR programs have been conducted by the United States, the United Kingdom, France, India, Japan, South Korea, and Russia. As I explained in Chapter 5, the American experience has been less than happy. While the technology had advanced well, the politics and economics were not seen in a favorable light and most of the work has been shut down. There remain only small efforts within General Electric and Argonne National Laboratory to develop super safe modular concepts.

The chief political issue is that if you are going to breed plutonium at some point in your fuel cycle, you are going to have separated plutonium. The concern was that this material represented a target for bad guys to divert to make nuclear weapons. Now, whether the bad guys were terrorists or nations currently not possessing nuclear weapons, the idea of either group getting access to plutonium was sufficiently troubling for the United States to step away from breeder reactors and to suggest the rest of the world do the same.

The rest of the world did not do that. France in particular worked pretty hard on LMR breeders. They built and operated a 350 MW thermal power demonstration plant, the Phenix. They scaled this up to the biggest LMR ever built, the 1200 MW electric Super Phenix. However, that latter plant ran only a short time and was shut down. While safety, environmental, and political concerns went into that decision, a lot of it was based on the fact that, as currently configured, it just didn't make economic sense. Uranium is currently pretty darn cheap and the chemical process to get the plutonium out of the fuel and targets of a LMR breeder is not cheap.

The other really large program is the Russian effort. They have progressed through a series of ever-larger units. At 135 MW electrical power, the BN-350 reactor the Soviet Union built in Kazakhstan operated for 27 years. The larger BN-600 at 560 MW electrical power has been running since 1981. They would like to finish a BN-800 reactor they started and have plans for a 1200 MW electric reactor.

In addition to these straightforward extensions of LMR technology, the Russians have advocated a different coolant. Rather than the usual liquid sodium they have explored the use of molten lead. It is harder to pump and not quite as efficient with neutrons, but it is chemically inert. They get away from the nasty little problem with sodium and water. Their argument is that the world picked sodium as *the* LMR coolant at a time when making as much plutonium as fast as you could was important. At that time, the early 1970s, the belief was that nuclear energy was going to expand rapidly. Therefore, we needed to make a lot of fuel for the new reactors before they came online. Therefore, they chose a LMR coolant that was the very best at leaving neutrons alone so they could get on with making plutonium. The Russians say that now we can be a little more relaxed about the whole thing. As long as we make more new plutonium than uranium-235 or recycled plutonium we use up, we will be ahead of the game and will get where we need to be. There is no rush about it. This makes a lot of sense, but once again you have all those pesky mechanical engineering things to work out on a concept that we have little experience with. Time will tell.

The last potential future reactor technology I want to talk about is proliferation-resistant reactors. As we discussed in Chapter 6, one of the unfortunate things about generating electricity with nuclear power is that the civilian nuclear power industry intersects with the production of nuclear weapons in common portions of their respective fuel cycles. We saw that there is no need to make weapons as you make electricity nor any need to make electricity if you are making weapons. The two can be completely separated. However, if you have a uranium-fueled reactor, you are going to make some plutonium during operation. If you have a military point of view and sufficient technology, you can get that plutonium out of your spent fuel and try to make nuclear weapons. Clearly that is not a simple thing to do. There is a lot of tough science and more than a little difficult engineering to complete the task. However, because it is possible, folks have been uncomfortable when nations who do not have nuclear weapons start up a civilian nuclear power program. Is that because they are trying to work toward nuclear weapons? Are they going to use their "civilian" facilities for military purposes? The worldwide consensus is that the North Koreans were doing just that. Both India and Pakistan gained military capabilities from the civilian side of their nuclear programs. This increase or proliferation of nuclear weapons capability is and should be troubling.

Now this creates a real problem if we are hoping to use nuclear power to offset a huge increase in energy production in developing countries that otherwise would be using fossil fuels. What can we do to allow those developing countries to employ clean nuclear energy if we are worried about the potential for them to divert the technology and/or materials to weapons purposes?

The first line of defense is the International Atomic Energy Agency (IAEA). This organization has teams of inspectors that travel the globe checking "civilian" nuclear facilities to ensure that the facilities are not being misused and that all the "civilian" nuclear material is properly accounted for. This provides a lot of comfort, but you need to recognize that you are expecting the host country to act in good faith. If they don't, people worry there might be ways to sneak around the inspectors or to totally hide the military fraction from prying eyes. There is also the fact that IAEA inspectors are only allowed in countries if those countries invite them in.

To address these concerns, several concepts for proliferation-resistant power reactors have been developed. While there are a variety of ideas, the central theme is a small modular reactor with a long-lived fuel load. The proliferation-resistant concept would probably use a uranium-235 enrichment above the normal 3 to 5% found in current light water reactors. It might be in the 10 to 20% range. This would be enough to allow the reactor to run a long time on that fuel load but not so high as to make the uranium itself attractive as a weapons material. You would also probably try to design the fuel to make and use plutonium efficiently. Whoa, I thought plutonium was just what we didn't want to make? Well, if we can be assured it is going to stay inside the reactor, it will be all right. The idea is that if we can efficiently make and use plutonium within the reactor, we can extend the fuel life by half or more.

The key to the proliferation-resistant reactor designs is that they are a welded shut integral unit. You provide them to the non-weapons developing country as a finished product. They hook up water pipes and power lines to it and turn it on. They run it for 10 to 20 years, and then they give it back to you. During that time, they just maintain the external machinery but never open the reactor vessel itself. There is no need. Just add water, electricity comes out. This would be cool for them, and cool for you. Now you still need the IAEA to inspect and make sure they don't try to break into it. However, because the reactor vessel is welded shut, it isn't going to come apart easily or quickly, and that makes the IAEA's job a lot easier.

Well, that is a review of reactor systems. Is that all that we can look forward to in the future of nuclear energy? No, there are a lot of other neat things coming down the pike.

Besides full-scale reactor concepts, there are many potential new technologies that will affect the various components that make up a nuclear power plant. Let's start with fuels and materials.

Current light water reactors are limited in their thermal efficiency by the limits on their operating temperature. Is that inherent in the nuclear physics? Is nuclear energy just not as hot as burning natural gas? Not hardly. The temperatures you can achieve with nuclear energy are limited only by the structure you use to contain the release of energy. If you don't contain it at all, as in a nuclear weapon, you can get millions of degrees. That is tough to contain, but clearly the reason we operate a light water reactor at temperatures lower than those achieved in a natural gas burner are not the limitations of the energy release process. It is the fuels and structural materials we use. In light water reactors, that is a uranium dioxide fuel and zirconium alloy tubes we put around the fuel. Now these are really useful materials. The uranium dioxide is very tough, retains fission products very well, and is very stable under irradiation. Zirconium alloys are almost magical in how they can take many displacements of the individual atoms in the metal matrix caused by neutrons banging into them and still find their way back into a useful, stable position.

However, as good as both these materials are, there are limitations. Uranium dioxide is a ceramic material. That is why it is so tough. Remember, a ceramic is material used in your coffee cup, provided it isn't plastic or metal. Tough stuff, but it is also an insulator of heat. That is a good thing in a coffee cup because you want to keep your coffee warm. That is not a good thing in a nuclear fuel in which you are trying to get the heat out of the fuel and into the coolant. Uranium dioxide is such a good insulator that in 0.3-inch-diameter fuel pin, the difference between the temperature at the center and at the outer surface can be over a thousand degrees during operation. But because the uranium dioxide can take those temperatures and it has all those other good properties, we put up with this low conductance of heat.

But uranium dioxide isn't the only type of tough fuel we can make. There are uranium nitrides and carbides that have interested nuclear fuels folks.

These are also tough and can take very high temperatures. Both have better heat conductance than oxides. Nitrides are especially good at that. There is a good deal more research to do to ensure that we get all the good properties of the uranium dioxide that we have so much experience with, as well as enhanced properties. Also we have to be sensitive to the question of cost and long-term operational stability.

The zirconium alloys are really magical. However, they too have a few unfortunate properties. One of the most irritating is that zirconium metal wants to be zirconium hydride. And zirconium hydride is a brittle material. Therefore, if you put zirconium in water and heat it at the right temperature, the hydrogen in the water reacts with the zirconium metal and your wonderful, strong, ductile fuel clad turns into easily breakable hydride. It turns out you can surface treat the zirconium alloys to slow that way, way down, but it does become an issue as to how long you can leave your fuel in the reactor.

If there was an accident in which you lost all normal cooling, the fuel would get very, very hot. If that happened, the zirconium would corrode rapidly. In doing that, it would strip the oxygen out of the water and leave hydrogen gas. If that hydrogen gas then found some more oxygen, it would either burn or explode. That is what happened at Chornobyl after the accident got rolling. In U.S. light water reactors, the design of the emergency cooling system and the operating power limits are established to ensure this doesn't happen. That is good, but it adds expense to your plant and limits the maximum operating temperature, and therefore, the thermal efficiency of your power plant.

Almost any metal you can think of to be a clad for fuel rods is going to have chemical properties like this that either limit the total lifetime of your fuel or limit your operating temperature. However, there are a number of neat ideas to resolve these problems based on the increasing understanding of what is going on at the molecular and atomic levels. Better surface treatments and micro-coatings show promise at reducing both hydriding and corrosion at accident temperatures. There is even a possibility that we can develop a silicon carbide or related material that could be used as a clad. That would remove all the metal-related chemistry problems, if we can make it strong enough and durable enough.

We are also learning a lot more about how other structural materials in a reactor behave over time. It is interesting to reflect that, in the history of the earth, never before have materials been subjected to the conditions found inside a nuclear reactor. The combination of heat, pressure, coolant chemistry, and radiation has never been seen before. But we now have had nearly 60 years to look at how materials behave, and recently we have been developing really good models that describe what we are observing. Once you have a model of what you think is going on, you can get a lot better at predicting what might happen under a range of circumstances. That makes it easier to come up with ways to mitigate problems. It is a pretty exciting time for the materials scientists.

For the overall reactor designer, what the enhancements in fuels and materials mean is it gives them opportunities to design reactors that can operate longer, more reliably, more safely, and at higher temperatures. This means greater thermal efficiency, which means more electricity for the same reactor power, which means cheaper electricity. I can recall as an undergraduate in the early 1970s recognizing that the true performance of nuclear energy was determined by materials, not by the nuclear physics. Build a better pot to boil the water in, and, as a physicist, I can get you any temperature you want or can imagine.

Beyond materials, there are many other areas where we could apply new technologies to the next generation of nuclear power plants. One of the most rapidly expanding areas is in instrumentation and control systems. Remember all the concern over the so-called Y2K problem that was going to cause all computers to go nuts at midnight January 1, 2000? Notice that despite all the flap about nuclear power plant safety absolutely nothing bad happened? Well, that has two parts to it. One, a lot of very smart people worked very hard to scrub a lot of systems and make sure nothing bad would happen. The other part is a very large fraction of the nuclear power plants in the world were designed in the days when Bill Gates was still a high school kid. The original control systems did not use a lot of digital-based controls and instruments. To a great degree, the Y2K effort had to do with checking on when and where updates had been done. But the older systems didn't have the problem.

What that means is we haven't fully taken advantage of the reliability and operability enhancements we could get out of modern digital systems in much of our nuclear power plants. Such upgrades would give us greater understanding of what our plant was doing as well as make it easier to maintain instruments. Beyond the normal instruments and controls, we have the ability to add new sensors that would tell us far in advance when systems and components needed maintenance. By monitoring flows, pressures, voltages, vibrations, and the like, we could predict just how long it would be before a pump or a valve or a sensor would be ready for replacement or tune-up. Again, this would improve the reliability of the plant and help ensure all safety systems were working at peak capability. It would also help us do the right maintenance at the right time rather than overdoing it, which can be a costly error on the side of caution.

Operation is not the only place we could employ modern computerized methods. The reason I get uncomfortable reporting the $1700/kilowatt number for construction of a new nuclear plant is that estimate is based on the way we used to build nuclear plants. Even if we didn't go to smaller modular plants that might have a lot of components efficiently built in a factory, we could still do a lot more to reduce construction costs. The current generation of nuclear plants was built at a time when no one had a PC on their desk and everyone worked with blueprints that a draftsperson drew by hand. No one does that anymore.

Using modern construction management systems and computer-aided drafting and design would work a lot of the bugs out of the construction process. It is a real drill to get a couple thousand folks together and put something as complicated as a nuclear power plant together. With sharp construction management, you could do a lot of pre-planning, staging of parts and pre-fabricated components, and get the most efficient use out of your construction crew. This would reduce time to build and the amount of time between when you bought your parts and when they were all put together and making money for you. The fact that recent Japanese and Korean reactors are getting construction times down to 4 years and reducing construction costs in parallel supports this view.

The most important new technology out there for nuclear energy is the recycling of spent fuel and the use of breeder reactors. This is the fuel cycle issue I promised you earlier.

It is really frustrating to me when I see articles dismissing nuclear energy because after all we only have 50 or 60 years of uranium left, so what is the point of building nuclear power plants since we are about out of uranium? No. No. No. Sure you can get numbers like that if you make the right assumptions: assumptions like you will only mine the easiest uranium from mines you already have in operation, assumptions like you will never look for any more nor will you pay even a little bit more once the highest grade ore is used up, or assumptions like you will continue to throw 95+% of the fuel value of the uranium you mine. Uranium is more common than silver, but I don't see anyone making a run on silver jewelry before it is all gone.

Let's look at some real numbers to understand the real situation. Table 9.1 is a collection of numbers related to uranium resources. Let me just provide these first, and then we will figure out what they mean and the real story in those numbers as we go along.

Now the first four numbers come from data collected across the world by the Organization for Economic Co-operation and Development. As you see they update the information periodically and provide it by the potential price. That is, if you are willing to pay more, then miners will dig harder and get more for you.

In 1997, their estimate at $80 per kilogram uranium was 2.3 million metric tons of uranium. That might sound like a lot, but we are using uranium at a rate of about 60,000 metric tons per year right now. That means we would go through the 2.3 million metric tons in only 38 years.

TABLE 9.1 Nuclear energy resources.

Data Set	Estimated Amount (metric tons)
1997 OECD @ $80/kg U	2.3 million
1999 OECD @ $80/kg U	3.34 million
1997 OECD @ $130/kg U	4.3 million
1998 OECD Total Conventional Resources	15.5 million
Estimated Thorium Reserves	4.1 million
Uranium in Oceans	170,000 to 1,300,000 million
Uranium in Earth's Crust	100,000,000 million
Pre-mined Uranium Tails	1.2 million

OECD = Organization for Economic Co-operation and Development

Recall, when we were looking at fossil fuel reserves, we tried real hard not to do math like that. The problem is that the estimated reserve on a given date at a given price is not "how much there is." It is more like how much people have found, that they will tell you about, at that time, at that price. There are a lot of caveats it that definition.

Looking at the OECD numbers just 2 years later we see the $80/kg uranium number has moved up to 3.34 million metric tons. Several big new and very rich mines were coming on in Canada between 1993 and 1999, which may have impacted this. The point there is that we may have a lot more high-grade uranium left to find.

Then there is the issue of price. In 1997, the OECD suggested that if we went to $130/kg uranium, we could get 4.3 million metric tons of uranium, which is 187% of the amount available at $80/kg uranium.

Before we go further, we need to do a little calibration here to see how impactful such higher prices might be on the cost of nuclear electricity. The Energy Information Administration's Annual Energy Outlook 2001 estimated a 2005 nuclear generation cost of 6.3¢ per kilowatt-hour. Of that, about 5.1¢ was due to capital costs—the cost of building the plant, 0.75¢ was for operations and maintenance, and about 0.45¢ was for fuel. But of the fuel cost only about a quarter is related to the cost of the ore itself. The rest of the cost is in enrichment and manufacture of the fuel rods. So the cost of the ore in their number is about 0.11¢ per kilowatt-hour.

Now let's assume the price of the ore they were using was the July 2001 spot market price of $10 per pound U_3O_8. That is equal to about $26/kg uranium. So, if the ore was instead $80/kg uranium, it would be over 3 times more expensive. That would take the ore cost up to 0.34¢ and the total generation cost up to 6.53¢ per kilowatt-hour or about a 3.6% generation cost increase for a tripling of the ore cost. If we did the $130/kg uranium case, we would get a 5 times increase in ore cost resulting in a 7% increase in generation cost. That is saying very strongly that the price of the ore isn't a big factor in the cost of nuclear power. For a natural gas plant where the cost of fuel is about two-thirds the cost of the power, a tripling of the fuel price would increase the cost of power by 237%.

But wait, you said the cost of construction of nuclear power was coming down. Won't that make you more vulnerable to ore prices? Ah, yes, that is a good question. Let's get real positive and assume we get the total cost of nuclear generation all the way down to 2.5¢/kilowatt-hour. Now that is a long way to go, but let's just try that. If we did that and we did it without decreasing the fuel cost, still set at 0.45¢ with an ore cost of 0.11¢ and then we jumped from $26/kg uranium ore to $130/ kg uranium ore, a factor of 5 increase, then the cost of generation would bump up to 2.94¢/kilowatt-hour, an 18% increase. Yes, a bigger bump but still a very stable price of power compared to even huge changes in the cost of the ore.

O.K., let's return to the numbers in the table. The point we just made is that there is more ore out there, if you can pay for it, and we saw that paying more for the ore isn't going to affect your electricity bill very much at all.

But still how much is out there? After all, the 4.3 million metric tons is only good for 70 years at our current rate, and you want to expand nuclear energy use. Well, the OECD put out a value of 15.5 million metric tons as "total conventional resources." This would appear to be getting uranium from current types of mines with relatively high-grade ores. That would work out to about 250 years at current rates.

But again, that is not all there is. Uranium is actually found almost everywhere. It is found in common rocks in concentrations of about 4 parts in every million. Those minerals have dissolved in a lot of seawater, and the oceans hold dissolved uranium at concentrations from 0.3 to 2.3×10^{-6} grams per liter. Using those concentrations you can come up with the estimates in the table for total uranium out there. Now it is unlikely you are going to strain the entirety of earth's oceans for uranium, but the point is that there is a lot of uranium out there, and remember we can pay quite a bit more for the ore than we do now and still not impact the final cost of the power much.

Now, the next thing to understand is that even with all that uranium potentially available doing the "how long will it last" math based on the way we currently use uranium isn't the only way it can be done. After all, the subject here is supposed to be new nuclear technologies.

The way we use uranium now is that we mine natural uranium, enrich it to something like 3.5% in uranium-235, run it through the reactor, and then pitch it out. When we enrich uranium what we are doing is taking the natural uranium, which has 0.7% uranium-235 and 99.3% uranium-238, and sifting it to get one stream with more uranium-235 and one with less. The nature of the enrichment process is that there is a balance between how hard you squeeze the ore and how much the ore costs. That is, you can get just about all the uranium-235 out but that takes a lot of energy, which costs money. The current processes work out that leaving about 0.3% behind optimizes the cost of enrichment against the cost of the ore. If you do all the math, what this works out to is that you will need about 8 kilograms of natural uranium to get one kilogram of uranium enriched to 3.5% uranium-235, and you get 7 kilograms of "tails," which is so-called depleted uranium, having 0.3% uranium-235.

O.K., after I have put one kilogram of that fuel through my typical light water reactor, I am going to have 950 grams of uranium-238, about 10 grams each of uranium-235 and plutonium-239, and 30 grams of fission products. Only the 30 grams of fission products are really waste, but in the United States current once-through fuel cycle, we pitch out the whole kilogram.

If now, we reprocessed the spent fuel and reused—recycled—the good stuff, then we could extend the usefulness quite a bit. What we are trying to get in the recycled fuel is something that looks like our original fresh fuel with 35 grams of uranium-235 in each kilogram of fuel. Well, we already have 10 grams of uranium-235 and 10 grams of plutonium-239. Plutonium-239 works about as good as uranium-235, so in rough terms, we are looking for another 15 grams of uranium-235 for each kilogram of new fuel. If we pitched out 430 grams of the uranium-238 from the spent fuel and replaced it with 430 grams of fresh fuel enriched to 3.5% in uranium-235, we would get those 15 grams. To get that much new fresh fuel, we would need 8 times 430 grams or 3.4 kilograms of natural uranium. What all that means is that with light water reactor recycling, I could get the equivalent of one kilogram of new fresh fuel with 3.4 kilograms instead of 8 kilograms of natural uranium.

Now, that is only with a single recycle step. You could recycle the recycle and so on. However, you can't go too far because you start to make too much of the higher plutonium isotopes and uranium-236. These are generally not real helpful to you. But even with a single recycle, you would extend the worth of your ore supply by 2.3 times. That would make our 250 years at the 15.5 million metric ton level stretch out to 575 years. If we increased the use of nuclear power by a factor of 8, it would still allow us about 72 years. Why do I suggest 8 times our current rate? Well, in rough numbers, that would be enough to provide all the global electricity we are forecasting for 2020.

But we are just getting started on extending the usefulness of our nuclear fuel sources. You might have noticed the line about thorium resources in Table 9.1. What is that? Well, thorium is a heavy element with an atomic number (that is, number of protons) just two less than uranium. The normal stable isotope of thorium is thorium-232. The really neat thing about thorium-232 is that, if you expose it to a bunch of neutrons, it will absorb some and become uranium-233. And the neat thing about uranium-233 is that it will fission with any energy neutron and hence looks a lot like uranium-235 as far as nuclear fuel goes. Therefore, if you fuel a reactor with a mixture of uranium-235 and thorium you get some uranium-233 made, very much like you get plutonium-239 made if you mix uranium-235 and uranium-238. In effect then, you could use all the thorium supplies just like you would the natural uranium supplies to make new fuel. After you got some uranium-233, you could fuel the next time with uranium-233 instead of uranium-235 and run your reactor essentially totally on thorium thereafter. This requires a little better neutron economy than current light water reactors in which you only get about 0.7 "new" fuel atoms created for each one you use up. But even if you could use as much as half that 4.1 million metric tons of thorium as fuel, you could run reactors for a very long time. How long?

If we are using around 60,000 metric tons of natural uranium per year now, then we are using about 240 metric tons of uranium-235. That is, in each kilogram of uranium we mine we have 7 grams of uranium-235, but in the enrichment tails, we leave about 3 grams, giving 4 grams uranium-235 out of the original 7 we really use. Therefore, if you could convert even half the thorium to uranium-233 and you used uranium-233 essentially the same as uranium-235, that would get you 8500 years at our current nuclear fuel use rate, or over a thousand years at 8 times the current rate.

But if we start to use thorium-232 to make new fuel, why don't we also use the uranium-238 as new fuel? Well, we could do that. That is what breeder reactors are all about. If we employed liquid metal fast breeder reactors with reprocessing of the spent fuel and recycling of the spent fuel and new plutonium created, we could make the uranium-238 fuel.

Now the last number in Table 9.1 is 1.2 million metric tons of uranium tails. In our example above, we saw that it took 8 kilograms of natural uranium to make one kilogram of uranium enriched to 3.5% in uranium-235. The tails is the other 7 kilograms that we have just put aside into storage. The piles of this we have built up come from both the civilian nuclear power program and the nuclear weapons and naval reactor enrichment efforts. In the United States, we have about half the world's reserves of depleted uranium. Now, following the same logic as we did above for thorium, if you could get a fleet of breeder reactors going, you could make a lot of good use of those tailings. You can't get every last bit, probably only 80% or so, but just to get a feel for the value of the resource, let's guess you could get at least half. Then with the uranium **we have already mined,** you could run nuclear power reactors at their current rate for about 2500 years. At 8 times the current rate, you could run for over 300 years.

If we went on to mine the total conventional resource at 15.5 million tons, got only half the value with recycling from our breeder reactors, and increased nuclear by 8 times, we would have fuel for 4000 years. Add another 1000 years for thorium at 8 times our current rate, and you get over 5000 years.

If you gave me a few thousand years to think about it, I bet you I could come up with an economical way to get at least 0.01% of the uranium out of the seawater and get you another 5000 years.

No, don't tell me there is no point in nuclear power because we are going to run out of uranium in 40 years.

If you use the technology available, nuclear energy can provide a great deal of energy that can last us a long time. Energy that comes without greenhouse gases and air pollution. Energy that can be affordable. Energy that can be had at a stable cost and be available whether the sun shines or not or whether the wind blows or not. It can buy us a lot of time to find other truly long-term solutions to the questions of how to be the technologic dependent species that reigns as good stewards of the planet.

However, nuclear energy can do that only if the technologies of the future are allowed to be employed. That is not a scientific question. We know that the science can work. There is a good bit of fun engineering to make it work as efficiently and inexpensively as we can consistent with good environmental stewardship. However, the real questions about how much we are going to use nuclear energy are political not technical.

The whole point of my effort here has been to show you that a lot of what you have been told about nuclear energy has been misrepresented by those who are politically pre-disposed against nuclear power. My mission has been to give you both a larger perspective to consider the questions and sufficient technical background to make your own balanced judgments. If I have succeeded and pulled you over to my side, then we can go to the last chapter and see what we ought to do with all this.

10

WHAT ALL THIS MEANS TO YOU AND WHAT YOU CAN DO

Well, here we are. We are almost done. It has taken you some hours and me some years to get to this point. What have we learned together?

We took a very long view of the small place that the human species has in the history of our planet and gained a little perspective on how fragile our hold is on that place. We saw how technology and the energy that make technology work are vital to maintaining humanity's current numbers. We saw that with the momentum built into our current population, we are likely to increase our numbers. Even if we didn't, we saw that we will need a lot more energy if we are going to raise the standard of living of much of our population out of grinding, hopeless poverty. We saw that continued dependence on fossil fuels as our primary energy source would lead us to either political or environmental disasters. We also saw that alternative energies were unlikely to, by themselves, leap in to save the day.

We walked through some of the science and engineering behind nuclear power. Hopefully, that demystified nuclear energy somewhat for you and gave you the ability to judge some of the things you have heard about nuclear energy. We spent some time directly addressing the "bad" things about nuclear energy and looked at those issues in a larger context. We also looked at the state of nuclear power today and took a peek at what it could do for us tomorrow. We saw that if given the chance, nuclear energy could go a long way toward delaying the disasters ensured by sole dependence on fossil fuels. That delay might give us time to solve the technical and social challenges of being energy and technology dependent creatures.

Although we finished the last chapter talking about thousands of years of energy available if we really turn on nuclear energy, we need to recognize that this doesn't mean we have got it made. It is going to be really hard to get to the state where those thousands of years are available to us.

Remember I talked about the race between the forces of our destruction and the forces of our salvation? Well, nuclear energy **can** give us thousands of years, **provided** we move fast enough to get enough of an offset from fossil fuel use to allow us to last long enough to get nuclear rolled out. As I said, the race is on and frankly it doesn't look real good. The thousands of years calculation is like the coach telling you all you need is two touchdowns and one field goal. The hard part is getting those touchdowns before the clock runs out. If you are still in the third quarter, it is only hard. If you have already had the two-minute warning, well, it is going to be really tough.

O.K., we have learned all these things and been given the pep talk by the coach. Is that it? What do you do with all that?

I recall somewhere in my training the admonition that you never write a business letter unless you recommend or request an action. If you don't want the recipient of the letter to do something, then don't bother them with the letter. You should be asking them to do something, refrain from doing something, buy something from you, sell something to you, or team with you on some project.

Well, this is a pretty darn long letter I have written to you, but the action I am requesting of you is to team with me in getting the largest possible deployment of nuclear power as fast as possible. Hopefully, if we work at that hard enough, we can use nuclear power to get us past the crisis point and save our species and the planet. Does this sound like a worthy cause? I think so.

But what can I do, you ask? You're the nuclear engineer, you say, not me. I can't design some new reactor system or invent a better fuel design. No, I am not asking you to do any of those things. What I am asking of you is a lot harder.

What I am asking of you is to join me in the effort to get all the political barriers out of the way. We have the scientists and engineers to do the easy part. The tough social and political issues take a lot more people and a lot more effort.

Now, one of the things you might be thinking is, ugh, he is talking about politics and we all know that is a bad thing. Well, like our tarnishing of technology, politics has gotten a bad name that it doesn't fully deserve.

Some years ago I found myself on an airplane coming back from several days of meetings in Washington, D.C. As I was going over the meetings in my mind, I realized, to my horror, that essentially everything we had done was politics. There was very little science or engineering in the discussions. It was all about how to convince people with the power to decide that our idea was better than the other guys' idea. Who do we have to convince, in what order, what do they want, how can we display our idea in the best light against those needs, who else can we get on our side of the debate, what do we need to trade to make coming on to our side attractive? Sure, there was some science in there someplace, but the name of the game was politics.

So I thought about this a bit. Was that a bad thing? Is politics inherently evil? What really is politics, at the heart of it? The more I thought about it the more I realized that what politics really comes out to be is a marketplace for ideas.

In a normal financial marketplace, you trade things for money. You set prices and work deals to transact for goods and services. In a political market, you do the same thing, but the currency is not money. It is influence, the ability to effect an outcome. The goods traded are ideas. Ideas that will become public policy. Will we have a king or an elected president? Will the speed limit be 70 mph or 55 mph? How much money will we spend on schools, how much on new highways? A lot of different people have a lot of different views on any of these and a million more topics. How do you get all those views heard, sifted, weighed, and make choices? How do you even decide what things you are going to decide about? Politics, that is how. Politics is that art and science of formulating public policy through the interaction of ideas and values.

There is nothing inherently evil about politics. Politics gets a bad name because some people do some pretty evil things to get politics to favor their preferred outcome. I once worked on a project that in the end was going to be worth billions of dollars to the folks whose idea won out. I have to admit I saw at least a little lying and cheating done along the way. I like to think that while we always looked for the best way to present our way, we never told untruths and fessed up honestly to any weakness in our proposal.

So while I was first dismayed that I had become as much a politician as an engineer, I came to the conclusion that if you really believed you had the better idea your engineering ethics (use science and technology to help) actually required you to become a politician. If you didn't, then your better idea would not happen.

So, yes, in that regard, I am asking you to become a politician. Now I am not asking you to sell your house and move to Washington, D.C. What I am asking of you is to become involved in that public policy marketplace of ideas called politics. You can do that right where you are. In fact, right where you are is the best place to do it from.

There are a few simple keys to engaging in this marketplace. The first key is simply to recognize that the marketplace exists. Hopefully, you are onboard on that. The second is to have an idea you want to support. After reading this far, you probably know the idea I am suggesting you support. Now the last key is to come up with the tactical actions, that is, specific things, you can do to help the effort.

As we mentioned in the first chapter, you have many people working for you whose jobs and future careers depend on doing what you want. Obviously, this includes elected officials like senators and congressmen. But it also includes appointed government officials, like members of your state's public utility commission or your local planning council. It includes your school board and teachers. It also includes corporations' officials, planners, and managers, and not just corporations that build or operate power plants, but corporations like banks that lend money to power companies.

All those people work for you. They owe their jobs to you. Clearly you can vote a senator out of office. You can also let those elected officials who hire and fire appointed officials know if you are happy or unhappy with any given appointed official. You can influence decisions made by corporations as either a stockholder or simply as a consumer. If we all decided tomorrow that any car with less than 45 miles per gallon was unacceptable, and started buying cars based on that, a lot of people in Detroit, Tokyo, and Frankfurt would start changing their worlds pretty darn fast.

Another really important key to all this is the disproportionate weight you have if you speak up. A senator might send out the occasional poll or have his or her staff look at polls others conduct on specific issues. However, few things influence elected officials like a direct comment. A letter or e-mail or telegram is a sign to them that there is an individual who really cares about the issue they are writing in about. That also means that it is an issue important enough that the elected official's performance relative to the writer's view on the issue is going to impact how the voter is going to vote in the next election. And it also means that if that one voter took the trouble to write in, that there are probably thousands or tens of thousands of voters who feel the same way.

I once spent 9 months working on a big project for the Department of Energy in Washington, D.C. Our little project office had about 10 people in it. One fellow's primary task was to answer letters that citizens had written to Congress about the project. Nearly 10% of our total effort was answering congressmen's mail from you. Why? Because we got our money from Congress and Congress got their jobs from you. No one in that food chain was confused about who held the real power. But the key was the people in Congress were only answering the letters they received. Only those folks who spoke out got heard.

This is how it works. Let's suppose there is a proposal to build a bridge and it takes $50 million of federal highways funds to make it happen. The problem is you have to tear down the county old folks' home to do that. Now the Acme bridge company might get the city officials on either side of the proposed bridge to write to congressmen telling them how grateful everyone would be to get the federal funds. Maybe they would get construction union officials to write in saying how badly the area needs the jobs. Add to that a number of letters from out-of-work steelworkers and underemployed grocery clerks who would benefit from the bridge, and Acme has a good shot at it. But if you flood the congressmen's offices with a few hundred letters from the old folks and their families, Acme isn't going to get the bridge.

Notice we are asking 280 million Americans to fund the bridge, and maybe 50 thousand people would be affected by the bridge in some way, but a few hundred people can make it happen or not, if they speak up.

Writing to congressmen might not be the only way to impact the bridge. Maybe there would be planning commission public hearings or an environmental review that asked for public comment. If you liked the idea of the bridge, you could get a new patio poured by one of the firms pitching the bridge and tell them what a great job they are doing by bringing jobs to the area. If you didn't want the bridge, you could go by that concrete firm and tell them you are taking your business elsewhere because they are being so mean to the old folks' home.

Maybe you own stock in Acme bridge construction or own stock in Ace bank that holds the construction loans to Acme. If you like what is going on, maybe you buy more stock. If you are aghast at their insensitivity to the old folks' home, maybe you sell your stock or better yet show up at the stockholders' meeting and ask a few "how dare you?" questions from the floor.

The two points are that there are a lot of ways to get your voice heard and that, since so few people will even try to make a voice heard, it will be those few who influence what happens to everyone else.

But won't those elected officials, appointed officials, business people, and other decision makers just do the "right thing" without us spouting off at them? Well, they might, but there are several problems in just setting back and expecting the "right thing" to happen.

First of all, it gets pretty tricky to say just what the "right thing" is. If you run a little grocery store on the west side of the river and they just built a big housing development on the east side of the river, then the bridge looks pretty good to you. If your grandmother lives in and loves the old folks' home and tearing in down would separate her from all her friends ruining these golden days of her life, well, the bridge isn't the "right thing" for you.

Another problem has to do with planning horizons. If you own land in Westberg and connecting to Eastberg will allow long-term growth in your area improving the value of your land, the bridge is a good thing in the long run to you. But to the congressman with a hundred angry old folks picketing his office in their wheelchairs, his long-term view is as far as today's six o'clock news.

There is a wonderful saying about elected officials and news media. It is said that if you understand two truths about them, then the behavior of those groups of people will make sense to you. If you don't understand these two truths, then almost nothing those groups do will make sense. The two truths are the purpose of elected officials is to get elected again, and the purpose of news media is to sell advertising space. Think about it.

What this does is really limit the planning horizon of elected officials. Congressmen are really in a jam. They go up for election every 2 years. They are always running for office. Something that won't come to fruition in 2 years or less just isn't on the list. If the "right thing" costs money and hardship in the near term but has long-term pay off, it is going to be tough to get the House of Representatives behind it. They aren't bad people; they are just driven by a 2-year window to make the voters happy.

I will tell you another story. Many years ago we had a congressman from a neighboring state in town. We showed him the idea we had for our government project and how it would save a lot of money and in the long term be the "right thing." He agreed completely. Cool, we had it made. We then asked if we could count on his support. He said, no, we could not. Politely, as you ask congressmen, we asked why that might be. He explained that it was his job to represent people from his district. Those people would not understand what we were trying to do and since it was somehow "nuclear" and they were against anything "nuclear" he couldn't support us. I was floored. Wasn't it this guy's job to do the right thing, make the government work for people? He agreed our project was the right thing and saved federal money by the basketful. But because his folks at home had not been given the 10 minutes of useful instruction on the topic that would clarify it for them, he would reflect that ignorance and let the "wrong thing" happen instead.

Once I learned about the two truths of elected officials and news media, it all made perfect sense to me.

So, no, you cannot count on elected officials to do the right thing. You have to let them know what you think the right thing is.

Coming back to the subject of nuclear power, getting some supporting voices out there is really, really important. Just as we have had this brainwashing that technology is bad, we have gotten a storm of nuclear is bad. A lot of this is coming from actually a relatively few but very vocal people. This is magnified by a media in which scary bad news sells more advertising space than rational balanced reporting.

I would like to believe that of those opposed to nuclear energy, a lot are sincere and most, I think, really mean well. Most, I think, however, don't appreciate the larger issues and don't understand the actual science involved and, more importantly, don't grasp the choices they are making. I expect a lot of that back-to-nature, love-the-planet wishful thinking would turn around quickly if those advocates appreciated that they were condemning five-sixths of the population to death and the remainder to grinding toil and poverty. Great-great grandpa might have gotten by with 40 acres, a mule, and a wood stove, but that was when there was only one-sixth as many of us. Furthermore, great-great grandpa could expect a hard life of maybe 50 years and half his children to die before adulthood. My guess is great-great grandpa didn't really have a very good time.

That is the choice the anti-technologists actually offer, if they would only understand it. Worse yet, even that bleak future would be allowed for only a very few, because we simply can't feed five-sixths of us without full tilt technological agriculture. And again, that technology is driven by energy. Some feel it is great to "keep big oil out of the national parks, protect the tidewater, and stop that nuclear plant." But where do you get the diesel to fuel the trucks that bring your groceries to the neighborhood store, or the electricity to run the factories or light the schools?

Every voice that is raised against technology or energy must be heard in the context of the things that don't happen if the technology or energy source is blocked. The way politics works is that the decision to build the bridge or not comes about by the balancing of the voices for and against. In the debate on nuclear energy, it has been a long time since the "voices for" have been heard over the din of "voices against." Especially as the media amplifies the negative anti-nuclear messages.

If we are going to have a balanced debate, I need you to join in the expression of support. We need to hear voices saying, "well, that is good, it would be nice if there were no nuclear waste, but what do you suggest instead? Coal-fired electricity with much more dangerous waste, solar with much greater energy cost that will bankrupt the nation, or do we just starve out over half the human race?" We need to hear challenges of statements like "horribly radioactive for millions of years." You understand the physics of it now. The uninformed or purposeful misleading can't tell you untruths anymore.

But what can you do to get your voice heard? Well, like the old joke about the elections in Chicago, vote early, vote often.

You get a lot of chances to vote for nuclear power, to offer a balancing voice. Let's look at a potential list.

- **Tell elected officials what you think**. Look for opportunities. Does your congressman send out polls? Has some elected official made some statement you either agree with or don't agree with? Write them a letter or an e-mail telling them which side you are on. Is there a vote coming up that impacts something you care about? Let them know. Can't think of an issue? Well, how about suggesting that the research and development budget for nuclear energy at least be as large as that for coal. That would be a factor of 10 increase for the nuclear team.

- **Tell non-elected officials what you think**. Now these people might not be quite as sensitive to the voters as elected officials, but you still hold their chain through elected officials who appoint them and through your influences on how tax dollars are spent (again, through elected officials and initiative processes). Look for actions of public officials you support and those you don't. Let them know. A state official whose gets a lot of press blocking a nuclear waste shipment that he knows full well is extremely low risk, tying up a lot of state resources that could be much better used keeping drunks off the road, needs to be told we don't endorse that grandstanding foolishness.
- **Make your voice heard in polls.** Now and then you will see polls asking for opinions on energy and/or nuclear matters. Join in. My buddies have an e-mail alert system that when such an Internet poll comes out, we jump on it. Not a scientific poll, but it is just one more way to collect the view that here are "X" many people who care enough about the topic to take the time to speak up.
- **Be a positive voice at public hearings.** There are a number of opportunities to make your views known to the government through public hearings. These can be environmental impact hearings, planning commission hearings, state legislative hearings, government advisory boards, and others as well. Can't get to the meeting? Don't like to speak in public? Fine, that is O.K., most of these venues have some way to submit written comments. It just takes a little work on your part to find the way.
- **Let the corporate world know what you think.** This has two forms. One as a stockholder, one as a customer. I own a little bit of stock in a power company. So I get one of those proxy letters every so often. It is just junk mail to me, I don't know anything about the issues at hand, and I know votes of my few shares are going to be a drop in the bucket. But once one of the issues had to do with the nuclear power policy of the company. That one I filled out and sent back in. If you don't own stock in a related company, you might think about buying some. If you find a company you like and it is investing in a future you want to leave for this generation's grandchildren, then maybe they should be rewarded with your financial expression of confidence.

  As a customer you have the ultimate vote for any corporation. You buy their product or not. Can't get any more yes/no than that. What you might have to do, however, is write in to tell the corporation why you are or are not buying what they sell. It is very much like training dogs or raising children. You reward good behavior and discourage negative behavior. The trick is getting the trainee to recognize that they are being trained.

- **Get engaged in education.** For those of you with children, do you know what your kids are being taught about science and technology? At one time there was a clear "technology is bad" slant to grade school textbooks. What is in your kids' books? Education should be about the provision of information and the development of the ability to think rationally and independently. I have tried in this volume to provide a balanced education on things nuclear. To an anti-nuclear advocate, my balanced view probably looks very pro-biased. The point is that it is hard to present any topic without some spin. You tell the story of the American Civil War appropriately condemning slavery, but do you also condemn the oppressive import duties the Northern states imposed on the Southern states? Do you present technology as the source of pollution and degradation in the human condition or as the unique identifier of the human species that allows us to exist? One, or the other, or some balance of both? Ask your kids what they think and then ask them who told them it was that way. If you like the answers, tell your school board, your kids' teachers, and the PTA. If you don't like the answers, then perhaps you might want to start a little dialogue with some or all of those people.
- **Don't let the media get away with biased or incorrect reporting.** This is hard, and it is also very important. I have given you the example of a national publication that used the phrase "renewable energy sources like wind and solar" and then went on to cite numbers for hydro and biomass giving the totally erroneous impression that wind and solar were big players. I doubt very many people caught that. I wrote and told them their implication was misleading and a disservice to their readership. I never saw a correction. Now if a few thousand of you would have joined me in writing in, we could have adjusted their frame of reference.

  Writing in to a newspaper or television station or magazine is a good way to provide feedback, both positive and negative when they get something right or wrong about energy. But it isn't the only way. Years ago, during the height of the clean energy excitement, a national magazine had an article on wind power. Associated with the article was a drawing of a Darrius wind generator (the eggbeater looking type) which I knew at that time topped out at about 250 kilowatts. In the illustration the wind generator was connected to power lines going off to a city skyline in the distance. The clear implication was that this little thing was going to power the city. Let's be generous and give them a 30% capacity factor. Doing the math you would get enough electricity for about 55 people. That is hardly the entire city as the illustration implied. I wrote them a letter of complaint and canceled my subscription. Well, I hardly broke the firm and last year they had a nice article on the revival of nuclear power. Perhaps I ought to subscribe again. The point is that the real power is the power of the pocketbook. Buying or not buying their publication or better yet buying or not buying the product of their advertisers is the thing that drives all media. If station XYZ evening news comes out with a nuclear scare story long on hyperbola and short on science and that segment is sponsored by Acme Snack Crackers, a polite but pointed letter to Acme Snack Crackers Inc. is likely to help station XYZ re-obtain journalistic balance.

- **Don't let public figures get away with uninformed public advocacy.** It has always been a mystery to me as to why the opinions of an actor, singer, or sports star about toothpaste or politics or anything but their own field of expertise should matter to anyone. However, mysterious or not, those opinions do seem to matter and have a lot of influence. For the nuclear power fan, the unfortunate thing about this phenomena is that, with the technology is bad and nuclear is evil undercurrent, a lot of public figures seem to feel it is fashionable to be anti-nuclear. Again, while accepting that at least some of these people are well meaning if misinformed, we are still left with their disproportionate impact on national feeling and policy due to their wide media access. If a major actor tells you that plant down the road is bad and marches in the parade protesting that, well, it must be bad, right? Well, he got his name in the paper. Now what if you wrote to his studio and told them you felt that this behavior was detrimental to the future of humanity and that all things considered it would be better if the major actor were to either educate himself properly or just keep quiet? What if you suggested your opinion of the actor and the studio was negatively impacted by this behavior and that you would be less inclined to buy the product of either in the future?

  I have been extremely impressed with the expressions of American patriotism in our public figures since September 11. Now I am sure a lot of that is honest and heart felt. But I would guess at least a little of that is the star's P.R. folks telling their client that patriotism is **IN** and mocking the military or federal government right now is not a good thing. If we could get across the message that cheap shots at things nuclear and uninformed opposition was not only **not cool** but in point of fact a **bad** thing, then we could perhaps get on with a balanced view of life.

- **Spread the good word to your co-workers, friends, and neighbors.** If you have come around to my point of view that nuclear energy deserves the opportunity to help us delay the day of reckoning from fossil fuel dependency, then you will have dozens of opportunities to share that view with people you meet everyday. I am not asking you to go get a reader board and stand on street corners, or snag everyone you run into and beat them with the message. What I am suggesting is that when the topic comes up, you offer the pro-nuclear view. What I have tried to do is arm you with enough information and perspective to do that rationally. Besides, wouldn't you sound cool if you could explain a positive void coefficient to your neighbor over the back fence as you explain that no reactor in the United State can have an accident like Chornobyl because we all have negative void coefficients?

Let me tell you one more story, and I will let you go. At the reactor where I worked years ago, we imported an industrial safety program originally developed by the DuPont Company. The training associated with this program was long, boring, extremely repetitive, and massively redundant. It was also very effective. The basic safety philosophy got pounded into you until in permeated your very bones. The single most important concept was that safety was everyone's responsibility. Along with that responsibility, everyone had the authority to act in the name of safety. Whether you were the CEO, a manager, an engineer, a technician, a craftsman, a janitor, or the boy who goes for coffee, if you saw an unsafe job, you had the authority to stop that job. And it was hammered home that not only could you stop work to correct unsafe conditions, you were required to do so. It was your responsibility. If oil got spilled on the floor and someone slipped in it, the people who saw the oil and walked by were every bit as responsible as the person who spilled the oil in the first place. The companion concept was that you needed to act every time you saw something wrong and as soon as you saw it. The key words were "every time," and "right away." You were not to think it is not my job, or someone else will do it, or I am busy, or maybe I will come back later and see if some else hasn't already fixed it. No, you did it every time and right away.

So I come back to vote early, vote often. Every time you have the opportunity to vote for nuclear power, do it. Do it every time, and right away. It is your future you are voting on. Who else is going to take the responsibility to provide you the future you would choose? A political decision to turn away from technology was made in China 600 years ago by relatively few people, and several billion Chinese have suffered for it since. Decisions to turn from technology and its vital and viable energy drivers will produce a dark future and a lot of suffering for the entire planet. You have lots of ways to make your views known, to cast your votes. Look for the opportunities to make your voice heard. When you see those opportunities, speak up. Do it every time, right away. I have done my part arming you with the basics and the mission. Let's get Little Nell off the tracks; let's give the generations to come a life we would be proud to leave them.

BIBLIOGRAPHY

Braggs, Melvyn, *On Giants' Shoulders.* Hoboken, New Jersey: John Wiley & Sons, 1998.

British Petroleum (BP), *BP Statistical Review of World Energy*, London: British Petroleum, June 2001.

Cohen, Joel E., *How Many People Can the Earth Support?* New York: W.W. Norton, 1995.

Deffeyes, Kenneth S., *Hubbert's Peak–The Impending World Oil Shortage.* Princeton: Princeton University Press, 2001.

Diamond, Jared, *Guns, Germs, and Steel–The Fates of Human Societies.* New York: W. W. Norton, 1999, 1997.

Driscoll, Charles, et al., "Acidic Deposition in the Northeastern Unites States," *Bio-Science*, March 2001.

Fairlie, Ian, *Dry Storage of Spent Nuclear Fuel: The Safer Alternative to Reprocessing*, Report to Greenpeace International In Response to Cogema Dossiers to the La Hague Public Inquiry, London, May 2000.

General Accounting Office, *Radiation Standards: Scientific Basis Inconclusive, and EPA and NRC Disagreement Continues*, GAO/RCED-00-152. Washington, D.C.: General Accounting Office, June 2000.

Hamza, Khidhir, *Saddam's Bombmarker*, New York: Scribner, 2000.

Holldobler, Bert, and Wilson, Edward, *Journey of the Ants: A Story of Scientific Exploration*, Cambridge, Massachusetts: Belknap Press, 1995.

International Energy Agency (IEA), *Key World Energy Statistics from the IEA 2001 Edition*, Paris: International Energy Agency, 2002.

Kirby, Richard; Withington, Sidney; Darling, Arthur; and Kilgour, Frederick, *Engineering in History*. New York: Dover Publications, 1990 (McGraw-Hill, 1956).

LeMay, Curtis, and Yenne, Bill, *Superfortress, The B-29 and American Air Power*, New York: McGraw-Hill, 1988.

Levathes, Louise, *When China Ruled the Seas–The Treasure Fleet of the Dragon Throne 1405–1433*. New York: Simon & Schuster, 1994.

Mokyr, Joel, *The Lever of Riches–Technological Creativity and Economic Progress*. New York: Oxford Press, 1990.

Pinentel, David, and Giampietro, Mrio. *Food, Land, Population and the U.S. Economy*, Washington, D.C.: Carrying Capacity Network, November 21, 1994.

Raup, David M., *Extinction–Bad Genes or Bad Luck?* New York: W. W. Norton, 1991.

Rist, Curtis. "Why We'll Never Run Out of Oil," *Discover*, June 1999, 20(6):80-87.

Rodigues, Baxter, McEachern, & McGill, *Transmutation of Nuclear Waste Using Thermal and Fast Neutron Energy Spectra*, Los Alamos, New Mexico: General Atomics.

"Silicon Valley Turbines Too Perilous for Condors," March 25, 2001, *Seattle Times*.

Tattersall, Ian, and Schwartz, Jeffrey, *Extinct Humans*. Boulder, Colorado: Westview Press, 2001.

U.S. Census Bureau, *Statistical Abstracts of the United States: 2000*, Washington, D.C.: U.S. Department of Commerce, 2000, Table 971.

U.S. Department of Energy, Energy Information Administration, *Electricity Prices in a Competitive Environment–Marginal Cost Pricing of Generation Services and Financial Status of Electrical Utilities*, DOE/EIA-0614. Washington, D.C.: U.S. Department of Energy, 1997 (August).

U.S. Department of Energy, Energy Information Administration, *Annual Energy Outlook 2001*, DOE/EIA-0383. Washington, D.C.: U.S. Department of Energy, 2001 (December).

U.S. Department of Energy, Energy Information Administration, *International Energy Outlook–2000*, DOE/EIA-0484. Washington, D.C.: U.S. Department of Energy, 2000 (March).

U.S. Department of Energy, Energy Information Administration, *International Energy Outlook–2001*, DOE/EIA-0484. Washington, D.C.: U.S. Department of Energy, 2001 (March).

Ward, Peter D., and Brownlee, Donald, *Rare Earth–Why Complex Life is Uncommon in the Universe*. New York: Copernicus, 2000.

World Almanac Books, *1960 World Almanac and Book of Facts 2001*, World Almanac Books.

A Few Interesting Web Sites and Additional Links

International Atomic Energy Agency links to information sources page: www.iaea.org/programmes/ne/qlinks.htm

International Energy Agency—global look at energy: www.iea.org/

Joseph Gonyeau's Virtual Nuclear Tourist—a delightful and eclectic private collection of information: www.nucleartourist.com/

Nuclear Energy Institute—nuclear energy industry trade organization: www.nei.org/

University of California-Berkeley Department of Nuclear Engineering—has interesting material you won't find elsewhere: www.nuc.berkeley.edu/

U.S. Department of Energy—Energy Information Administration—source of a lot of information on energy use and applications—constantly updated: www.eia.doe.gov/

U.S. Department of Energy—Office of Nuclear Energy, Science and Technology—focus point of U.S. government research and development on nuclear energy: www.ne.doe.gov/

U.S. Nuclear Regulatory Commission: www.nrc.gov/

World Nuclear Association—Good general reference, up-to-date global view, wealth of information: www.world-nuclear.org/index.htm

World Nuclear Association, links page—truly huge resource of site links: www.world-nuclear.org/portal/index.htm

INDEX

F

O

P